Weakened Ramsey Theory

Weakened Ramsey Theory provides readers with an overview of weakened generalizations of various parameters studied in Ramsey theory. Rather than determining how large a structure must be to guarantee the existence of monochromatic substructures in a t-coloring, weakened Ramsey theory considers how large a structure must be to guarantee the existence of a substructure that uses at most a given number of colors. Weakened generalizations of Ramsey numbers, Gallai-Ramsey numbers, Schur numbers, Rado numbers, rainbow Schur numbers, and others have all been studied in the literature, and this book offers a thorough overview of the known results, with complete proofs.

This book should prove a valuable resource to advanced undergraduates, graduate students, and researchers working in Ramsey theory. Only a basic background in graph theory, algebra, and number theory is assumed, although a concise review of important definitions and background results is included.

Features

- Self-contained treatment of the subject, requiring only a minimal background in graph theory, algebra, and number theory
- Complete proofs of major results in weakened Ramsey theory
- Numerous color figures to assist the reader
- Comprehensive list of references
- Tables of known weakened parameters in Ramsey theory, providing a useful reference for researchers.

Discrete Mathematics and Its Applications
Series Editors: Miklós Bóna, Donald L. Kreher, Douglas B. West

Computational Complexity of Counting and Sampling
Istvan Miklos

Volumetric Discrete Geometry
Karoly Bezdek, Zsolt Langi

The Art of Proving Binomial Identities
Michael Z. Spivey

Combinatorics and Number Theory of Counting Sequences
Istvan Mezo

Applied Mathematical Modeling
A Multidisciplinary Approach
Douglas R. Shier, K.T. Wallenius

Analytic Combinatorics
A Multidimensional Approach
Marni Mishna

50 years of Combinatorics, Graph Theory, and Computing
Edited By Fan Chung, Ron Graham, Frederick Hoffman, Ronald C. Mullin, Leslie Hogben, Douglas B. West

Fundamentals of Ramsey Theory
Aaron Robertson

Methods for the Summation of Series
Tian-Xiao He

The Lambert W Function
Its Generalizations and Applications
István Mező

Combinatorics of Permutations, Third Edition
Miklos Bona

Topics in Graph Theory
Mark Anderson, Jonathan Gross and Jay Yellen

Applied Combinatorics, Third Edition
Fred S. Roberts and Barry Tesman

Discrete Mathematics
An Open Introduction, Fourth Edition
Oscar Levin

Introduction to Enumerative and Analytic Combinatorics, Third Edition
Miklós Bóna

Handbook of Combinatorial Algebraic Geometry
Subvarieties of the Flag Variety
Edited by Erik Insko, Martha Precup, and Edward Richmond

Weakened Ramsey Theory
Mark R. Budden

https://www.routledge.com/Discrete-Mathematics-and-Its-Applications/book-series/CHDISMTHAPP

Weakened Ramsey Theory

Mark R. Budden

CRC Press
Taylor & Francis Group
Boca Raton London New York

CRC Press is an imprint of the
Taylor & Francis Group, an **informa** business
A CHAPMAN & HALL BOOK

Designed cover image: Shutterstock

First edition published 2026
by CRC Press
2385 NW Executive Center Drive, Suite 320, Boca Raton FL 33431

and by CRC Press
4 Park Square, Milton Park, Abingdon, Oxon, OX14 4RN

CRC Press is an imprint of Taylor & Francis Group, LLC

ISBN: 978-1-032-80702-7 (hbk)
ISBN: 978-1-003-49885-8 (pbk)
ISBN: 978-1-003-49886-5 (ebk)

DOI: 10.1201/9781003498865

Typeset in Latin Modern font
by KnowledgeWorks Global Ltd.

Publisher's note: This book has been prepared from camera-ready copy provided by the authors.

This book is dedicated to my mother, Joelle Budden, and my late father, John Budden, for instilling in me a love of learning.

Contents

Preface

Combinatorial objects usually have multiple parameters associated with them, which we use to describe their structure and connectivity. A characteristic theme in Ramsey theory (and more generally, extremal combinatorics) is to fix all but one parameter, then to try to determine the threshold of the variable parameter at which a given property must occur.

Features of combinatorial objects are often studied by considering partitions of their parts. In classical Ramsey theory, one seeks to determine how large a structure must be for certain properties to be guaranteed of its partitions. Variations of this idea have arisen in the study of Ramsey numbers, Schur numbers, rainbow numbers, and Rado numbers, among others. As partitions can be described via t-colorings, such numbers usually focus on the existence of monochromatic substructures. In weakened Ramsey theory, this concept is generalized to guarantee the existence of substructures that use at most s colors, where $1 \leq s < t$.

The first time this concept was formally studied was in the work of K. Chung, M. Chung, and C. Liu [62] (see also [63] and [164]) in 1977, when they defined the concept of a chromatic Ramsey number. Their work was soon extended by M. Jacobson [132, 133], and in 1999, H. Harborth and M. Möller [123] introduced the name "weakened Ramsey number" for the diagonal case of a chromatic Ramsey number. Over the past few decades, weakened variations of other Ramsey-theoretic parameters have been considered within the study of graphs, hypergraphs, and combinatorial number theory.

This book seeks to consolidate the known results in weakened Ramsey theory, along with relevant background and detailed proofs. Mistakes in several of the original sources are corrected, and some results are generalized. Numerous open problems in weakened Ramsey theory are thoroughly discussed with the goal of stimulating future research in this area.

Variations in Ramsey theory, like the ones presented here, are excellent topics for young researchers to investigate. The problems are relatively simple to describe, and students can quickly become engaged in working out special cases by hand. Undergraduate students can become involved in meaningful research with only a background in proof-writing and a basic course in discrete mathematics. I am very grateful to have had the opportunity to collaborate with many talented students over the years, a number of whom are cited in the bibliography. While this book certainly showcases their work, it is intended to serve as a valuable resource for future students as well as seasoned researchers that work in Ramsey theory.

Mark R. Budden
Cullowhee, NC

Biography

Mark R. Budden completed his Ph.D. in Mathematics in 2003 at the University of Missouri-Columbia. He was a faculty member at Armstrong Atlantic State University (now Georgia Southern University's Armstrong campus) until 2010, when he joined the faculty at Western Carolina University, where he is now a professor. He has published 47 research articles and is the author of the book *Star-Critical Ramsey Numbers for Graphs*. Most of his research is in Ramsey theory or other related topics in combinatorial number theory, and the majority of his publications have included student coauthors.

List of Figures

List of Tables

Symbols

$\|S\|$	The cardinality of the set S.
$\overline{G}$	The complement of the graph G.
$\deg_G(x)$	The degree of $x \in V(G)$.
$N_G(x)$	The neighborhood of $x \in V(G)$.
$G_1 \cup G_2$	The union of the graphs G_1 and G_2.
$G_1 + G_2$	The join of the graphs G_1 and G_2.
$G_1[G_2]$	The lexicographic product of the graphs G_1 with G_2.
$\chi(G)$	The chromatic number of the graph G.
$s(G)$	The chromatic surplus of the graph G.
$\omega(G)$	The clique number of the graph G.
$\alpha(G)$	The independence number of the graph G.
P_n	The path of order n.
C_n	The cycle of order n.
K_n	The complete graph of order n.
F_n	The fan $K_1 + nK_2$.
$F_{k,n}$	The generalized fan $K_1 + nK_k$.
B_n	The book $K_2 + nK_1$.
$K_{m,n}$	The complete bipartite graph having partite sets with orders m and n.
$\mathrm{Cay}(G; S)$	The Cayley graph for a group G and generating set S.
$\mathcal{P}_q$	The Payley graph of order q
$\mathbb{N}$	The natural numbers.
$\mathbb{Z}$	The integers.
$\mathbb{Q}$	The rational numbers.
$\mathbb{R}$	The real numbers.
$\mathbb{C}$	The complex numbers.
$[s, t]$	The set of integers $\{s, s+1, \ldots, t\}$, where $s \leq t$.
$\mathbb{F}_q$	The finite field with q elements.
$\mathbb{F}^{\times}$	The multiplicative group of the field $\mathbb{F}$.

1

Foundations in Ramsey Theory

The predominant theme in Ramsey theory is the realization that various properties of the partitions of a structure can be guaranteed if parameters associated with the structure reach a certain threshold. The goal of Ramsey theorists is to determine the threshold of a given parameter that forces the existence of a desired property. This subject has many flavors (graph theory, number theory, geometry, etc...), but the theme remains the same. Partitions can be visualized using colors, and the properties most commonly sought entail the existence of certain monochromatic substructures. Weakened Ramsey theory softens the expectations by seeking substructures that use at most a given number of colors.

This chapter provides an overview of the definitions and notations that will be used throughout this book, as well as a brief overview of relevant results. For notations and basic results in graph theory, and in particular, those that concern the coloring of graphs, the reader is referred to the books by Chartrand and Zhang [59], Harary [120], König [151], and Soifer [196].

1.1 Combinatorial Prerequisites

Let $\mathbb{N}$, $\mathbb{Z}$, $\mathbb{Q}$, $\mathbb{R}$, and $\mathbb{C}$ denote the natural numbers, integers, rational numbers, real numbers, and complex numbers, respectively. Since our focus is on a topic in discrete mathematics, most of the parameters we will encounter assume values in $\mathbb{Z}$. For $x \in \mathbb{R}$, denote by $\lfloor x \rfloor$ the greatest integer less than or equal to x, and denote by $\lceil x \rceil$ the least integer greater than or equal to x. These are referred to as the *floor* and *ceiling* functions of x, respectively.

The following axiom plays an important role in number theory and combinatorics. While we will not reference it directly, it is fundamental to the existence of most of the graphical parameters we study.

The Well-Ordering Principle *Every nonempty subset of* $\mathbb{N}$ *contains a least element.*

Of course, assuming the well-ordering principle is true, if S is any nonempty subset of $\mathbb{Z}$ that has a least element, then every nonempty subset of S contains a least element.

A classic combinatorial tool that we will use frequently is the pigeonhole principle.

Theorem 1.1 (The Pigeonhole Principle) *Let* $m, n, r \in \mathbb{N}$ *satisfy* $n > rm$. *If* n *distinct elements are placed into* m *sets, then some set has cardinality at least* $r + 1$.

Proof Assume that n distinct elements are placed into m sets, with each set containing at most r elements. Then there are at most rm elements in total. However, the assumption $n > rm$ leads to a contradiction. Therefore, some set contains at least $r + 1$ elements. □

DOI: 10.1201/9781003498865-1

If S is any finite set, then we denote by $|S|$ the cardinality of S. A *partition* of a set S is a collection of nonempty subsets $S_1, S_2, \ldots, S_k \subseteq S$ such that

$$S = \bigcup_{i=1}^{k} S_i \quad \text{and} \quad S_i \cap S_j = \emptyset \text{ when } i \neq j.$$

Partitions of sets will usually be described by relating the elements in a given subset with a unique color corresponding to that subset. A *t-coloring* of a set S is a map $f : S \longrightarrow \{1, 2, \ldots, t\}$. When f is surjective, it is called an *exact t-coloring* of S.

A *graph* $G = (V, E)$ consists of a nonempty finite set $V = V(G)$ of *vertices* and a set $E = E(G)$ of *edges*, each of which is a 2-element subset of V. We will denote the edge consisting of distinct vertices x and y by xy. A graph is called *simple* if the elements in its edge set E are distinct. Throughout this book, we only consider graphs that are simple. A graph is called *empty* if its edge set is the empty set. The *order* and *size* of a graph G are given by $|V(G)|$ and $|E(G)|$, respectively. Two distinct vertices are called *adjacent* if they form an edge. Two distinct edges are called *adjacent* if they have a vertex in common. If xy is an edge, then we say that x and xy are *incident.* If $x \in V(G)$, the *degree* of x, denoted $\deg(x)$ (or $\deg_G(x)$ when one wishes to indicate the graph being considered), is the number of edges in $E(G)$ that are incident with x. The set

$$N(x) := \{y \mid xy \in E(G)\}$$

is called the *neighborhood* of x in G. Note that $|N(x)| = \deg(x)$.

The following theorem is sometimes called the "First Theorem of Graph Theory" as it describes a fundamental principle that applies to all graphs.

Theorem 1.2 (First Theorem of Graph Theory) *If G is any graph, then*

$$\sum_{x \in V(G)} \deg(x) = 2|E(G)|.$$

Proof When summing up the degrees of the vertices in a graph, each edge is counted twice since it has two distinct endpoints. □

Given any graph G, we can partition its vertex set $V(G)$ into the sets

$$U =: \{x \in V(G) \mid \deg(x) \text{ is even}\} \quad \text{and} \quad W =: \{x \in V(G) \mid \deg(x) \text{ is odd}\}.$$

Theorem 1.2 implies that

$$\sum_{x \in U} \deg(x) + \sum_{x \in W} \deg(x) = 2|E(G)|.$$

Since the first sum is even, the second sum must also be even. The only way that a sum of odd numbers can be even is if there are an even number of terms in the sum. Hence, we obtain the following corollary to Theorem 1.2.

Corollary 1.1 *Every graph contains an even number of vertices of odd degree.*

Two graphs G and G' are *isomorphic* if there exists a bijection $\phi : V(G) \longrightarrow V(G')$ such that $xy \in E(G)$ if and only if $\phi(x)\phi(y) \in E(G')$. When such a bijection exists, we call the map ϕ an *isomorphism* and write $G \cong G'$. Given a graph G, the *complement* of G, denoted $\overline{G}$, is the graph with vertex set $V(\overline{G}) = V(G)$ and edge set

$$E(\overline{G}) := \{xy \mid x, y \in V(\overline{G}) \text{ and } xy \notin E(G)\}.$$

A graph is called *self-complementary* if it is isomorphic to its complement.

The *complete graph* K_n is the graph of order n in which every distinct pair of vertices is an edge. Note that $|E(K_n)| = \binom{n}{2} = \frac{n(n-1)}{2}$. A *path* is a sequence of distinct vertices $x_1x_2\cdots x_n$ such that every consecutive pair of vertices form an edge. We refer to such a path as an $x_1 - x_n$ path. When x_1x_n is also an edge, we call the sequence $x_1x_2\cdots x_nx_1$ a *cycle*. We write P_n for a path of order n and C_n for a cycle of order n. The *length* of a path or cycle is the number of edges contained in that path or cycle. So, P_n has length $n-1$ while C_n has length n.

Given a graph G, a *subgraph* H is a graph such that $V(H) \subseteq V(G)$ and $E(H) \subseteq E(G)$. An *H-subgraph* of a graph G is a subgraph of G that is isomorphic to H. For any nonempty subset $S \subseteq E(G)$, the subgraph of G *spanned* by the edges in S is the subgraph with vertex set $V(G)$ and edge set S. The subgraph $G - S$ is the graph with vertex set $V(G)$ and edge set $E(G) \setminus S$. When $S = \{e\}$, we simplify this notation to $G - e$. A *spanning subgraph* of a graph G is a subgraph H that satisfies $V(H) = V(G)$. In this sense, a spanning cycle (path) in a graph G is a cycle (or path) that spans G. This is also called a *Hamiltonian cycle (path)*.

Given a graph G and a nonempty subset $S \subseteq V(G)$, the subgraph *induced* by G, denoted $G[S]$, has vertex set $V(G[S]) = S$ and edge set

$$E(G[S]) = \{xy \in E(G) \mid x, y \in S\}.$$

The graph $G - S$ is the subgraph of G induced by $V(G) \setminus S$. If $S = \{x\}$, we simplify this notation to $G - x$.

The main partitions of graphs that we will consider are paritions of their edge sets. So, a *t-coloring of a graph* G is a t-coloring of its edge set $E(G)$. Given a t-coloring of a graph G and $x \in V(G)$, let G_i be the subgraph of G of order $|V(G)|$, spanned by the edges in color i, write $\deg_i(x)$ for the degree of x in G_i, and write $N_i(x)$ for the neighborhood of x in G_i. For small values of t, we will replace 1, 2, 3, 4, and 5 with red, blue, green, purple, and orange, respectively. In this situation, we replace the indices 1, 2, 3, 4, and 5 with r, b, g, p, and o, respectively. When we consider subgraphs spanned by edges in more than one color, we use similar notations. For example, we write G_{rb} and $\deg_{rb}(x)$ for the subgraph of G of order $|V(G)|$, spanned by the red and blue edges, and the degree of x in G_{rb}, respectively.

If for every distinct pair of vertices x and y in a graph G, there exits an $x - y$ path, then G is called *connected*. Otherwise, G is *disconnected* and its largest connected subgraphs are called its *connected components*. If x and y are distinct vertices in a graph G, then the minimum length among all $x - y$ paths is called the *distance* from x and y, and is denoted $d_G(x, y)$. If no such path exists, we write $d_G(x, y) = \infty$. By convention, we set $d_G(x, x) = 0$. A path from x to y that has length $d_G(x, y)$ is called an $x - y$ *geodesic*. The following theorem concerning the connectivity of a graph and its complement is usually credited to Erdős and Rado (e.g., see Theorem 1.1 of [22]).

Theorem 1.3 (Erdős and Rado) *If a graph G is disconnected, then $\overline{G}$ is connected.*

Proof Suppose that G is disconnected and its connected components are labeled $G_1, G_2, \ldots, G_k$, where $k \geq 2$. Select two distinct vertices x and y. If x and y are in the same component G_i, then pick a vertex z from a different component G_j. Then xzy is an $x - y$ path in $E(\overline{G})$. If x and y are in different components, then the edge xy is an $x - y$ path in $E(\overline{G})$. Since x and y were chosen arbitrarily, it follows that $\overline{G}$ is connected. □

From the proof of Theorem 1.3, we can observe that if G is disconnected, then $d_{\overline{G}}(x, y) \in \{1, 2\}$ for every distinct pair of vertices x and y.

A graph G is called *bipartite* if $V(G)$ can be partitioned into two sets V_1 and V_2 such that every edge in $E(G)$ joins a vertex in V_1 to a vertex in V_2. Here, V_1 and V_2 are called the *partite sets* of G. The *compete bipartite graph* $K_{m,n}$ is the bipartite graph with partite sets satisfying $|V_1| = m$, $|V_2| = n$, and in which every edge that joins a vertex in V_1 with a vertex in V_2 is included in its edge set (e.g., see the first image in Figure 1.1). Note that

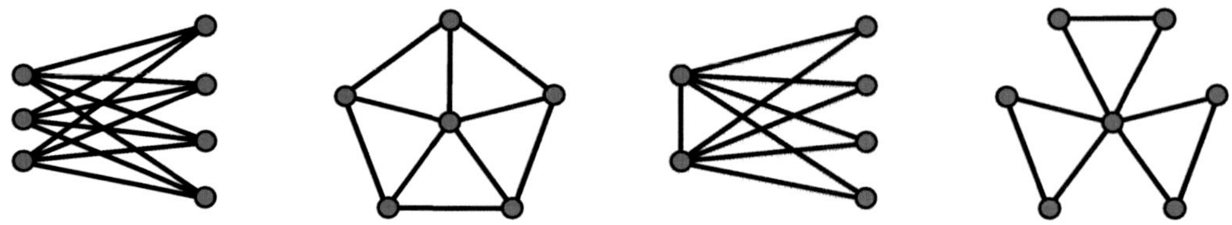

FIGURE 1.1: The graphs $K_{3,4}$, W_6, B_4, and F_3.

$|E(K_{m,n})| = mn$. A complete bipartite graph of the form $K_{1,n}$, where $n \geq 1$, is called a *star*.

The following well-known theorem (e.g., see Theorem 2.4 of [120]) describes a useful property of bipartite graphs. From this result, it follows that wheels, books, and fans are not bipartite since they all contain cycles of odd length.

Theorem 1.4 *A graph is bipartite if and only if all of its cycles have even length.*

Proof First, assume that G is bipartite and $x_1x_2 \cdots x_nx_1$ is a cycle in G. Without loss of generality, assume that $x_1 \in V_1$. Then $x_2 \in V_2$, $x_3 \in V_1$, ... Every vertex with an odd index is in V_1 and every vertex with an even index is in V_2. Since x_n is adjacent to x_1, it follows that $x_n \in V_2$, implying that n is even.

Now, assume that G is a graph in which every cycle has even length. It suffices to prove that G is bipartite when G is connected since a graph is connected if and only if all of its connected components are connected. Select a vertex x and define the sets

$$V_1 := \{y \in V(G) \mid d_G(x, y) \text{ is even}\} \quad \text{and} \quad V_2 := \{y \in V(G) \mid d_G(x, y) \text{ is odd}\}.$$

Note that $x \in V_1$. We must prove that no two vertices in V_1 are adjacent and no two vertices in V_2 are adjacent. Suppose that two vertices y and z are in V_1 and $yz \in E(G)$. Let P be an $x - y$ geodesic and let Q be an $x - z$ geodesic. Let w be the last vertex that P and Q have in common (which could be x). The length of the $w - y$ subpath P' of P and the length of the $w - z$ subpath Q' of Q must have the same parity, otherwise, P and Q would have different parities. Now a cycle of odd length can be formed using P', Q', and edge yz, which is a contradiction. Hence, no two vertices in V_1 are adjacent. A similar proof can be used to show that no two vertices in V_2 are adjacent, and hence, G is bipartite. □

Given two graphs G_1 and G_2, the *union* of G_1 and G_2, denoted $G_1 \cup G_2$, is the graph with vertex set $V(G_1 \cup G_2) = V(G_1) \cup V(G_2)$ and edge set $E(G_1 \cup G_2) = E(G_1) \cup E(G_2)$. If G is any graph, then nG denotes the union of n disjoint copies of G. The *join* of G_1 and G_2, denoted $G_1 + G_2$, is the graph with vertex set $V(G_1 + G_2) = V(G_1) \cup V(G_2)$ and edge set

$$E(G_1 + G_2) := E(G_1) \cup E(G_2) \cup \{xy \mid x \in V(G_1) \text{ and } y \in V(G_2)\}.$$

For each $n \geq 4$, the *wheel* of order n is defined by $W_n := K_1 + C_{n-1}$ (e.g., the second image in Figure 1.1 shows W_6). In this definition, the K_1-subgraph is called the *hub* of the wheel. For $n \geq 1$, define the *book* $B_n := K_2 + \overline{K_n} = K_2 + nK_1$ (e.g., the third image in Figure 1.1 shows B_4). In this definition, the K_2-subgraph is called the *spine* of the book and the K_1-subgraphs are called the *pages* of the book. For $n \geq 1$, define the *fan* $F_n := K_1 + nK_2$

(e.g., the fourth image in Figure 1.1 shows F_3). In this definition, the K_1-subgraph is called the *center* of the fan and the K_2-subgraphs are called the *blades* of the fan. These graphs satisfy $|V(W_n)| = n$, $|E(W_n)| = 2n$, $|V(B_n)| = n+2$, $|E(B_n)| = 2n+1$, $|V(F_n)| = 2n+1$, and $|E(F_n)| = 3n$. Note that some authors define W_n to be K_1+C_n rather than K_1+C_{n-1}, but we prefer for the index to indicate the order of the wheel.

A graph G is called *n-regular* if all of its vertices have degree n. More generally, a *regular graph* is one in which all of the vertices have the same degree. Note that cycles are 2-regular and the complete graph K_n is $(n-1)$-regular.

A *tree* is a connected graph that does not contain any cycles (e.g., paths and stars are both examples of trees). A tree is *minimally connected* in the sense that it is connected, but removing any edge (while retaining all of its vertices) results in a disconnected graph. A vertex of degree 1 in a tree is called a *leaf.* It is a simple exercise to prove that every tree contains at least one leaf. Also, if a tree has order n, then it has size $n-1$. Since trees lack cycles of odd length, Theorem 1.4 implies that every tree with order at least 2 is bipartite.

An *independent set of vertices* in a graph G is a nonempty subset $I \subseteq V(G)$ such that no two distinct vertices in I are adjacent. The maximum cardinality of an independent set of vertices in a graph G is called the *independence number* of G and is denoted $\alpha(G)$. The order of a largest complete subgraph of a graph G is called the *clique number* of G, and is denoted $\omega(G)$. Observe that $\alpha(G) = \omega(\overline{G})$ and $\omega(G) = \alpha(\overline{G})$.

An *independent set of edges* in a graph G is a subset $I \subseteq E(G)$ such that no two edges in I are adjacent. An independent set of edges is also called a *matching.* If a mathcing has size $n \geq 1$, then it is usually denoted nK_2. A *perfect matching* (also known as a *1-factor*) in a graph G is a matching that spans G. In order for G to have a perfect matching, $|V(G)|$ must be even.

A *factor* of a nonempty graph G is a spanning subgraph H of G (i.e., $V(H) = V(G)$) that is not an empty graph. A *factorization of* G is a collection of factors $G_1, G_2, \ldots, G_n$ such that

$$E(G) = \bigcup_{i=1}^{n} E(G_i) \quad \text{and} \quad E(G_i) \cap E(G_j) = \emptyset \text{ when } i \neq j.$$

In other words, a factorization of G corresponds with a partition of $E(G)$. Here, we say that G can be *factored* into $G_1, G_2, \ldots, G_n$. We now consider two well-known theorems that describe useful factorizations of complete graphs (e.g., see [120] and [151]).

Theorem 1.5 *For all $n \in \mathbb{N}$, the complete graph K_{2n} can be factored into $2n-1$ perfect matchings.*

Proof Denote the vertices in K_{2n} by $x, v_0, v_1, \ldots, v_{2n-2}$. We will partition the edge set into $2n-1$ pairwise disjoint subsets $S_0, S_1, \ldots, S_{2n-2}$, each of which forms a perfect matching. Note that every edge in K_{2n} has one of the forms

$$\begin{aligned} &v_iv_j \quad \text{where } i,j \in \{0,1,\ldots,2n-2\} \text{ and } i \neq j, \text{ and} \\ &xv_i \quad \text{where } i \in \{0,1,\ldots,2n-2\}. \end{aligned}$$

Let $v_iv_j \in S_r$ if and only if $i+j \equiv r \pmod{2n-1}$ and let $xv_i \in S_r$ if and only if $2i \equiv r \pmod{2n-1}$. Note that every edge is assigned to one and only one of the sets S_r. We now show that the edges in S_r form a perfect matching for every r (i.e., we show that every vertex is the endpoint of one and only one edge in S_r). For a vertex v_i, either v_iv_j or xv_i is the only edge in S_r that includes v_i, based on whether the only solution j $(0 \leq j \leq 2n-2)$ to $i+j \equiv r \pmod{2n-1}$ satisfies $i \equiv j \pmod{2n-1}$ or $i \not\equiv j \pmod{2n-1}$. Also, if we consider x, then xv_i is the only edge in S_r that includes the vertex x. □

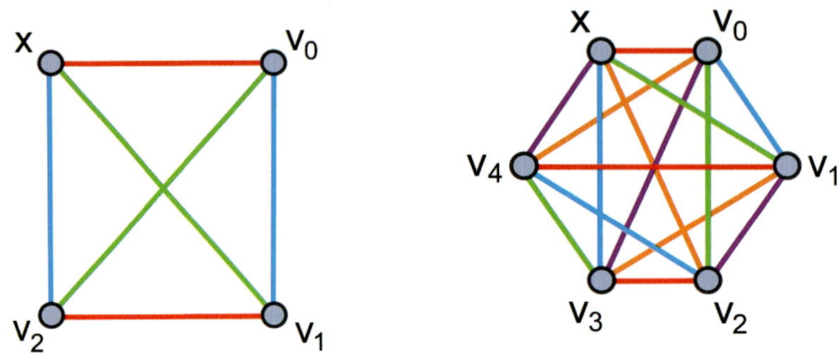

FIGURE 1.2: Factoring K_4 and K_6 into perfect matchings.

For example, in Figure 1.2, the factorizations of K_4 and K_6 into perfect matchings is shown, following the construction given in the proof of Theorem 1.5.

Theorem 1.6 *For all $n \in \mathbb{N}$, the complete graph K_{2n+1} factors into n Hamiltonian cycles.*

Proof Denote the vertices in K_{2n+1} by $x, v_0, v_1, \ldots, v_{2n-1}$ and construct n paths P_i $(1 \leq i \leq n)$ on $\{v_0, v_1, \ldots, v_{2n-1}\}$ in the following way:

$$P_i : v_i v_{i-1} v_{i+1} v_{i-2} \cdots v_{i+n-1} v_{i-n}.$$

Here, the jth vertex in P_i is v_k, where $k = i + (-1)^{j+1} \lfloor \frac{j}{2} \rfloor$, with all subscripts reduced to one of $0, 1, \ldots, 2n-1$ modulo $2n$. For example, when $n = 3$, we obtain the 3-coloring of K_6 shown in Figure 1.3. The Hamiltonian cycle C_i can then be formed by joining x to the

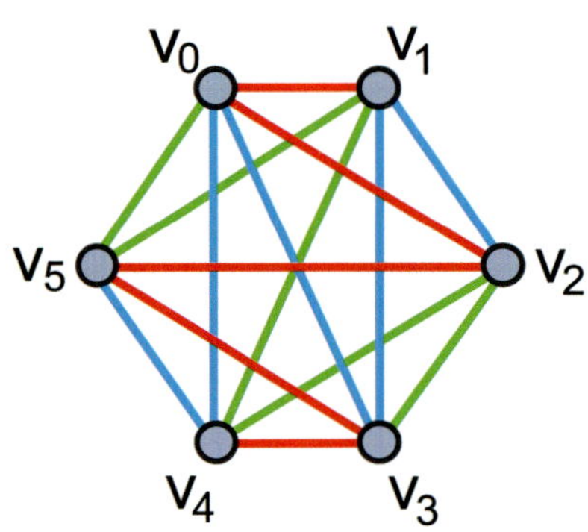

FIGURE 1.3: Factoring K_6 into Hamiltonian paths.

endpoints of P_i with edges in color i. □

In the process of proving Theorem 1.6, we showed the following corollary.

Corollary 1.2 *For all $n \in \mathbb{N}$, the complete graph K_{2n} can be factored into n Hamiltonian paths.*

Given two graphs G_1 and G_2 and a vertex $x \in V(G_1)$, we must introduce the concept of *replacing x with G_2*. The result is a graph with vertex set $(V(G_1) \setminus \{x\}) \cup V(G_2)$ and edge set

$$\{yz \mid y, z \in V(G_1) \setminus \{x\} \text{ and } yz \in E(G_1)\} \cup \{uv \mid u, v \in V(G_2) \text{ and } uv \in E(G_2)\}$$
$$\cup\{uy \mid y \in V(G_1) \setminus \{x\},\ u \in V(G_2), \text{ and } xy \in E(G_1)\}.$$

Here, G_1 is called the *base graph* and G_2 is called a *block*. We will often find ourselves replacing a vertex (or multiple vertices) in a t-coloring of a graph G_1 with a t-coloring of G_2. Edges occurring in the first set of edges are colored according to their colors in G_1, edges occurring in the second set of edges are colored according to their colors in G_2, and the edge uy in the third set of edges is colored according to the color of xy in G_1.

If one replaces all of the vertices in G_1 with copies of G_2, then the resulting graph is called the *lexicographic product* of G_1 with G_2, and is denoted $G_1[G_2]$. It has order $|V(G_1)|\cdot|V(G_2)|$ and size $|V(G_1)| \cdot |E(G_2)| + |E(G_1)| \cdot |V(G_2)|^2$. In general, the lexicographic product is not commutative: $G_1[G_2] \neq G_2[G_1]$. It is a particularly useful operation in Ramsey theory as it satisfies the properties

$$\alpha(G_1[G_2]) = \alpha(G_1)\alpha(G_2) \quad \text{and} \quad \omega(G_1[G_2]) = \omega(G_1)\omega(G_2).$$

In 1941, Pál Turán [203] proved the following theorem, leading to the introduction of Turán numbers.

Theorem 1.7 (Turán's Theorem [203]) *If a graph G satisfies $\omega(G) = p \geq 2$, then*

$$|E(G)| \leq \left\lfloor \left(1 - \frac{1}{p}\right) \frac{|V(G)|^2}{2} \right\rfloor.$$

Proof We proceed by induction on $n = |V(G)|$. If $n \leq p$, then it is not possible for G to contain a K_{p+1}-subgraph. The maximum number of edges in such a graph G is

$$\frac{n(n-1)}{2} = \frac{n^2}{2} - \frac{n}{2} \leq \frac{n^2}{2} - \frac{n}{p} \cdot \frac{n}{2} = \left(1 - \frac{1}{p}\right) \frac{n^2}{2}.$$

Now suppose that the theorem is true for all graphs having fewer than n vertices and let G be a graph of order n satisfying $\omega(G) = p < n$. Then G contains a K_p-subgraph, but not a K_{p+1}-subgraph. Let X be the vertex set for a K_p-subgraph in G and note that

$$|E(G[X])| = \frac{p(p-1)}{2}.$$

Let $Y = V(G) - X$ and apply the inductive hypothesis to obtain

$$E(G[Y])| \leq \left(1 - \frac{1}{p}\right) \frac{(n-p)^2}{2}.$$

Denote by $E_{X,Y}$ the set of edges that join vertices in X to vertices in Y. In order to avoid a K_{p+1}-subgraph, each vertex in Y joins to X using at most $p-1$ edges. Hence,

$$|E_{X,Y}| \leq (n-p)(p-1).$$

Thus, the total number of edges in G satisfies

$$\begin{aligned}
|E(G)| &\leq \frac{p(p-1)}{2} + \left(1 - \frac{1}{p}\right) \frac{(n-p)^2}{2} + (n-p)(p-1) \\
&= (p-1)\left(\frac{p}{2} + \frac{(n-p)^2}{2p} + n - p\right) \\
&= (p-1)\left(\frac{p^2 + n^2 - 2np + p^2 + 2p(n-p)}{2p}\right) \\
&= \left(1 - \frac{1}{p}\right) \frac{n^2}{2}.
\end{aligned}$$

Since $|E(G)| \in \mathbb{N}$, we can apply the floor function to obtain the statement given in the theorem. □

Given graphs G and H, we say that G is *H-free* if H is not a subgraph of G. The *Turán number $ex(n,H)$* is defined to be the maximum number of edges in any H-free graph of order n. From Theorem 1.7, it follows that

$$ex(n, K_p) \leq \left\lfloor \left(1 - \frac{1}{p-1}\right) \frac{n^2}{2} \right\rfloor.$$

Although we will usually consider partitions of the edge sets of graphs, at times, we will also consider partitions of vertex sets. A *proper vertex coloring* of a graph G is a map $f : V(G) \longrightarrow \{1, 2, \ldots, t\}$ such that $f(x) \neq f(y)$ whenever $xy \in E(G)$. Here, we identify the elements in $\{1, 2, \ldots, t\}$ with colors. The minimum number of colors needed to properly vertex color G is called the *chromatic number* of G, and is denoted by $\chi(G)$. Among all proper vertex colorings of G that use exactly $\chi(G)$ colors, denote by $s(G)$ the minimum number of vertices in a color class. Here, $s(G)$ is called the *chromatic surplus* of G.

It will be necessary for us to be able to argue that a connected graph contains a path of a certain length. The following theorem is Exercise 7 of Chapter 1 in Diestel's book [78]. It has a proof similar to that of Dirac's theorem [79].

Theorem 1.8 ([78]) *If G is a connected graph, then it has a path of length* $\min\{2\delta(G), |V(G)| - 1\}$.

Proof Let $P : x_0x_1 \cdots x_k$ be a path of maximal length in a connected graph G. If the length of P is $|V(G)| - 1$, then P is a Hamiltonian path and we are done. So, assume that the length of P is less than $|V(G)| - 1$ (i.e., not all of the vertices in G are in the path P). Then $N_G(x_0)$ and $N_G(x_k)$ are both subsets of $V(P)$, otherwise, a longer path could be formed. Consider the set of ordered pairs $S = \{(x_0, x_1), (x_1, x_2), \ldots, (x_{k-1}, x_k)\}$ and note that $|S| = k$.

Suppose that $k < 2\delta(G)$. Since x_0 and x_k both have at least $\delta(G)$ neighbors in $V(P)$, the pigeonhole principle implies that there exists $(x_i, x_{i+1}) \in S$ such that x_0 is adjacent to x_{i+1} and x_k is adjacent to x_i. Then $x_0x_1 \cdots x_ix_kx_{k-1} \cdots x_{i+1}x_0$ forms a cycle of length k. Since G is connected, there exists a vertex y such that yx_ℓ is an edge for some $\ell \in \{0, 1, \ldots, k\}$. The cycle of length k, along with vertex y, form a path of length $k+1$, which is a contradiction. Hence, $k \geq 2\delta(G)$. □

One way of generalizing the concept of a graph is to allow for edges to contain any number of vertices (not just two), leading to the concept of a hypergraph. In order to maintain a suitable level of structure, we restrict our attention to uniform hypergraphs. A *k-uniform hypergraph* $H = (V, E)$ consists of a nonempty set $V = V(H)$ of vertices and a set $E = E(H)$ of *hyperedges*, each of which is a distinct k-element subset of V. When $k = 2$, this definition coincides with that of a graph. In general, we assume that $k \geq 2$, however, many results often still hold when $k = 1$. Denote by $K_n^{(k)}$ the *complete k-uniform hypergraph of order n*. Its hyperedge set consists of all k-element subsets of its n-element vertex set. As a general reference on hypergraphs, it is recommended that the reader consult the text by Claude Berge [20].

Most of the terminology and concepts surrounding Ramsey numbers for graphs extend to hypergraphs in a natural way. Given k-uniform hypergraphs $H_1, H_2, \ldots, H_t$, the *hypergraph Ramsey number* $r(H_1, H_2, \ldots, H_t; k)$ is defined to be the least $n \in \mathbb{N}$ such that every t-coloring of the hyperedges of $K_n^{(k)}$ contains a copy of H_i in color i, for some $i \in \{1, 2, \ldots, t\}$. When $H_1 = H_2 = \cdots = H_t$, we simplify the notation to $r^t(H; k)$. Generalizations of some known Ramsey theorems to the setting of hypergraphs can be found in [42].

1.2 Algebraic and Number-Theoretic Prerequisites

In this section, we review some of the standard definitons and concepts from algebra and number theory that will be used in the constructions of various lower bounds for chromatic and weakened Ramsey numbers. Since this book focuses on a topic in discrete mathematics, we refrain from proving the well-known theorems given in this section. More details can be found in the texts by Gallian [103] and Artin [9]. For background that is more number-theoretic, the reader is referred to the texts by Fröhlich and Taylor [98], Ireland and Rosen [129], and Lemmermeyer [158].

A *group* $(\mathfrak{G}, *)$ consists of a non-empty set $\mathfrak{G}$ and an operation $*$ such that the following statements are true:

1. $a * b \in \mathfrak{G}$, for all $a, b \in \mathfrak{G}$,
2. $a * (b * c) = (a * b) * c$, for all $a, b, c \in \mathfrak{G}$,
3. there exists $e \in \mathfrak{G}$ such that $e * g = g = g * e$, for all $g \in \mathfrak{G}$, and
4. for every $g \in \mathfrak{G}$, there exists $g^{-1} \in \mathfrak{G}$ such that $g * g^{-1} = g^{-1} * g = e$.

In this definition, Statement 1 is referred to as *closure*, Statement 2 is referred to as *associativity*, the element e in Statement 3 is called an *identity*, and the element g^{-1} in Statement 4 is called an *inverse* of g. It is a simple exercise to prove that the identity of a group is unique, as is the inverse of any element $g \in \mathfrak{G}$. For a general group $(\mathfrak{G}, *)$, we usually replace the operation with multiplication (i.e., we write ab in place of $a * b$). When the operation is understood, we may refer to a group $(\mathfrak{G}, *)$ by just $\mathfrak{G}$. The sets $\mathbb{Z}$, $\mathbb{Q}$, $\mathbb{R}$, and $\mathbb{C}$ all form groups under the operation of addition, as does the set $\mathbb{Z}_m := \{0, 1, \ldots, m-1\}$ under the operation of addition modulo m.

The *order of a group* $\mathfrak{G}$, denoted $|\mathfrak{G}|$, is the number of elements in $\mathfrak{G}$ (which may be infinite). If $g \in \mathfrak{G}$, then the *order of* g, denoted $|g|$, is the least natural number n such that $g^n = e$. If no such n exists, then we write $|g| = \infty$.

If $\mathfrak{G}$ is a group and $\mathfrak{H} \subseteq \mathfrak{G}$ that forms a group under the same operation as that of $\mathfrak{G}$, then $\mathfrak{H}$ is a *subgroup* of $\mathfrak{G}$. For $g \in \mathfrak{G}$, define

$$\langle g \rangle := \{g^n \mid n \in \mathbb{Z}\}.$$

It is easily confirmed that $\langle g \rangle$ is a subgroup of $\mathfrak{G}$, called the *cyclic subgroup of* $\mathfrak{G}$ *generated by* g. If a group $\mathfrak{G}$ satisfies $\mathfrak{G} = \langle g \rangle$, for some $g \in \mathfrak{G}$, then we say that $\mathfrak{G}$ is a *cyclic group with generator* g. The generator of a cyclic group need not be unique.

The relationship between the order of a group and the order of a subgraph is described by Lagrange's theorem.

Theorem 1.9 (Lagrange's theorem) *If $\mathfrak{H}$ is a subgroup of a finite group $\mathfrak{G}$, then $|\mathfrak{H}|$ divides $|\mathfrak{G}|$.*

If $\mathfrak{G}_1$ and $\mathfrak{G}_2$ are groups, then a *group homomorphism* $\psi : \mathfrak{G}_1 \longrightarrow \mathfrak{G}_2$ is a function that satisfies $\psi(ab) = \psi(a)\psi(b)$, for all $a, b \in \mathfrak{G}_1$. The *kernel* of ψ is the set

$$\mathrm{Ker}(\psi) := \{g \in \mathfrak{G}_1 \mid \psi(g) = e_{\mathfrak{G}_2}\}.$$

That is, the kernel is the preimage of the identity element in $\mathfrak{G}_2$. It is a simple exercise to check that $\mathrm{Ker}(\psi)$ is a subgroup of $\mathfrak{G}_1$. If a group homomorphism ψ is bijective, then it is called a *group isomorphism*. If there exists a group isomorphism $\psi : \mathfrak{G}_1 \longrightarrow \mathfrak{G}_2$, then we say that $\mathfrak{G}_1$ and $\mathfrak{G}_2$ are *isomorphic* and we write $\mathfrak{G}_1 \cong \mathfrak{G}_2$.

A *ring* $(\mathfrak{R}, +, \cdot)$ consists of a nonempty set $\mathfrak{R}$ and operations $+$ and $\cdot$, called *addition* and *multiplication*, respectively, such that the following conditions hold:

1. $(\mathfrak{R}, +)$ forms a group,
2. $a \cdot (b \cdot c) = (a \cdot b) \cdot c$, for all $a, b, c \in \mathfrak{R}$, and
3. $a \cdot (b + c) = a \cdot b + a \cdot c$ and $(a + b) \cdot c = a \cdot c + b \cdot c$, for all $a, b, c \in \mathfrak{R}$.

If $a \cdot b = b \cdot a$ for all $a, b \in \mathfrak{R}$, then we say that $\mathfrak{R}$ is a *commutative ring*. The identity in the group $(\mathfrak{R}, +)$ is denoted by 0. If a ring $\mathfrak{R}$ has a multiplicative identity, then we denote it by 1 (i.e., $1 \cdot a = a \cdot 1 = a$, for all $a \in \mathfrak{R}$). When a ring has a 1, we can consider whther or not an element $a \in \mathfrak{R}$ has a multiplicative inverse (i.e., an element $a^{-1} \in \mathfrak{R}$ such that $a \cdot a^{-1} = a^{-1} \cdot a = 1$). If an element has a mutiplicative inverse, then it is called a *unit*. To simplify notation, we usually write ab in place of $a \cdot b$, and $\mathfrak{R}$ in place of $(\mathfrak{R}, +, \cdot)$.

The canonical example of a ring is $\mathbb{Z}$, along with regular addition and multiplication. Another fundamental property of $\mathbb{Z}$ is described by the division algorithm.

Theorem 1.10 (The Division Algorithm) *If $m, n \in \mathbb{Z}$ such that $n \neq 0$, then there exists unique $q, r \in \mathbb{Z}$ such that $m = nq + r$, where $0 \leq r < n$.*

The set $\mathbb{Z}_m$ also forms a ring under the operations of addition and multiplication modulo m. It is a simple exercise to check that the units in $\mathbb{Z}_m$ are the nonzero elements that are relatively prime to m. *Euler's totient function* $\varphi(m)$ is the number of units in $\mathbb{Z}_m$. In particular, when p is a prime number, $\varphi(p) = p - 1$.

If $\mathfrak{R}_1$ and $\mathfrak{R}_2$ are rings, then a *ring homomorphism* is a map $\psi : \mathfrak{R}_1 \longrightarrow \mathfrak{R}_2$ such that

$$\psi(a + b) = \psi(a) + \psi(b) \quad \text{and} \quad \psi(ab) = \psi(a)\psi(b),$$

for all $a, b \in \mathfrak{R}_1$. If ψ is bijective, then it is called a *ring isomorphism*. If there exists a ring isomorphism $\psi : \mathfrak{R}_1 \longrightarrow \mathfrak{R}_2$, then we say that $\mathfrak{R}_1$ and $\mathfrak{R}_2$ are *isomorphic* and we write $\mathfrak{R}_1 \cong \mathfrak{R}_2$.

A *field* is a commutative ring with a 1 in which every nonzero element has a multiplicative inverse (i.e., for every $a \neq 0$, there exists a^{-1} such that $aa^{-1} = a^{-1}a = 1$). Note that certain classical texts, in particular, many that were written by French mathematicians, do not assume that fields are commutative (e.g., see [209]). If $\mathbb{F}$ is a field, then it is easily checked that $\mathbb{F}^{\times} := \mathbb{F} - \{0\}$ forms a group under the operation of multiplication. The group $\mathbb{F}^{\times}$ is called the *multiplicatve group of* $\mathbb{F}$. Examples of fields include $\mathbb{Q}$, $\mathbb{R}$, and $\mathbb{C}$. When p is a prime number, $\mathbb{Z}_p$ forms a field under the operations of addition and multiplication modulo p.

The following two theorems are standard, so we refrain from proving them here. They can be found in Section 7.1 of Ireland and Rosen's text [129].

Theorem 1.11 *Every finite field has an order that is a prime power. For every prime p and every $f \in \mathbb{N}$, there exists a unique (up to isomorphism) finite field with p^f elements.*

Due to this theorem, we may refer to "the" finite field $\mathbb{F}_q$, where $q = p^f$ for some prime p and $f \in \mathbb{N}$. When p is a prime number, $\mathbb{F}_p \cong \mathbb{Z}_p$. When $q = p^f$ for $f > 1$, the finite field $\mathbb{F}_q$ is constructed by considering the factor ring of polynomials with coefficients in $\mathbb{F}_p$ modulo an ideal generated by a monic irreducible polynomial of degree f. This construction goes beyond the scope of this text.

Theorem 1.12 *If $\mathbb{F}$ is a finite field, then its multiplicative group $\mathbb{F}^{\times}$ is cyclic.*

A *character* of a group $\mathfrak{G}$ is a group homomorphism $\psi : \mathfrak{G} \longrightarrow \mathbb{C}^\times$. When the order of $\mathfrak{G}$ is n, $\psi(g)$ is an nth root of unity in $\mathbb{C}^\times$, for all $g \in \mathfrak{G}$ since

$$(\psi(g))^n = \psi(g^n) = \psi(e) = 1.$$

The main characters that we will consider are characters of $\mathbb{F}_q^\times$. The set of all characters of $\mathbb{F}_q^\times$ form a group under the operation $\psi_1\psi_2(g) = \psi_1(g)\psi_2(g)$ (see Section 8.1 of [129]). Denote by $\widehat{\mathbb{F}}_q^\times$ the *character group* of $\mathbb{F}_q^\times$. It well-known that $\mathbb{F}_q^\times \cong \widehat{\mathbb{F}}_q^\times$ (e.g., see Proposition 4.18 of [158]). Given a character $\psi : \mathbb{F}_q^\times \longrightarrow \mathbb{C}^\times$, let r_ψ denote the *order* of ψ as an element of $\widehat{\mathbb{F}}_q^\times$.

When q is odd, the group $\mathbb{F}_q^\times$ has a unique element of order 2. It follows that there exists a unique character of order 2 (called the *quadratic character*). When p is an odd prime, the quadratic character for $\mathbb{F}_p^\times$ is called the *Legendre symbol.* If p and p' are distinct odd primes, and ψ_p and $\psi_{p'}$ are their respective Legendre symbols, then they are related by the ubiquitous *law of quadratic reciprocity*:

$$\psi_p(p')\psi_{p'}(p) = (-1)^{(p-1)(p'-1)/4}$$

(see [158] and Section 5.2 of [129]). First proved by Gauss [104] in 1801, this law is perhaps the most proved result in all of mathematics in that there are several hundred "different" proofs that have been published (e.g., see Appendix B of [158]). Along with the supplementary laws

$$\psi_p(-1) = (-1)^{(p-1)/2} \quad \text{and} \quad \psi_p(2) = (-1)^{(p^2-1)/8}, \tag{1.1}$$

and the unique factorization of integers, one obtains an algoirthm for evaluating the Legendre symbol on any element in $\mathbb{F}_p^\times$.

Let $\mathfrak{G}$ be a finite group with identity e. Let $\mathcal{S}$ be a subset of $\mathfrak{G}$ such that $e \notin \mathcal{S}$ and $\mathcal{S} = \mathcal{S}^{-1}$ (i.e., $s \in \mathcal{S}$ if and only if $s^{-1} \in \mathcal{S}$). Define the *Cayley graph* Cay$(\mathfrak{G}; \mathcal{S})$ to be the graph with vertex set $\mathfrak{G}$, and two distinct vertices g and h are joined by an edge if and only if $h = gs$ for some $s \in \mathcal{S}$ (see [6]). Cayley graphs on the cyclic group $\mathbb{Z}_m$ are called *circulant graphs.* For example, Figure 1.4 shows the circulant graphs Cay$(\mathbb{Z}_8; \{\pm1, \pm3\})$ and Cay$(\mathbb{Z}_8; \{\pm2, 4\})$.

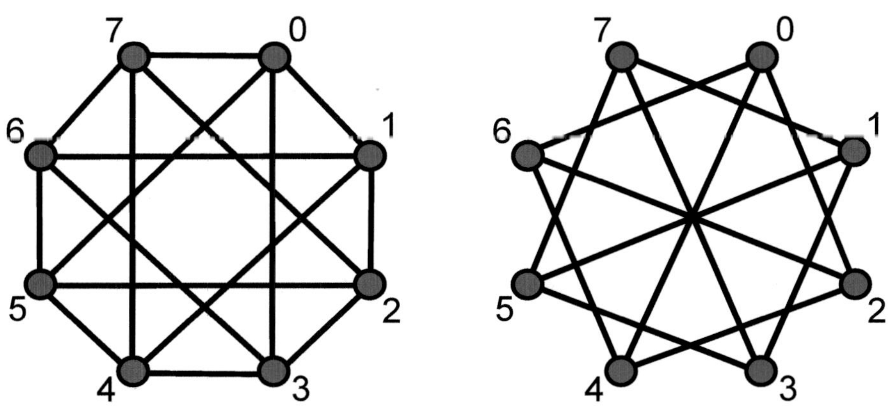

FIGURE 1.4: The Cayley graphs Cay$(\mathbb{Z}_8; \{\pm1, \pm3\})$ and Cay$(\mathbb{Z}_8; \{\pm2, 4\})$.

One important type of Cayley graph encountered in Ramsey theory is known as a Paley graph. In order to define a Paley graph, first note that for any odd prime power q, the set

$$\mathbb{F}_q^{\times 2} := \{g^2 \mid g \in \mathbb{F}_q^\times\}$$

forms a subgroup of $\mathbb{F}_q^\times$. Let $q \equiv 1 \pmod 4$ be a prime power and $\psi_q : \mathbb{F}_q^\times \longrightarrow \mathbb{C}^\times$ be the unique quadratic character on $\mathbb{F}_q^\times$. The *Paley graph* $\mathcal{P}_q$ is the graph with vertex set $V(\mathcal{P}_q) := \mathbb{F}_q$ and edge set

$$E(\mathcal{P}_q) := \{xy \mid \psi_q(y - x) = 1\} = \{xy \mid y - x \in \mathbb{F}_q^{\times 2}\}.$$

The assumption $q \equiv 1 \pmod 4$ guarantees that $-1 \in \mathbb{F}_q^{\times 2}$ (cf., the first supplementary law given in (1.1)), making $\mathcal{P}_q$ well-defined. Here, we have that $\mathcal{P}_q \cong \mathrm{Cay}(\mathbb{F}_q; \mathbb{F}_q^{\times 2})$. If g is any quadratic nonresidue in $\mathbb{F}_q^\times$ (i.e., $g \in \mathbb{F}_q^\times \setminus \mathbb{F}_q^{\times 2}$), then the map $x \mapsto gx$ defines an isomorphism from $\mathcal{P}_q$ to $\overline{\mathcal{P}_q}$. In other words, $\mathcal{P}_q$ is self-complementary.

The Paley graphs $\mathcal{P}_3$ and $\mathcal{P}_5$ are isomorphic to C_3 and C_5, respectively. Figure 1.5 shows the Paley graphs $\mathcal{P}_{13}$ and $\mathcal{P}_{17}$.

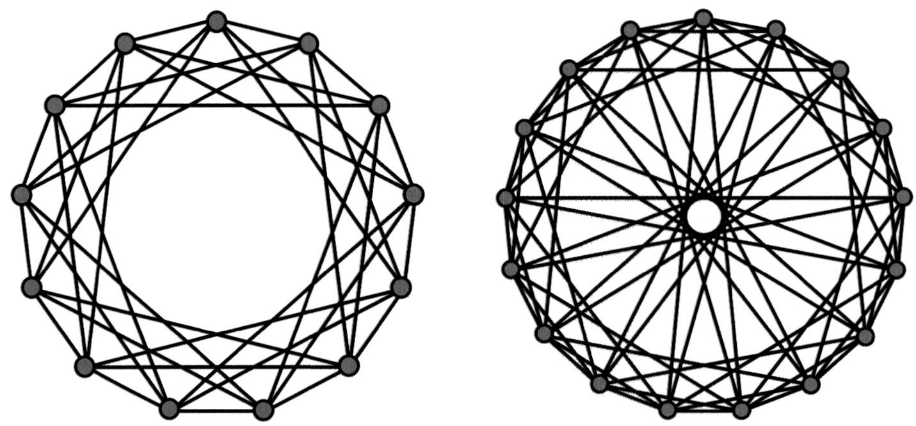

FIGURE 1.5: The Paley graphs $\mathcal{P}_{13}$ and $\mathcal{P}_{17}$.

A *strongly-regular graph* (specifically, an $\mathrm{srg}(n, k, \lambda, \mu)$ graph) is a k-regular graph G of order n such that the following conditions hold:

1. every pair of adjacent vertices have exactly λ common neighbors, and
2. every pair of non-adjacent vertices have exactly μ common neighbors.

It is known that the Paley graph $\mathcal{P}_q$ is an $\mathrm{srg}(q, \frac{q-1}{2}, \frac{q-5}{4}, \frac{q-1}{4})$ graph (e.g., see Chapter VII, Section 11 of [73], which was written by A. Brouwer).

Paley graphs can be further generalized by usng characters with orders greater than 2. Suppose that $q \equiv 1 \pmod{2r_\psi}$, where $\psi : \mathbb{F}_q^\times \longrightarrow \mathbb{C}^\times$ is a character of order r_ψ. Then

$$\mathbb{F}_q^{\times r_\psi} := \{g^{r_\psi} \mid g \in \mathbb{F}_q^\times\}$$

is a subgroup of $\mathbb{F}_q^\times$ that contains -1. Here, the Cayley graph $\mathrm{Cay}(\mathbb{F}_q; \mathbb{F}_q^{\times r_\psi})$ is called a *character difference graph.* One benefit of working with character difference graphs is that enumeration can be carried out using character sums (e.g., see [37], [169], and Chapter 8 of [129]).

1.3 Ramsey Theory for Graphs

There are many great books on Ramsey theory (e.g., see [108], [109], [140], [162], [182], [192], [197], and [210]), so it is not the goal of this text to reproduce an existing exposition on

Ramsey numbers. That said, we attempt to be as comprehensive as possible in developing the background needed to prove the main results in weakened Ramsey theory. In this section, we review some fundamental results concerning Ramsey numbers.

1.3.1 Multicolor Ramsey Numbers

Given graphs $G_1, G_2, \ldots, G_t$, the *t-color Ramsey number* $r(G_1, G_2, \ldots, G_t)$ is the least $n \in \mathbb{N}$ such that every t-coloring of K_n contains a monochromatic subgraph isomorphic to G_i in color i, for some $i \in \{1, 2, \ldots, t\}$. When $G_1 = G_2 = \cdots = G_t$ (the "diagonal" case), we write $r^t(G_1)$ in place of $r(G_1, G_2, \ldots, G_t)$. A t-coloring of $K_{r(G_1,G_2,\ldots,G_t)-1}$ that avoids a monochromatic copy of G_i in color i, for all $i \in \{1, 2, \ldots, t\}$ is called a *critical coloring for* $r(G_1, G_2, \ldots, G_t)$. The determination of $r(G_1, G_2, \ldots, G_t)$ can be very difficult in general, but for specific graphs, many cases have been evaluated. Radziszowski's dynamic survey [189] offers a current overview of the known values.

Since $|V(G)|$ vertices are required to have a G-subgraph, the 1-color Ramsey number is given by $r^1(G) = |V(G)|$. The naming of the "colors" used in a t-coloring of K_n is arbitrary, so it follows that permuting the colors does not affect the value of $r(G_1, G_2, \ldots, G_t)$. For example, when $t = 2$, we have that $r(G_1, G_2) = r(G_2, G_1)$. In the special case where $G_1 = K_1$, the trivial Ramsey number

$$r(K_1, G_2, \ldots, G_t) = 1$$

since the existence of a single vertex is a K_1. Also, a t-coloring of K_n either contains an edge in color 1 or it avoids color 1 altogether. It follows that when $G_1 = K_2$, we have

$$r(K_2, G_2, \ldots, G_t) = r(G_2, \ldots, G_t).$$

In particular, $r(K_2, G_2) = |V(G_2)|$.

From the definiton, it is not immediately clear that the Ramsey number $r(G_1, G_2, \ldots, G_t)$ exists for all selections of graphs $G_1, G_2, \ldots, G_t$. In 1929, Ramsey's ubiquitous paper [190] proved the existence of Ramsey numbers for graphs (and hypergraphs), as well as many other related numbers that we will encounter in this text. In the following theorem, we provide a proof of Ramsey's theorem for graphs.

Theorem 1.13 (Ramsey's Theorem for Graphs [190]) *For* $t, n_i \geq 1$, *the Ramsey number* $r(K_{n_1}, K_{n_2}, \ldots, K_{n_t})$ *is finite.*

Proof When $t = 1$, it is trivial that $r(K_{n_1}) = n_1$. To prove the $t = 2$ case, we first show that

$$r(K_{n_1}, K_{n_2}) \leq r(K_{n_1-1}, K_{n_2}) + r(K_{n_1}, K_{n_2-1}). \tag{1.2}$$

Let

$$m = r(K_{n_1-1}, K_{n_2}) + r(K_{n_1}, K_{n_2-1})$$

and consider a 2-coloring of K_m. If a vertex x satisfies $\deg_r(x) \geq r(K_{n_1-1}, K_{n_2})$, then $N_r(x)$ contains a red K_{n_1-1} or a blue K_{n_2}. In the latter case, we are done, so assume the former case. Including vertex x, along with the red K_{n_1-1} produces a red K_{n_1}. If vertex x satisfies $\deg_r(x) < r(K_{n_1-1}, K_{n_2})$, then

$$\begin{aligned}
\deg_b(x) &= m - 1 - \deg_r(x) \\
&\geq m - 1 - (r(K_{n_1-1}, K_{n_2}) - 1) \\
&= r(K_{n_1}, K_{n_2-1}).
\end{aligned}$$

Hence, $N_b(x)$ either contains a red K_{n_1} or a blue K_{n_2-1}. In the former case, we are done, so assume the latter case. Including vertex x, along with the blue K_{n_2-1} produces a blue K_{n_2}. It follows that $r(K_{n_1}, K_{n_2}) \leq m$, confirming Inequality (1.2). The existence of $r(K_{n_1}, K_{n_2})$ now follows by induction on $n_1 + n_2$ with $r(K_1, K_{n_2}) = 1 = r(K_{n_1}, K_1)$ serving as the base cases.

To see that the theorem is true for $t > 2$, we use induction on $t \geq 2$, with the existence of $r(K_{n_1}, K_{n_2})$ serving as the base case. Suppose that all $(t-1)$-color Ramsey numbers for complete graphs are finite and consider $r(K_{n_1}, K_{n_2}, \ldots, K_{n_t})$. If we group together the first two colors, we find that

$$r(K_{n_1}, K_{n_2}, \ldots, K_{n_t}) \leq r(K_{r(K_{n_1}, K_{n_2})}, K_{n_3}, \ldots, K_{n_t})$$

since having a $K_{r(K_{n_1}, K_{n_2})}$ in colors 1 and 2 guarantees the existence of a K_{n_1} in color 1 or a K_{n_2} in color 2. It follows that all t-color Ramsey numbers are finite. □

Ramsey's theorem may be viewed as a generalization of the pigeonhole principle, which corresponds to the case $t = 1$. When H is a subgraph of G_1, the observation

$$r(H, G_2, \ldots, G_t) \leq r(G_1, G_2, \ldots, G_t)$$

shows that all t-color Ramsey numbers exist, not just those with complete graphs as arguments.

In 1955, Greenwood and Gleason [111] determined some of the first nontrivial 2 and 3-color Ramsey numbers for complete graphs. Interestingly, it appears as though they were unaware of Ramsey's work at the time, as they did not cite him. Using Inequality (1.2) and lower bounds constructed using character difference graphs, Greenwood and Gleason showed that $r^2(K_3) = 6$, $r(K_3, K_4) = 9$, $r^2(K_4) = 18$, $r(K_3, K_5) = 14$, and $r^3(K_3) = 17$. Currently, the only other known Ramsey numbers for complete graphs (i.e., the "classical" Ramsey numbers) are $r(K_3, K_6) = 18$, $r(K_3, K_7) = 23$, $r(K_3, K_8) = 28$, $r(K_3, K_9) = 36$, $r(K_4, K_5) = 25$, and $r(K_3, K_3, K_4) = 30$ (see [189]). In the case of $r^2(K_3)$, the lower bound follows from the critical coloring given in Figure 1.6. It is well-known that up to

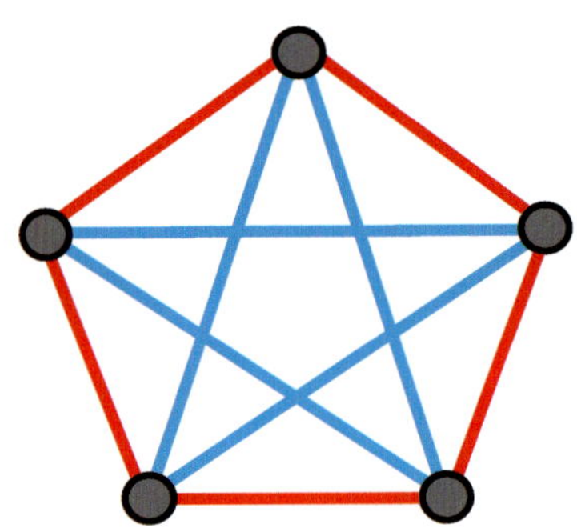

FIGURE 1.6: The critical coloring for $r(K_3, K_3)$.

isomorphism, the critical coloring for $r^2(K_3)$ given in Figure 1.6 is unique (e.g., see [110]). For the sake of completeness, we provide a proof of this fact, incorporating it into the proof that $r^2(K_3) = 6$.

Theorem 1.14 ([111]) *If a 2-coloring of K_5 avoids a monochromatic K_3, then the subgraphs spanned by each of the two colors are C_5-subgraphs. Furthermore, $r^2(K_3) = 6$.*

Proof Consider a 2-coloring of K_5 that avoids a monochromatic K_3. Suppose that a vertex x is incident with at least three edges of the same color. Without loss of generality, assume that xa, xb, and xc are red. If any edge in the subgraph induced by $\{a, b, c\}$ is red, then a red K_3 is formed. The only other possibility is that the subgraph induced by $\{a, b, c\}$ forms a blue K_3. Hence, every vertex must be incident with exactly two red edges and two blue edges. Assume that xa and xb are red and xc and xd are blue. Then ab is blue and cd is red. Now consider two cases.

Case 1: If ac is red, then ad must be blue, bd must be red, and bc must be blue. It follows that $xacdbx$ is a red C_5 and $xcbadx$ is a blue C_5.

Case 2: If ac is blue, then bc must be red, bd must be blue, and ad must be red. It follows that $xadcbx$ is a red C_5 and $xcabdx$ is a blue C_5.

Now, the 2-coloring given in Figure 1.6 implies that $r^2(K_3) \geq 6$. To prove the reverse inequality, consider a 2-coloring of K_6 and select a vertex x. By the pigeonhole principle, at least three of the edges incident with x have the same color. Without loss of generality, suppose that x is incident with at least three red edges. If the subgraph induced by $N_r(x)$ contains a red edge, then that edge, along with vertex x, forms a red K_3. If all edges in this subgraph are blue, then they form a blue K_3. It follows that $r^2(K_3) \leq 6$. □

Theorem 1.15 ([111]) $r(K_3, K_4) = 9$.

Proof Let $S_r := \{\pm 3, \pm 4\}$ and $S_b := \{\pm 1, \pm 2\}$ and factor K_8 into the Cayley graphs $\mathrm{Cay}(\mathbb{Z}_8; S_r)$ and $\mathrm{Cay}(\mathbb{Z}_8; S_b)$, with edges colored red and blue, respectively. The resulting 2-coloring of K_8 is shown in Figure 1.7. Since it avoids a red K_3 and a green K_4, it follows

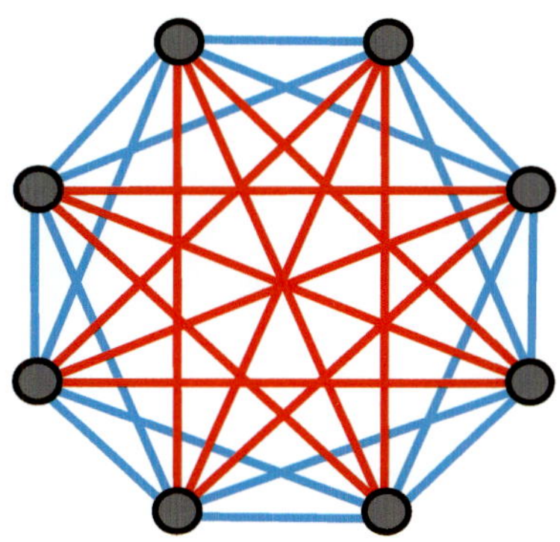

FIGURE 1.7: A critical coloring for $r(K_3, K_4)$.

that $r(K_3, K_4) \geq 9$. To prove the reverse inequality, consider a 2-coloring of K_9. We break the proof into the following cases.

Case 1: Suppose that some vertex x satisfies $\deg_r(x) \geq 4$. If the subgraph induced by $N_r(x)$ contains a red edge, then that edge, along with vertex x forms a red K_3. If no such edge exists, thenthe subgraph induced by $N_r(x)$ is a blue K_4.

Case 2: Suppose that some vertex y satisfies $\deg_b(y) \geq 6$. By Theorem 1.14, the subgraph induced by $N_b(y)$ either contains a red K_3 or a blue K_3. In the latter case, the blue K_3, along with vertex y, forms a blue K_4

Case 3: Suppose that Cases 1 and 2 do not occur. Since $\deg_r(z) + \deg_b(z) = 8$ for every vertex z, every vertex must satisfy $\deg_r(z) = 3$ and $\deg_b(z) = 5$. In this case, the subgraph spanned by the red edges has order 9, with every vertex having degree 3. This contradicts Corollary 1.1, so no such 2-coloring of K_9 exists.

We have shown that every 2-coloring of K_9 contains a red K_3 or a blue K_4. It follows that $r(K_3, K_4) \leq 9$. □

Theorem 1.16 ([111]) $r(K_3, K_5) = 14$.

Proof Let $S_r := \{\pm 1, \pm 5\}$ and $S_b := \{\pm 2, \pm 3, \pm 4, \pm 6\}$ and factor K_{13} into the Cayley graphs $\text{Cay}(\mathbb{Z}_{13}; S_r)$ and $\text{Cay}(\mathbb{Z}_{13}; S_b)$, with edges colored red and blue, respectively. The resulting 2-coloring of K_{13} is shown in Figure 1.8. The red subgraph has clique number 2,

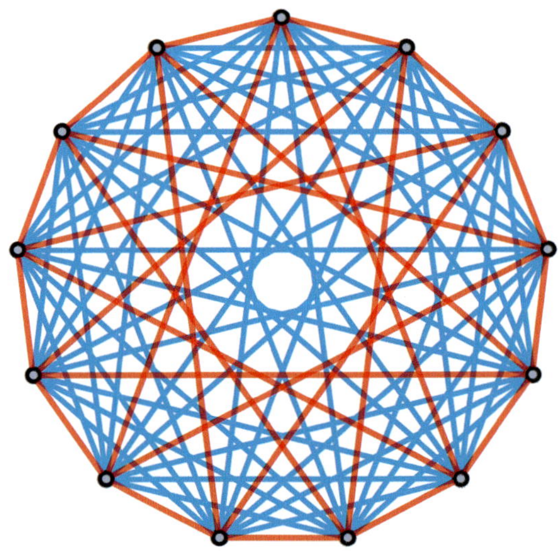

FIGURE 1.8: A critical coloring for $r(K_3, K_5)$.

while the blue subgraph has clique number 4. It follows that $r(K_3, K_5) \geq 14$. To prove the reverse inequality, consider a 2-coloring of K_{14}. We separate the proof into two cases.

Case 1: Assume that some vertex x satisfies $\deg_r(x) \geq 5$. If the subgraph induced by $N_r(x)$ contains a red edge, then that edge, along with vertex x, forms a red K_3. If the subgraph induced by $N_r(x)$ does not contain any red edges, then all of its edges are blue and it contains a blue K_5.

Case 2: Assume that some vertex y satisfies $\deg_b(y) \geq 9$. By Theorem 1.15, the subgraph induced by $N_b(y)$ contains a red K_3 or a blue K_4. In the latter case, the blue K_4, along with vertex y, forms a blue K_5.

Since $\deg_r(z) + \deg_b(z) = 13$ for every vertex z, one of the above cases must occur. It follows that $r(K_3, K_5) \leq 14$. □

Theorem 1.17 ([111]) $r^2(K_4) = 18$.

Proof Figure 1.9 shows the Paley graph $\mathcal{P}_{17}$ in red, and its complement (which is isomorphic to $\mathcal{P}_{17}$) in blue. Although it is not easy to determine by hand, $\omega(P_{17}) = 3$ (this can be shown by enumerating the K_4-subgraphs in Paley graphs, as was done in Corollary 6 of [10]). Thus, no monochromatic K_4 exists in the 2-coloring given in Figure 1.9. It follows that $r^2(K_4) \geq 18$.

To prove the reverse inequality, consider a 2-coloring of K_{18} and select a vertex x. By the pigeonhole principle, x must be incident with at least 9 edges that are the same color. Without loss of generality, assume that x is incident with at least 9 red edges. By Theorem 1.15, $N_r(x)$ contains a red K_3 or a blue K_4. In the former case, the red K_3, along with vertex x, forms a red K_4. It follows that $r^2(K_4) \leq 18$. □

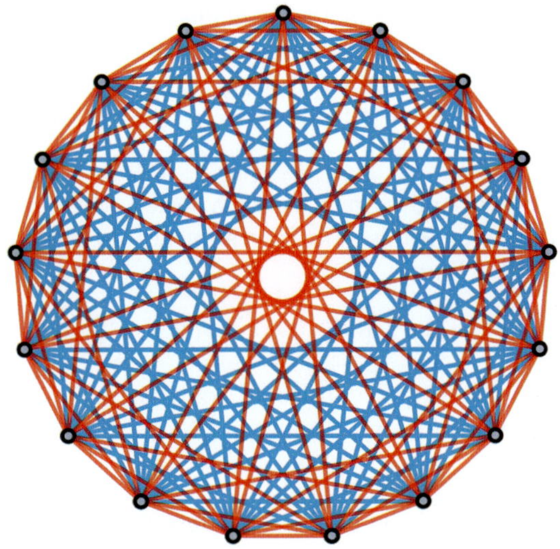

FIGURE 1.9: A critical coloring for $r^2(K_4)$.

Theorem 1.18 ([111]) $r^3(K_3) = 17$.

Proof Let $\psi : \mathbb{F}_{16}^{\times} \longrightarrow \mathbb{C}^{\times}$ be a character of order 3 (i.e., a *cubic character*) and suppose that $\mathbb{F}_{16}^{\times} = \langle g \rangle$. The character difference graph $\mathrm{Cay}(\mathbb{F}_{16}; \mathbb{F}_{16}^{\times 3})$ is often referred to as the *Greenwood and Gleason graph* (see Figure 1.10). One can confirm that $\omega(\mathrm{Cay}(\mathbb{F}_{16}; \mathbb{F}_{16}^{\times 3})) = 2$.

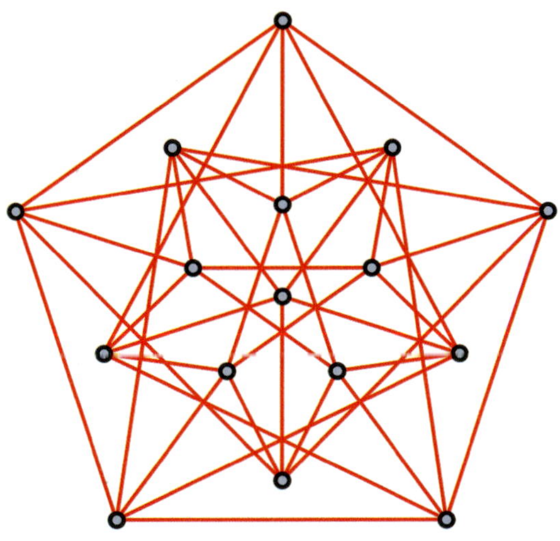

FIGURE 1.10: The Greenwood and Gleason graph $\mathrm{Cay}(\mathbb{F}_{16}; \mathbb{F}_{16}^{\times 3})$.

Along with $\mathrm{Cay}(\mathbb{F}_{16}; \mathbb{F}_{16}^{\times 3})$, the isomorphisms $x \mapsto gx$ and $x \mapsto g^2x$ define two more subgraphs of K_{16} that are isomorphic to $\mathrm{Cay}(\mathbb{F}_{16}; \mathbb{F}_{16}^{\times 3})$. These three subgraph form a factorization of K_{16} into three copies of $\mathrm{Cay}(\mathbb{F}_{16}; \mathbb{F}_{16}^{\times 3})$, each of which is assigned its own color. Since we have constructed a 3-coloroing of K_{16} that avoids a monochromatic K_3, it follows that $r^3(K_3) \geq 17$.

To prove the reverse inequality, consider a 3-coloring of K_{17} and select a vertex x. By the pigeonhole principle, at least six of the edges incident with x must have the same color.

Without loss of generality, suppose that x is incident with at least six red edges. If the subgraph induced by $N_r(x)$ contains a red edge, then that edge, along with vertex x, forms a red K_3. If all of the edges in the subgraph induced by $N_r(x)$ are blue and green, then it contains a blue/green K_6. Theorem 1.14 then implies that it contains a blue K_3 or a green K_3. Therefore, $r^3(K_3) \leq 17$. □

Beginning in 1972, Chvátal and Harary wrote a series of papers [66], [67], [68], and [69], in which they considered 2-color Ramsey numbers for graphs that are not complete. In [68], they proved the general lower bound

$$r(G_1, G_2) \geq (c(G_1) - 1)(\chi(G_2) - 1) + 1,$$

where $c(G_1)$ is the order of the largest connected component in G_1. When G_1 is connected, $c(G_1) = |V(G_1)|$. This lower bound was improved by Burr [53] in 1981, where he proved the following theorem.

Theorem 1.19 ([53]) *For all graphs G_1 and G_2 satisfying $c(G_1) \geq s(G_2)$,*

$$r(G_1, G_2) \geq (c(G_1) - 1)(\chi(G_2) - 1) + s(G_2).$$

Proof Begin with a blue $K_{\chi(G_2)}$ and replace $\chi(G_2)-1$ of the vertices with red $K_{c(G_1)-1}$-subgraphs and replace the last vertex with a red $K_{s(G_2)-1}$-subgraph. The resulting 2-coloring of $K_{(c(G_1)-1)(\chi(G_2)-1)+s(G_2)-1}$ does not have a large enough red connected component to contain a red G_1. To see that it also lacks a blue G_2, note that we can color the vertices according to which red $K_{c(G_1)-1}$-subgraph they are contained in to obtain a proper vertex coloring for every blue subgraph. If $s(G_2) = 1$, then there are only $\chi(G_2) - 1$ colors used, from which it follows that no blue G_2 exists. If $s(G_2) > 1$, then $\chi(G_2)$ colors are used, but one color class only has $s(G_2) - 1$ vertices, implying that no blue G_2 exists. □

In fact, Theorem 1.19 can easily be generalized to more than two colors. If $G_1, G_2, \ldots, G_t$ are graphs such that $t \geq 3$ and $r(G_1, G_2, \ldots, G_{t-1}) \geq s(G_t)$, then

$$r(G_1, G_2, \ldots, G_t) \geq (r(G_1, G_2, \ldots, G_{t-1}) - 1)(\chi(G_t) - 1) + s(G_t)$$

(see Theorem 5 of [41]).

In 1977, Chvátal [65] considered the Ramsey number for a tree versus a complete graph, proving that the lower bound given in Theorem 1.19 is exact.

Theorem 1.20 ([65]) *Let $m, n \in \mathbb{N}$. If T_m is any tree of order m, then*

$$r(T_m, K_n) = (m - 1)(n - 1) + 1.$$

Proof When T_m is a tree of order m, $c(T_m) = m$. Also, $\chi(K_n) = n$ and $s(K_n) = 1$. Theorem 1.19 implies that

$$r(T_m, K_n) \geq (m - 1)(n - 1) + 1.$$

To prove the reverse inequality, we proceed by (strong) induction on $m+n$. For the base cases, we have the following trivial values:

$$r(T_1, K_n) = 1 = r(T_m, K_1), \quad r(T_2, K_n) = n, \quad \text{and} \quad r(T_m, K_2) = m.$$

Assume that for any $m', n' \in \mathbb{N}$ such that $m' + n' < m + n$,

$$r(T_{m'}, K_{n'}) \leq (m' - 1)(n' - 1) + 1,$$

and consider a 2-coloring of $K_{(m-1)(n-1)+1}$. Let T_m be a tree of order m and define T' to be the tree formed by deleting a leaf from T_m. Let x denote the vertex in T' that was adjacent to the deleted leaf and note that T' has order $m-1$. By the inductive hypothesis,

$$r(T', K_n) \le (m-2)(n-1)+1 < (m-1)(n-1)+1.$$

So, there exists a red T' or a blue K_n. In the latter case, we are done, so assume that there exists a red T'.

If we delete the vertices corresponding with the red T' from the $K_{(m-1)(n-1)+1}$, we obtain a 2-coloring of $K_{(m-1)(n-2)+1}$. Applying the inductive hypothesis to this complete graph, we find that

$$r(T_m, K_{n-1}) \le (m-1)(n-2)+1.$$

Hence, there exists a red T_m or a blue K_{n-1}. In the former case, we are done, so assume that there exists a blue K_{n-1}. At this point, we have argued that the original 2-coloring of $K_{(m-1)(n-1)+1}$ contains a red T' that is disjoint from a blue K_{n-1}. Consider the edges that join x to the blue K_{n-1}. If any such edge is red, then a red T_m is formed. Otherwise, all such edges are blue, and x, along with the blue K_{n-1} form a blue K_n. It follows that

$$r(T_m, K_n) \le (m-1)(n-1)+1,$$

completing the proof of the theorem. □

Theorem 1.20 can easily be extended to a multicolor result, leading to the following corollary (cf., Theorem 1 of [33]).

Corollary 1.3 *Let $m, n_i \in \mathbb{N}$ for all $i \in \{1,2,\ldots,t\}$. If T_m is any tree of order m, then*

$$r(T_m, K_{n_1}, K_{n_2}, \ldots, K_{n_t}) = (m-1)(r(K_{n_1}, K_{n_2}, \ldots, K_{n_t}) - 1) + 1.$$

Proof Let $p = r(K_{n_1}, K_{n_2}, \ldots, K_{n_t})$ and start with a t-coloring of K_{p-1} that lacks a copy of K_{n_i} in color i, for all $i \in \{1,2,\ldots,t\}$ (i.e., a critical coloring for $r(K_{n_1}, K_{n_2}, \ldots, K_{n_t})$). Replace each of its vertices with red K_{m-1}-subgraphs (where the colors $1,2,\ldots,t$ are different from red). The resulting $K_{(m-1)(p-1)}$ still avoids a copy of K_{n_i} in color i, for all $i \in \{1,2,\ldots,t\}$. The largest red component has order $m-1$, so no red T_m exists. It follows that

$$r(T_m, K_{n_1}, K_{n_2}, \ldots, K_{n_t}) \ge (m-1)(p-1)+1.$$

To prove the reverse inequality, consider a $(t+1)$-coloring of $K_{(m-1)(p-1)+1}$. By Theorem 1.20, there exists a red T_m (and we are done) or a copy of K_p in colors $1,2,\ldots,t$ (here, we consider red to be different from colors $1,2,\ldots,t$). In the latter case, we find that there exists a copy K_{n_i} in color i, for some $i \in \{1,2,\ldots,t\}$. It follows that

$$r(T_m, K_{n_1}, K_{n_2}, \ldots, K_{n_t}) \le (m-1)(p-1)+1.$$

□

Table 1.1 shows the exact values of $r(T_m, K_{n_1}, K_{n_2}, \ldots, K_{n_t})$ that follow from Corollary 1.3 and known nontrival Ramsey numbers for complete graphs.

1.3.2 Star-Critical Ramsey Numbers

Star-critical Ramsey numbers were first introduced by Jonelle Hook in her dissertation [125] in 2010 (see also [35], [126], and [127]). They serve as a refinement of Ramsey numbers.

TABLE 1.1: Known nontrivial multicolor Ramsey numbers for a tree versus complete graphs.

$r(K_{n_1}, K_{n_2}, \ldots, K_{n_t})$	$r(T_m, K_{n_1}, K_{n_2}, \ldots, K_{n_t})$
$r(K_3, K_3) = 6$ [111]	$r(T_m, K_3, K_3) = 5m - 4$
$r(K_3, K_4) = 9$ [111]	$r(T_m, K_3, K_4) = 8m - 7$
$r(K_3, K_5) = 14$ [111]	$r(T_m, K_3, K_5) = 13m - 12$
$r(K_3, K_6) = 18$ [144]	$r(T_m, K_3, K_6) = 17m - 16$
$r(K_3, K_7) = 23$ [110], [142]	$r(T_m, K_3, K_7) = 22m - 21$
$r(K_3, K_8) = 28$ [112], [178]	$r(T_m, K_3, K_8) = 27m - 26$
$r(K_3, K_9) = 36$ [112], [142]	$r(T_m, K_3, K_9) = 35m - 34$
$r(K_4, K_4) = 18$ [111]	$r(T_m, K_4, K_4) = 17m - 16$
$r(K_4, K_5) = 25$ [141], [176]	$r(T_m, K_4, K_5) = 24m - 23$
$r(K_3, K_3, K_3) = 17$ [111]	$r(T_m, K_3, K_3, K_3) = 16m - 15$
$r(K_3, K_3, K_4) = 30$ [72], [142]	$r(T_m, K_3, K_3, K_4) = 29m - 28$

Writing $r(G_1, G_2, \ldots, G_t) = n$ has two meanings. First, it tells us that every t-coloring of K_n has a monochromatic copy of G_i in color i, for some $i \in \{1, 2, \ldots, t\}$. Second, it tells us that there exists a t-coloring of K_{n-1} that avoids a monochromatic copy of G_i in color i, for all $i \in \{1, 2, \ldots, t\}$. So, the addition of a single vertex (and all of the edges incident with that vertex) forces the Ramsey property to kick in. With star-critical Ramsey numbers, we add in a vertex, then we determine the exact number of edges incident with that vertex that will force the Ramsey property to hold.

Denote by $K_n \sqcup K_{1,k}$ the graph formed by taking K_n and a vertex v, and joining v with edges to exactly k of the vertices in the K_n, where $k \in \{0, 1, \ldots, n\}$. For example, Figure 1.11 shows the graph $K_5 \sqcup K_{1,2}$. For a collection of graphs $G_1, G_2, \ldots, G_t$, the

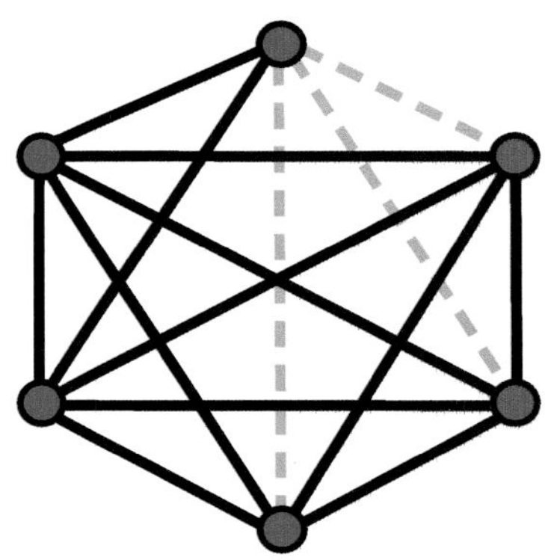

FIGURE 1.11: The graph $K_5 \sqcup K_{1,2}$.

star-critical Ramsey number $r_*(G_1, G_2, \ldots, G_t)$ is the least $k \in \{0, 1, \ldots, r(G_1, G_2, \ldots, G_t) - 1\}$ such that every t-coloring of $K_{r(G_1,G_2,\ldots,G_t)-1} \sqcup K_{1,k}$ contains a subgraph isomorphic to G_i in color i, for some $i \in \{1, 2, \ldots, t\}$. As long as none of the G_i contain isolated vertices, $r_*(G_1, G_2, \ldots, G_t) \geq 1$ (see [43]). In the case of a 1-color star-critical Ramsey number, every vertex of a graph must have minimum degree $\delta(G)$ in order for a G-subgraph to be present. It follows that $r_*^1(G) = \delta(G)$.

Similar to the determination of Ramsey numbers, evaluating a star-critical Ramsey number requires two steps. That is, $r_*(G_1, G_2, \ldots, G_t) = k$ if and only if both of the following statements are true:

1. There exists a t-coloring of $K_{r(G_1,G_2,\ldots,G_t)-1} \sqcup K_{1,k-1}$ that lacks a monochromatic copy of G_i in color i, for all $i \in \{1,2,\ldots,t\}$.
2. Every t-coloring of $K_{r(G_1,G_2,\ldots,G_t)-1} \sqcup K_{1,k}$ contains a monochromatic copy of G_i in color i, for some $i \in \{1,2,\ldots,t\}$.

Note that if Statement 1 is true (and none of the G_i contain isolated vertices), then there exists a t-coloring of $K_{r(G_1,G_2,\ldots,G_t)-1}$ that lacks a monochromatic copy of G_i in color i, for all $i \in \{1,2,\ldots,t\}$. If Statement 2 is true, then every t-coloring of $K_{r(G_1,G_2,\ldots,G_t)}$ contains a monochromatic copy of G_i in color i, for some $i \in \{1,2,\ldots,t\}$. So, the steps required to prove an exact value for a star-critical Ramsey number imply the steps needed to prove the value of the corresponding Ramsey number. For this reason, we will sometimes group the proofs of both numbers together.

Although star-critical Ramsey numbers had not yet been defined, the earliest evaluation of such a number was a proof that was equivalent to

$$r_*(K_{n_1}, K_{n_2}) = r(K_{n_1}, K_{n_2}) - 1$$

by Erdős, Faudree, Rousseau, and Schelp [84] in 1978. In the following theorem, we prove the multicolor verision of their result (see Theorem 1.1 of [35]).

Theorem 1.21 **([84])** *If $n_i \geq 2$ for all $i \in \{1,2,\ldots,t\}$, then*

$$r_*(K_{n_1}, K_{n_2}, \ldots, K_{n_t}) = r(K_{n_1}, K_{n_2}, \ldots, K_{n_t}) - 1.$$

Proof Let $\mathcal{G}$ be a critical coloring for $r(K_{n_1}, K_{n_2}, \ldots, K_{n_t})$ and select a vertex $x \in V(\mathcal{G})$. Introduce a new vertex y and for all $z \in V(\mathcal{G}) \setminus \{x\}$, color edge yz the same color as edge xz. Let edge xy be the "missing" edge. So, the subgraph induced by $(V(\mathcal{G}) \setminus \{x\}) \cup \{y\}$ is isomorphic to $\mathcal{G}$. Since no complete subgraph can include both x and y, it follows that the resulting

$$K_{r(K_{n_1},K_{n_2},\ldots,K_{n_t})-1} \sqcup K_{1,r(K_{n_1},K_{n_2},\ldots,K_{n_t})-2} = K_{r(K_{n_1},K_{n_2},\ldots,K_{n_t})} - e$$

lacks a copy of K_{n_i} in color i, for all $i \in \{1,2,\ldots,t\}$. □

1.3.3 Gallai-Ramsey Theory

A t-coloring $f : E(G) \longrightarrow \{1,2,\ldots,t\}$ of a graph G is called a *Gallai t-coloring* if it does not contain any rainbow triangles. That is, for all distinct $x, y, z \in V(G)$,

$$|\{f(xy), f(yz), f(xz)\}| \leq 2.$$

For a collection of graphs $G_1, G_2, \ldots, G_t$, the *Gallai-Ramsey number* $gr(G_1, G_2, \ldots, G_t)$ is the least $n \in \mathbb{N}$ such that every Gallai t-coloring of K_n contains a subgraph isomorphic to G_i in color i, for some $i \in \{1,2,\ldots,t\}$. Since at least three colors are required in order to have a rainbow triangle, this definition is equivalent to that of a Ramsey number when $t < 3$. When $G_1 = G_2 = \cdots = G_t$, we write $gr^t(G_1)$ for the corresponding Gallai-Ramsey number. Overviews of known results in Gallai-Ramsey theory can be found in [101] and [167]. Before we introduce weakened Gallai-Ramsey numbers, we first review some properties of Gallai colorings.

Lemma 1.1 **([86])** *For $n \geq 2$, every Gallai coloring of K_n uses at most $n-1$ colors.*

Proof We proceed by induction on $n \geq 2$. When $n = 2$, K_n only has a single edge, and hence, uses at most one color. Let $n \geq 3$ and assume that every Gallai coloring of K_{n-1} uses at most $n-2$ colors. Consider a Gallai coloring of K_n and select a vertex x. If we delete x (and all of the edges that are incident with x), then we obtain a Gallai coloring of K_{n-1} which, by the inductive hypothesis, uses at most $n-2$ colors. When reintroducing vertex x, if there exist two edges xy and xz that have two distinct colors not used in the K_{n-1}, then the subgraph induced by $\{x, y, z\}$ is a rainbow triangle. Since this cannot occur, it follows that there is at most one new color introduced with the addition of the vertex x, and the edges that are incident with it. Hence, the K_n uses at most $n-1$ colors. □

The following structure theorem provides a fundamental tool for proving results involving Gallai colorings. This theorem is a reinterpretation of a classic result of Gallai [102] (see Chapter 3 of [4] for an English translation of [102] by F. Maffray and M. Preissmann). The following reinterpretation of Gallai's theorem was first presented by Gyárfás and Simonyi [116] in 2004, where they proved two conjectures of Bialostocki and Voxman [24] concerning the existence of certain spanning trees in Gallai colorings.

Theorem 1.22 ([116]) *Every Gallai coloring of a complete graph can be formed by replacing the vertices in a* 2*-colored complete graph of order at least* 2 *with Gallai-colored complete graphs.*

In Theorem 1.22, the 2-colored complete graph is called the *base graph*, while the Gallai-colored complete graphs that replace the base graph's vertices are called the *blocks*. The partitioning of the vertex set of the original complete graph into the vertex sets of the blocks is called a *Gallai partition*. In order to prove Theorem 1.22, we must first prove the following lemma.

Lemma 1.2 ([116]) *Every Gallai-coloring of a complete graph that uses at least three colors has a color that spans a disconnected graph.*

Proof Let $\mathcal{G}$ be a counterexample to Lemma 1.2 with minimal order among all possible counterexamples. Suppose the Gallai coloring $\mathcal{G}$ uses $t \geq 3$ colors. Then every vertex must be incident with an edge in each of the t colors. Let $x \in V(\mathcal{G})$ and define $\mathcal{H}$ to be the Gallai-coloring induced by the vertex set $V(\mathcal{G}) \setminus \{x\}$. Then $\mathcal{H}$ cannot be 2-colored since any other color would have then spanned a disconnected subgraph in $\mathcal{G}$. Since $\mathcal{G}$ was chosen to be a minimal counterexample, some color in $\mathcal{H}$ (say, color 1) must span a disconnected subgraph. Let $G_1, G_2, \ldots, G_k$ be the connected components in the subgraph of $\mathcal{H}$ spanned by edges in color 1.

For $i \neq j$, suppose that the connected components G_i and G_j are joined by an edge yz in color 2, where $y \in V(G_i)$ and $z \in V(G_2)$. Consider a vertex $w \in V(G_i) \setminus \{y\}$. Since G_i is a connected graph in color 1, there exists an $y - w$ path $yv_1v_2 \ldots v_\ell w$ in color 1. As no edge in color 1 can join G_i and G_j, avoiding a rainbow triangle requires edge v_1z to have color 2. This argument can be repeated to see that $v_2z, v_3z, \ldots, v_\ell z, wz$ must all have color 2. A similar argment can then be made to show that if $u \in V(G_j) \setminus \{z\}$, then wu must have color 2. We obtain the following observation.

Observation: All edges between any distinct pair of components G_i and G_j must be the same color (other than color 1).

Since we assumed that $\mathcal{G}$ does not have a color that spans a disconnected graph, there must exist an edge xy_i in color 1, where $y_i \in V(G_i)$, for every $i \in \{1, 2, \ldots, k\}$. We also noted before that x must be incident with an edge in every color. Without loss of generality, suppose that xu and xv are edges in colors 2 and 3, respectively. Consider two cases.

<u>Case 1:</u> Suppose that u and v are in the same component G_i. For $i \neq j$, edge uy_j must have color 2 and edge vy_j must have color 3 in order to avoid rainbow triangles. However,

this contradicts the observation since it allows for edges in colors 2 and 3 to join G_i and G_j.

Case 2: Suppose that u and v and in different components: $u \in G_i$ and $v \in G_j$. Then uy_j must have color 2 and vy_i must have color 3 in order to avoid rainbow triangles. Once again, this contradicts the observation since it allows for edges in colors 2 and 3 to join G_i and G_j.

In both cases, we reach a contradiction. Hence, a minimal counterexample $\mathcal{G}$ does not exist. □

Proof of Theorem 1.22 We will proceed by using (strong) induction on $n \geq 1$. For $n \leq 3$, the statement in Theorem 1.22 is trivial since at most two colors can be used by Lemma 1.1. Suppose that Theorem 1.22 is true for all Gallai colorings of complete graphs of order at most $n-1$ and let $\mathcal{G}_n$ be a Gallai coloring of K_n, where $n \geq 4$. If $\mathcal{G}_n$ has at most two colors, then the theorem is trivial, so assume that it has at least $t \geq 3$ colors. For each $i \in \{1, 2, \ldots, t\}$, let G_i be the subgraph of $\mathcal{G}_n$ spanned by the edges in color i.

By Lemma 1.2, there exists an $i \in \{1, 2, \ldots, t\}$ such that G_i is disconnected. Without loss of generality, suppose that G_1 is disconnected and let $H_1, H_2, \ldots, H_k$ be its connected components. Also, assume that

$$1 \leq |V(H_1)| \leq |V(H_2)| \leq \cdots \leq |V(H_k)|.$$

Since color 1 appears in $\mathcal{G}_n$, it follows that $|V(H_k)| \geq 2$. By Observation 1 in the proof of Lemma 1.2, all edges joining any distinct pair of components must have the same color. So, we are able to collapse H_t down into a single vertex x to obtain a Gallai-coloring $\mathcal{G}'$ of a complete graph with order less than n. The inductive hypothesis can now be applied to $\mathcal{G}'$. Since $\mathcal{G}_n$ can be formed by replacing x in $\mathcal{G}'$ with H_t, it follows that the Gallai coloring $\mathcal{G}_n$ can be constructed by substituting the vertices of a 2-colored complete graph with Gallai-colored complete graphs. □

Since in every 2-coloring of a complete graph, the subgraph spanned by edges in one of the two colors is connected by Theorem 1.3, applying this observation to the base graph of a Gallai coloring results in the following corollary.

Corollary 1.4 *In any Gallai-coloring of a complete graph, there exists a monochromatic spanning subgraph.*

The main way in which we will use Theorem 1.22 is in the determination of upper bounds for various Gallai-Ramsey numbers. The theorem will allow us to break the proof of upper bounds into cases, based on the order of the base graph in a Gallai partition. The following lemma will further simplify some of our work.

Lemma 1.3 ([167]) *In any Gallai coloring of K_n, if the base graph is chosen to have minimal order, then its order is not* 3.

Proof Consider a Gallai coloring of K_n in which the base graph, which is chosen to have minimal order, has order 3. Since rainbow triangles are avoided, at least two of the edges in the base graph receive the same color. Denote the vertex sets for the blocks by X, Y, and Z and suppose, without loss of generality, that the edges joining X to Y and the edges joining X to Z are all red. Then this Gallai coloring can be viewed as having a Gallai partition with base graph of order 2 in which one block is X and the other block is $Y \cup Z$, contradicting the assumption that the base graph had minimal order. □

In 1983, Chung and Graham [61] proved a result that is equivalent to the following theorem (also, see Theorem 3 of [115]). We include its proof as an example of the utility of Theorem 1.22.

Theorem 1.23 ([61]) *For all $t \geq 1$,*

$$gr^t(K_3) = \begin{cases} 5^{t/2} + 1 & \text{if } t \text{ is even} \\ 2 \cdot 5^{(t-1)/2} + 1 & \text{if } t \text{ is odd.} \end{cases}$$

Proof To prove the lower bounds, first define $\mathcal{G}_1$ to be a K_2 in color 1. Define $\mathcal{G}_2$ to be the critical coloring for $r^2(K_3)$ in colors 1 and 2 as described in Theorem 1.14. For $t \geq 3$ odd, define $\mathcal{G}_t$ to be the t-coloring of $K_{2 \cdot 5^{(t-1)/2}}$ formed by taking a K_2 in color t and replacing each of its vertices with copies of $\mathcal{G}_{t-1}$ in colors $1, 2, \ldots, t-1$. If $t \geq 4$ is even, then define $\mathcal{G}_t$ to be the t-coloring of $K_{5^{t/2}}$ formed by taking a critical coloring for $r^2(K_3)$ in colors $t-1$ and t (see Theorem 1.14), and replacing each of its verticees with copies of $\mathcal{G}_{t-2}$ in colors $1, 2, \ldots, t-2$. Figure 1.12 shows the $t = 3$ and $t = 4$ cases. These constructions avoid both

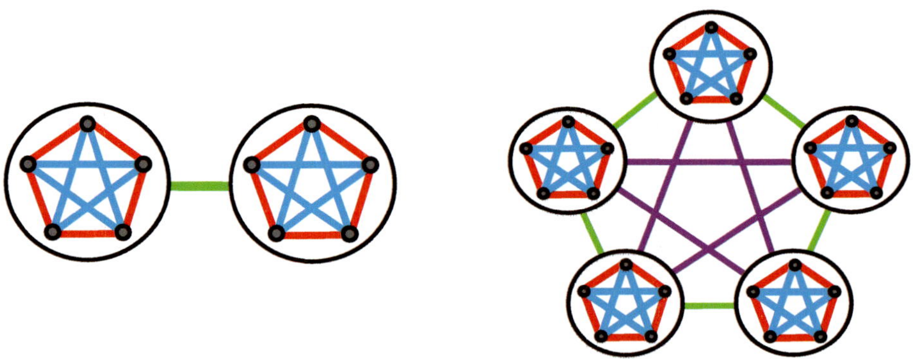

FIGURE 1.12: Critical colorings for $gr^3(K_3)$ and $gr^4(K_3)$.

rainbow and monochromatic triangles. It follows that $gr^t(K_3) \geq 2 \cdot 5^{(t-1)/2} + 1$ when t is odd and $gr^t(K_3) \geq 5^{t/2} + 1$ when t is even.

The upper bound will be proved by using (strong) induction on $t \geq 1$. The $t = 1$ case is the trivial Gallai-Ramsey number $gr^1(K_3) = r^1(K_3) = 3$, while the $t = 2$ case $gr^2(K_3) = r^2(K_3) = 6$ follows from Theorem 1.14. For all $k \in \{1, 2, \ldots, t-1\}$, assume that

$$gr^k(K_3) = \begin{cases} 5^{k/2} + 1 & \text{if } k \text{ is even} \\ 2 \cdot 5^{(k-1)/2} + 1 & \text{if } k \text{ is odd.} \end{cases}$$

Consider two cases.

Case 1: Assume that $t \geq 3$ is odd. Consider a Gallai t-coloring of $K_{2 \cdot 5^{(t-1)/2}+1}$ that avoids a monochromatic K_3, and let $\mathcal{B}$ be the base graph for this Gallai coloring, chosen to have minimal order. By Theorems 1.22 and 1.14, $2 \leq |V(\mathcal{B})| \leq 5$. Lemma 1.3 implies that $|V(\mathcal{B})| \neq 3$. We obtain the following subcases.

Subcase 1.1: Suppose that $|V(\mathcal{B})| = 2$. Without loss of generality, assume that the edge in $\mathcal{B}$ has color 1. Then neither block can contain an edge in color 1 without forming a K_3 in color 1. So, the blocks only contain edges in colors $2, 3, \ldots, t$. By the pigeonhole principle, some block has order at least $5^{(t-1)/2} + 1$. Since $t - 1$ is even, the inductive hypothesis implies that $gr^{t-1}(K_3) = 5^{(t-1)/2} + 1$. So, the larger of the two blocks must contain a monochromatic K_3 in one of the colors $2, 3, \ldots, t$, which is a contradiction.

Subcase 1.2: Suppose that $|V(\mathcal{B})| = 4$. Without loss of generality, assume that $\mathcal{B}$ has edges in colors 1 and 2. If any vertex in $\mathcal{B}$ is incident with edges in only one color, then it is possible to form a base graph with order smaller than that of $\mathcal{B}$, contradicting the minimality of $\mathcal{B}$. Hence, every vertex in $\mathcal{B}$ is incident with edges in both of the colors 1 and 2. If any block contains an edge in colors 1 or 2, then a monochromatic K_3 can be formed in that color. Hence, the only colors that appear within the blocks are colors $3, 4, \ldots, t$. By the pigeonhole principle, some block has order at least

$$\left\lceil \frac{2 \cdot 5^{(t-1)/2} + 1}{4} \right\rceil = \frac{2 \cdot 5^{(t-1)/2} + 2}{4},$$

since $5^{(t-1)/2} \equiv 1 \pmod 4$ for all odd $t \geq 3$. It can be checked that

$$\frac{2 \cdot 5^{(t-1)/2} + 2}{4} \geq 2 \cdot 5^{(t-3)/2} + 1 = gr^{t-2}(K_3),$$

for all odd $t \geq 3$. It follows that the largest block contains a monochromatic K_3, which is a contradiction.

Subcase 1.3: Suppose that $|V(\mathcal{B})| = 5$. Without loss of generality, assume that $\mathcal{B}$ has edges in colors 1 and 2. As we saw in the proof of Subcase 1.2, every vertex in $\mathcal{B}$ is incident with edges in both of the colors 1 and 2. If any block contains an edge in colors 1 or 2, then a monochromatic K_3 can be formed in that color. Hence, the only colors that appear within the blocks are colors $3, 4, \ldots, t$. By the pigeonhole principle, some block has order at least

$$\left\lceil \frac{2 \cdot 5^{(t-1)/2} + 1}{5} \right\rceil = 2 \cdot 5^{(t-3)/2} + 1 = gr^{t-2}(K_3).$$

It follows that the largest block contains a monochromatic K_3.

Case 2: Assume that $t \geq 4$ is even. Consider a Gallai t-coloring of $K_{5^{t/2}+1}$ that avoids a monochromatic K_3, and let $\mathcal{B}$ be the base graph for this Gallai coloring, chosen to have minimal order. By Theorems 1.22 and 1.14, $2 \leq |V(\mathcal{B})| \leq 5$. Lemma 1.3 implies that $|V(\mathcal{B})| \neq 3$. We obtain the following subcases.

Subcase 2.1: Suppose that $|V(\mathcal{B})| = 2$. Without loss of generality, assume that the edge in $\mathcal{B}$ has color 1. Then neither block can contain an edge in color 1 without forming a K_3 in color 1. So, the blocks only contain edges in colors $2, 3, \ldots, t$. By the pigeonhole principle, some block has order at least $\frac{5^{t/2}+1}{2}$. Since $t-1$ is odd, the inductive hypothesis implies that $gr^{t-1}(K_3) = 2 \cdot 5^{(t-2)/2} + 1$. One can check that $\frac{5^{t/2}+1}{2} \geq 2 \cdot 5^{(t-2)/2} + 1$ for all even $t \geq 4$. So, the larger of the two blocks must contain a monochromatic K_3 in one of the colors $2, 3, \ldots, t$, which is a contradiction.

Subcase 2.2: Suppose that $|V(\mathcal{B})| = 4$. Without loss of generality, assume that $\mathcal{B}$ has edges in colors 1 and 2. As we saw in the proof of Subcase 1.2, every vertex in $\mathcal{B}$ is incident with edges in both of the colors 1 and 2. If any block contains an edge in colors 1 or 2, then a monochromatic K_3 can be formed in that color. Hence, the only colors that appear within the blocks are colors $3, 4, \ldots, t$. By the pigeonhole principle, some block has order at least

$$\left\lceil \frac{5^{t/2} + 1}{4} \right\rceil = \frac{5^{t/2} + 3}{4},$$

since $5^{t/2} \equiv 1 \pmod 4$ for all even $t \geq 4$. It can be checked that

$$\frac{5^{t/2} + 3}{4} \geq 5^{(t-2)/2} + 1 = gr^{t-2}(K_3),$$

for all even $t \geq 4$. It follows that the largest block contains a monochromatic K_3, which is a contradiction.

Subcase 2.3: Suppose that $|V(\mathcal{B})| = 5$. Without loss of generality, assume that $\mathcal{B}$ has edges in colors 1 and 2. As we saw in the proof of Subcase 1.2, every vertex in $\mathcal{B}$ is incident with edges in both of the colors 1 and 2. If any block contains an edge in colors 1 or 2, then a monochromatic K_3 can be formed in that color. Hence, the only colors that appear within the blocks are colors $3, 4, \ldots, t$. By the pigeonhole principle, some block has order at least

$$\left\lceil \frac{5^{(t)/2} + 1}{5} \right\rceil = 5^{(t-2)/2} + 1 = gr^{t-2}(K_3).$$

It follows that the largest block contains a monochromatic K_3.

In all cases, we have shown that there exists a monochromatic K_3. It follows that

$$gr^t(K_3) \leq \begin{cases} 5^{t/2} + 1 & \text{if } t \text{ is even} \\ 2 \cdot 5^{(t-1)/2} + 1 & \text{if } t \text{ is odd.} \end{cases}$$

□

In 2015, Fox, Grinshpun, and Pach [96] posed the following conjecture.

Conjecture 1.1 ([96]) *For all $t \geq 1$ and $n \geq 3$,*

$$gr^t(K_n) = \begin{cases} (r^2(K_n) - 1)^{t/2} + 1 & \textit{if } t \textit{ is even} \\ (n-1)(r^2(K_n) - 1)^{(t-1)/2} + 1 & \textit{if } t \textit{ is odd.} \end{cases}$$

This conjecture was verified for $n = 4$ by Liu, Magnant, Saito, Schiermeyer and Shi [165] in 2020, when they proved that for all $t \geq 1$,

$$gr^t(K_4) = \begin{cases} 17^{t/2} + 1 & \text{if } t \text{ is even} \\ 3 \cdot 17^{(t-1)/2} + 1 & \text{if } t \text{ is odd.} \end{cases}$$

In 2022, Magnant and Schiermeyer [168] considered the Gallai-Ramsey number $gr^t(K_5)$, proving the following theorem.

Theorem 1.24 ([168]) *Let $t \geq 2$. If $r^2(K_5) \geq 44$, then*

$$gr^t(K_5) = \begin{cases} (r^2(K_5) - 1)^{t/2} + 1 & \textit{if } t \textit{ is even} \\ 4 \cdot (r^2(K_5) - 1)^{(t-1)/2} + 1 & \textit{if } t \textit{ is odd.} \end{cases}$$

Otherwise, $r^2(K_5) = 43$ and $gr^t(K_5)$ satisfies

$$\begin{cases} 42^{t/2} + 1 \leq gr^t(K_5) \leq 43^{t/2} + 1 & \textit{if } t \geq 4 \textit{ is even} \\ 169 \cdot 42^{(t-3)/2} + 1 \leq gr^t(K_5) \leq 4 \cdot 43^{(t-1)/2} + 1 & \textit{if } t \geq 3 \textit{ is odd.} \end{cases}$$

While the value of $r^2(K_5)$ is not known, it is known that $43 \leq r^2(K_5) \leq 46$ (the lower bound is due to Exoo [89] and the upper bound is due to Angeltveit and McKay [7]). In 1997, McKay and Radziszowski [177] posed the following conjecture.

Conjecture 1.2 ([177]) $r^2(K_5) = 43$.

By Theorem 1.24, it follows that exactly one of Conjectures 1.1 and 1.2 is true (see also Lemma 3.7 in [167]). Determining which of these conjectures is true would be a groundbreaking result in Ramsey theory.

1.3.4 Size Ramsey Numbers

In 1978, Erdős, Faudree, Rousseau, and Schelp [84] introduced the concept of a size Ramsey number. In order to define it, let $G_1, G_2, \ldots, G_t$ be graphs and define $\mathcal{C}(G_1, G_2, \ldots, G_t)$ to be the set of all graphs H such that every t-coloring of H contains a monochromatic copy of G_i in color i, for some $i \in \{1, 2, \ldots, t\}$. When $G_1 = G_2 = \cdots = G_t$, we shorten the notation to $\mathcal{C}^t(G_1)$. The *size Ramsey number* $\widehat{r}(G_1, G_2, \ldots, G_t)$ is then given by

$$\widehat{r}(G_1, G_2, \ldots, G_t) := \min\{|E(H)| \mid H \in \mathcal{C}(G_1, G_2, \ldots, G_t)\}.$$

When $G_1 = G_2 = \cdots = G_t$, we write $\widehat{r}^t(G_1)$.

The finiteness of size Ramsey numbers follows from the observation that

$$\widehat{r}(G_1, G_2, \ldots, G_t) \leq \binom{r(G_1, G_2, \ldots, G_t)}{2} = |E(K_{r(G_1, G_2, \ldots, G_t)})|,$$

since every t-coloring of $K_{r(G_1, G_2, \ldots, G_t)}$ is guaranteed to contain a copy of G_i in color i, for some $i \in \{1, 2, \ldots, t\}$. In Theorem 1 of [84], it was shown that

$$\widehat{r}(K_{n_1}, K_{n_2}) = \binom{r(K_{n_1}, K_{n_2})}{2}.$$

In Theorem 3.31 of Section 3.5, a generalization of this result will be proved for weakened size Ramsey numbers.

If $k = r_*(G_1, G_2, \ldots, G_t)$, then every t-coloring of $K_{r(G_1, G_2, \ldots, G_t)-1} \sqcup K_{1,k}$ contains a copy of G_i in color i, for some $i \in \{1, 2, \ldots, t\}$. If follows that

$$K_{r(G_1, G_2, \ldots, G_t)-1} \sqcup K_{1,k} \in \mathcal{C}(G_1, G_2, \ldots, G_t).$$

Hence, the size Ramsey number is related to the star-critical Ramsey number by the following inequality:

$$\widehat{r}(G_1, G_2, \ldots, G_t) \leq \binom{r(G_1, G_2 \ldots, G_t) - 1}{2} + r_*(G_1, G_2 \ldots, G_t).$$

1.4 Ramsey Theory on the Integers

When working over the integers, we will use the standard notation $[1, n] := \{1, 2, \ldots, n\}$. One of the first Ramsey-theoretic results involving subsets of integers was proved by Issai Schur [195] in 1916. His focus was on a topic in number theory: proving that the Fermat congruence

$$x^n + y^n \equiv z^n \pmod{p}$$

has a non-trivial integer solution when $n \geq 3$ and p is a sufficiently large prime number (see Theorem 8.1 of [156]). In the process of proving this result, Schur first proved a theorem concerning the existence of monochromatic solutions to the equation $x+y=z$ in t-colorings of $[1, n]$. His theorem guaranteed the existence of what we now refer to as "Schur numbers."

1.4.1 Schur and Rado Numbers

An m-tuple $(a_1, a_2, \ldots, a_m)$ that satisfies the equation

$$\mathcal{E}_m : \; x_1 + x_2 + \cdots + x_{m-1} = x_m$$

is called an *m-Schur solution.* Given a t-coloring of $[1,n]$, an m-Schur solution is monochromatic if all of its components receive the same color, and it is *rainbow* if no two of its components receive the same color. The $m=3$ case of the following theorem was proved by Schur [195], while this more general theorem is Corollary 8.19 of [44].

Theorem 1.25 (Generalized Schur's Theorem [44]) *For every $t \in \mathbb{N}$ and $m \geq 3$, there exists an $n \in \mathbb{N}$ such that every t-coloring of $\{1,2,\ldots,n\}$ contains a monochromatic m-Schur solution.*

Proof For any $t \in \mathbb{N}$ and $m \geq 3$, The Ramsey number $r^t(K_m)$ exists by Theorem 1.13. Let $p = r^t(K_m)$ and number the vertices in K_p by 1, 2, ..., p. Arbitrarily partition the set $\{1,2,\ldots,p-1\}$ into t subsets, corresponding to the t colors. If i and j are two distinct vertices, then color the edge connecting them according to the color of

$$|j-i| \in \{1,2,\ldots,p-1\}.$$

Since $p = r^t(K_m)$, it follows that the t-coloring of K_p that we just constructed necessarily contains a monochromatic K_m-subgraph. Suppose that $b_1 < b_2 < \cdots < b_m$ are the vertices of such a K_m-subgraph. If we let $a_1 = b_2 - b_1$, $a_2 = b_3 - b_2$, ..., $a_{m-1} = b_m - b_{m-1}$, and $a_m = b_m - b_1$, then it follows that

$$a_1 + a_2 + \cdots + a_{m-1} = (b_2 - b_1) + (b_3 - b_2) + \cdots + (b_m - b_{m-1}) = b_m - b_1 = a_m$$

is a monochromatic m-Schur solution. □

The *Schur number* $S(t)$ is the least $n \in \mathbb{N}$ such that every t-coloring of $\{1,2,\ldots,n\}$ contains a monochromatic 3-Schur solution. Currently, the only known Schur numbers are $S(1)=2$, $S(2)=5$, $S(3)=14$ [106], $S(4)=45$ [3], and $S(5)=161$ [124]. Schur numbers were extended to $\mathcal{E}_m$, where $m>3$, in 1982 by Beutelspacher and Brestovansky [21]. Define the *generalized Schur number* $S^t(m)$ to be the least $n \in \mathbb{N}$ such that every t-coloring of $\{1,2,\ldots,n\}$ contains a monochromatic solution to $\mathcal{E}_m$. Theorem 1.25 implies that $S^t(m) \leq r^t(K_m) - 1$. In [21], it was shown that $S^2(m) = m^2 - m - 1$, for all $m \geq 3$. Generalized Schur numbers were also studied in [3], [130], and [193].

A variation on the definition of a Schur number that has been introduced is the *strict Schur number* $\widehat{S}^t(m)$, defined to be the least $n \in \mathbb{N}$ such that every t-coloring of $\{1,2,\ldots,n\}$ contains a monochromatic solution to the system

$$\mathcal{E}_m^* : \begin{cases} x_1 + x_2 + \cdots + x_{m-1} = x_m \\ x_1 < x_2 < \cdots < x_m. \end{cases}$$

These numbers were studied in [1], [2], [25], [28], and [81]. They are sometimes referred to as "weak Schur numbers," but we will avoid this name to avoid confusion with our use of the word "weakened." It is known that $\widehat{S}^2(3) = 9$ [2], $\widehat{S}^2(4) = 24$ [28, 194], $\widehat{S}^2(5) = 52$ [28, 194], $\widehat{S}^2(6) = 101$ [81], $\widehat{S}^2(7) = 166$ [1], $\widehat{S}^2(8) = 253$ [1], $\widehat{S}^3(3) = 24$ [25], $\widehat{S}^3(4) = 94$ [1], $\widehat{S}^3(5) = 259$ [1], and $\widehat{S}^4(3) = 67$ [25]. An m-tuple $(a_1, a_2, \ldots, a_m)$ such that $a_1 + a_2 + \cdots + a_{m-1} = a_m$, where $a_1 < a_2 < \cdots < a_m$ is called a *strict m-Schur solution.*

In 1933, Richard Rado [188] (who was a Ph.D. student of Issai Schur), determined which systems of linear equations have monochromatic solutions under t-colorings of $\{1,2,\ldots,n\}$. For this reason, when Bialostocki and Schaal [23] considered another variation of Schur numbers for the systems

$$\mathcal{E}_m^\dagger : x_1 + x_2 + \cdots + x_{m-1} < x_m \qquad \text{and} \qquad \mathcal{E}_m^\ddagger : \begin{cases} x_1 + x_2 + \cdots + x_{m-1} < x_m \\ x_1 < x_2 < \cdots x_m, \end{cases}$$

they referred to them as "Rado numbers." That is, given a system $\mathcal{E}$ of equations/inequalities, define the *Rado number* $R^t(\mathcal{E})$ to be the least $n \in \mathbb{N}$ such that every t-coloring of $\{1, 2, \ldots, n\}$ contains a monochromatic solution to $\mathcal{E}$. Observe that $R^t(\mathcal{E}_m) = S^t(m)$ and $R^t(\mathcal{E}^*_m) = \widehat{S}^t(m)$. In [23], it was shown that for all $m \geq 3$, $R^2(\mathcal{E}^\dagger_m) = m^2 - m + 1$ and

$$R^2(\mathcal{E}^\ddagger_m) = \begin{cases} \frac{9}{16}m^3 - m^2 + m + 1 & \text{if } m \equiv 0 \pmod 4 \\ \frac{9}{16}m^3 - m^2 + \frac{13}{16}m + \frac{13}{8} & \text{if } m \equiv 1 \pmod 4 \\ \frac{9}{16}m^3 - m^2 + m + \frac{1}{2} & \text{if } m \equiv 2 \pmod 4 \\ \frac{9}{16}m^3 - m^2 + \frac{17}{16}m + \frac{5}{8} & \text{if } m \equiv 3 \pmod 4. \end{cases}$$

1.4.2 Rainbow Schur Numbers

The last type of number that we define in this section follows the ideas of Erdős, Simonovits, and Sós [86], where anti-Ramsey theory was introduced. Rather than fix the number of colors t and search for monochromatic solutions to a system of equations/inequalities, we fix the set $\{1, 2, \ldots, n\}$ and search for a rainbow solution to a given system. For $m \geq 3$ and $n \geq \frac{m(m-1)}{2}$, define the *rainbow Schur number* $RS_m(n)$ to be the minimum number of colors such that every exact $RS_m(n)$-coloring of $\{1, 2, \ldots, n\}$ contains a rainbow solution to $\mathcal{E}_m$. The assumption that $n \geq \frac{m(m-1)}{2}$ guarantees that $\mathcal{E}_m$ is solvable using distinct elements from $\{1, 2, \ldots, n\}$. If we color $\{1, 2, \ldots, n\}$ with n distinct colors, then a rainbow solution to $\mathcal{E}_m$ will exist. It follows that $RS_m(n) \leq n$. Note that this statment is not true if we consider an n-coloring that is not exact.

The rainbow Schur number is closely related to the *anti-Schur number* $AS_m(n)$, defined to be the maximum number of colors such that some exact $AS_m(n)$-coloring of $\{1, 2, \ldots, n\}$ lacks a rainbow solution to $\mathcal{E}_m$. Note that $RS_m(n) = AS_m(n) + 1$, for all $n \geq 3$. As these concepts are equivalent, we will focus on rainbow Schur numbers.

In [32], it was shown that

$$RS_3(n) = \lfloor \log_2(n) \rfloor + 2, \quad \text{for all } n \geq 3.$$

This result was independently proved by Fallon, Giles, Rehm, Wagner, and Warnberg [90], who also proved that

$$RS_4(n) = \left\lceil \frac{n+6}{2} \right\rceil, \quad \text{for all } n \geq 6.$$

The evaluation of $RS_m(n)$ was completed in 2024 in [44], where it was shown that for all $m \geq 4$ and $n \geq \frac{m(m-1)}{2}$,

$$RS_m(n) = \left\lceil \frac{(m-3)n + \frac{m(m-1)}{2}}{m-2} \right\rceil. \tag{1.3}$$

2

Chromatic Ramsey Numbers

Weakened Ramsey numbers have their origins in the foundational work of K. Chung, M. Chung, and C. Liu [62, 63, 164] starting in 1977, where they introduced the chromatic Ramsey number. Although it is unclear whether or not Chung, Chung, and Liu were aware of it, the numbers they considered were closely related to a problem posed by Erdős and Shelah (see Problem V of [82]). In this chapter, we give an overview of the known results involving chromatic Ramsey numbers. Note that our use of the name "chromatic Ramsey number" is different from that of Burr, Erdős, and Lovasz [54] and Chartrand and Polimeni [58], which depend upon the chromatic number of the graph being colored, and the chromatic numbers of the arguments, respectively.

2.1 Background and General Bounds

Given $t \geq 3$ colors, let $n, s_1, s_2, \ldots, s_n \in \mathbb{N}$ satisfy $1 \leq s_1 < s_2 < \cdots < s_n < t$. Order the $\binom{t}{s_1}$ subsets of s_1 of the colors, the $\binom{t}{s_2}$ subsets of s_2 of the colors, ..., the $\binom{t}{s_n}$ subsets of s_n of the colors, and let $b := \sum_{i=1}^{n} \binom{t}{s_i}$. For graphs $G_1, G_2, \ldots, G_b$, define the *t-color* $(s_1, s_2, \ldots, s_n)$-*chromatic Ramsey number* $r^t_{s_1,s_2,\ldots,s_n}(G_1, G_2, \ldots, G_b)$ to be the least $p \in \mathbb{N}$ such that every t-coloring of K_p contains a subgraph isomorphic to G_i whose edges only use colors in the ith subset of colors, for some $i \in \{1, 2, \ldots, b\}$. A t-coloring of $K_{r^t_{s_1,s_2,\ldots,s_n}(G_1,G_2,\ldots,G_b)-1}$ that lacks a subgraph isomorphic to G_i whose edges only use colors in the ith subset of colors, for all $i \in \{1, 2, \ldots, b\}$, is called a *critical coloring for* $r^t_{s_1,s_2,\ldots,s_n}(G_1, G_2, \ldots, G_b)$. In the special case where $G_1 = K_1$, we obtain the trivial value

$$r^t_{s_1,s_2,\ldots,s_n}(K_1, G_2, G_3, \ldots, G_b) = 1,$$

for all graphs $G_2, G_3, \ldots, G_b$.

Note that the defintion of $r^t_{s_1,s_2,\ldots,s_n}(G_1, G_2, \ldots, G_b)$ is dependent upon the chosen order of the subsets of colors. If $n = 1$ and $s_1 = t-1$, every ordering of the subsets can be achieved by permuting the colors, making the value of $r^t_{t-1}(G_1, G_2, \ldots, G_t)$ independent of the order of the arguments. These numbers are equivalent to the *complementary Ramsey numbers* studied in [85], [88], and [181]. In this case, we order the $(t-1)$-element subsets by letting the ith subset be

$$S_i := \{1, 2, \ldots, t\} \setminus \{t - i + 1\}.$$

That is, S_1 is missing color t, S_2 is missing color $t-1$, ..., S_t is missing color 1. For small values of t, we refer to the colors 1, 2, 3, 4, and 5 as red, blue, green, purple, and orange, respectively. When $G_1 = K_2$, avoiding a K_2 in colors $1, 2, \ldots, t-1$ forces all of the edges in a t-coloring of K_p to have color t. It follows that

$$r^t_{t-1}(K_2, G_2, G_3, \ldots, G_t) = \min\{|V(G_i)| \mid 2 \leq i \leq t\}.$$

DOI: 10.1201/9781003498865-2

Besides the $n = 1$ and $s_1 = t - 1$ case, we will also consider chromatic Ramsey numbers of the forms

$$r_{1,2}^3(G_1, G_2, G_3, G_{12}, G_{13}, G_{23}),$$

$$r_2^4(G_{12}, G_{34}, G_{13}, G_{24}, G_{14}, G_{23}),$$

$$r_{1,2}^4(G_1, G_2, G_3, G_4, G_{12}, G_{34}, G_{13}, G_{24}, G_{14}, G_{23}),$$

and

$$r_{1,3}^4(G_1, G_2, G_3, G_4, G_{123}, G_{124}, G_{134}, G_{234}).$$

Here, G_i is a graph with edges in color i, G_{ij} is a graph with edges in colors i and j, and G_{ijk} is a graph with edges in colors i, j, and k. Throughout this book, the orders of these subsets of colors will be fixed.

When $G_1 = G_2 = \cdots = G_b$, we write $r_s^t(G_1)$ in place of $r_s^t(G_1, G_2, \ldots, G_b)$ and call the resulting s-chromatic Ramsey number a *weakened Ramsey number*. This name was first introduced by Harborth and Möller [123] in 1999. Of course, the values of weakened Ramsey numbers are independent of the order of the $\binom{t}{s}$ subsets of colors.

In the remainder of this section, we prove some general upper and lower bounds for various chromatic Ramsey numbers. First, we prove a 3-color 2-chromatic Ramsey number lower bound that follows from a lexicographic product of colored complete graphs (cf. Corollary 1.3).

Theorem 2.1 *For $s, t \geq 2$, let $m = r(K_s, K_t)$. For any tree T_m of order m and any $p \geq 2$,*

$$r_2^3(T_m, K_{p(s-1)+1}, K_{p(t-1)+1}) \geq p(m-1) + 1.$$

Proof Let $\mathcal{G}$ be a critical coloring for $r(K_s, K_t)$ in colors red and blue. That is, $\mathcal{G}$ is a 2-coloring of K_{m-1} that avoids a red K_s and a blue K_t. Replace each of the vertices in a green K_p with copies of $\mathcal{G}$ (i.e., form the lexicographic product $K_p[\mathcal{G}]$). The resulting 3-coloring of $K_{p(m-1)}$ lacks a red/blue T_m since the subgraph G_{rb} spanned by the red and blue edges consists of p connected components of order $m - 1$. Defining G_{rg} and G_{bg} similarly, we find that

$$\omega(G_{rg}) = p(s-1) \quad \text{and} \quad \omega(G_{bg}) = p(t-1).$$

It follows that

$$r_2^3(T_m, K_{p(s-1)+1}, K_{p(t-1)+1}) > p(m-1),$$

which is equivalent to the inequality given in the statement of the theorem. □

The approach used to prove Theorem 2.1 can also be used to prove an analogous lower bound for 4-color 2-chromatic Ramsey numbers.

Theorem 2.2 *For $k, \ell, s, t \geq 2$, let $m = r(K_k, K_\ell)$ and $n = r(K_s, K_t)$. Then*

$$r_2^4(T_m, K_n, K_{(k-1)(s-1)+1}, K_{(\ell-1)(t-1)+1}, K_{(k-1)(t-1)+1}, K_{(\ell-1)(s-1)+1}) \geq (m-1)(n-1)+1.$$

Proof Suppose that $\mathcal{G}$ is a critical coloring for $r(K_k, K_\ell)$ in colors red and blue and $\mathcal{G}'$ is a critical coloring for $r(K_s, K_t)$ in colors green and purple. That is, $\mathcal{G}$ is a red/blue K_{m-1} that lacks a red K_k and a blue K_ℓ, and $\mathcal{G}'$ is a green/purple K_{n-1} that lacks a green K_s and a purple K_t. Replace each of the vertices in $\mathcal{G}'$ with a copy of $\mathcal{G}$ (i.e., form the lexicographic product $\mathcal{G}'[\mathcal{G}]$). The subgraph G_{rb} spanned by the red and blue edges lacks a T_m since it consists of $n - 1$ connected components of order $m - 1$. Also,

$$\omega(G_{rg}) = (k-1)(s-1), \quad \omega(G_{rp}) = (k-1)(t-1), \quad \omega(G_{bg}) = (\ell-1)(s-1),$$

$$\omega(G_{bp}) = (\ell-1)(t-1), \quad \text{and} \quad \omega(G_{gp}) = n-1.$$

It follows that

$$r_2^4(T_m, K_n, K_{(k-1)(s-1)+1}, K_{(\ell-1)(t-1)+1}, K_{(k-1)(t-1)+1}, K_{(\ell-1)(s-1)+1}) > (m-1)(n-1),$$

completing the proof of the theorem. □

For 3-color 2-chromatic Ramsey numbers for complete graphs, Chung and Liu [63] proved that if x is any vertex in a 3-coloring of a complete graph that lacks a red/blue K_i, a red/green K_j, and a blue/green K_k, then

$$\begin{aligned} \deg_r(x) + \deg_b(x) &\le r_2^3(K_{i-1}, K_j, K_l) - 1, \\ \deg_r(x) + \deg_g(x) &\le r_2^3(K_i, K_{j-1}, K_k) - 1, \quad \text{and} \\ \deg_b(x) + \deg_g(x) &\le r_2^3(K_i, K_j, K_{k-1}) - 1. \end{aligned}$$

The following lemma generalizes this result to $t \ge 3$ colors.

Lemma 2.1 *If x is any vertex in a t-coloring of a complete graph that lacks a copy of G_i that only uses colors from S_i, for all $i \in \{1, 2, \ldots, t\}$, then*

$$\sum_{\substack{1\le j\le t \\ j\ne i}} \deg_j(x) \le r_{t-1}^t(K_{n_1}, \ldots, K_{n_{i-1}}, K_{n_i-1}, K_{n_{i+1}}, \ldots, K_{n_t}) - 1.$$

Proof Consider a critical coloring for $r_{t-1}^t(K_{n_1}, K_{n_2}, \ldots, K_{n_t})$ (i.e., a t-coloring of $K_{r_{t-1}^t(K_{n_1}, K_{n_2}, \ldots, K_{n_t})-1}$ that lacks a copy of K_{n_i} spanned by edges in the colors from S_i). Suppose that a vertex x satisfies

$$\sum_{\substack{1\le j\le t \\ j\ne i}} \deg_j(x) \ge r_{t-1}^t(K_{n_1}, \ldots, K_{n_{i-1}}, K_{n_i-1}, K_{n_{i+1}}, \ldots, K_{n_t}) \tag{2.1}$$

for some color $i \in \{1, 2, \ldots, t\}$. Then the subgraph induced by $\bigcup_{\substack{1\le j\le t \\ j\ne i}} N_i(x)$ contains a K_{n_i-1} spanned by the edges in colors from S_i, or for some $j \ne i$, it contains a K_{n_j} spanned by the edges in colors from S_j. In the latter cases, we are done. In the former case, including vertex x results in a K_{n_i} spanned by edges in colors from S_i. Either way, it follows that Inequality (2.1) is not true. □

The next two theorems provide general lower and upper bounds for the chromatic Ramsey number $r_2^3(K_i, K_j, K_k)$.

Theorem 2.3 ([63]) *For all $i \le j \le k$, $r_2^3(K_i, K_j, K_k) \ge (i-1)(j-1) + 1$.*

Proof We will form a 3-coloring of $K_{(i-1)(j-1)}$ that avoids a red/blue K_i, a red/green K_j, and a blue/green K_k. Begin by indexing the vertices with the set of ordered pairs

$$\{(a, b) \mid 1 \le a \le i-1 \text{ and } 1 \le b \le j-1\}.$$

Color an edge red if its vertices correspond with (a, b) and (a, b') for $b \ne b'$, and color an edge blue if its vertices correspond with (a, b) and (a', b) for $a \ne a'$ (see Figure 2.1 for the case where $i = j = 4$). Color all remaining edges green. The largest red/blue complete subgraph uses at most $i-1$ vertices if a green edge is avoided. Similarly, the largest red/blue complete subgraph uses at most $j-1$ vertices if a green edge is avoided. Finally, the largest

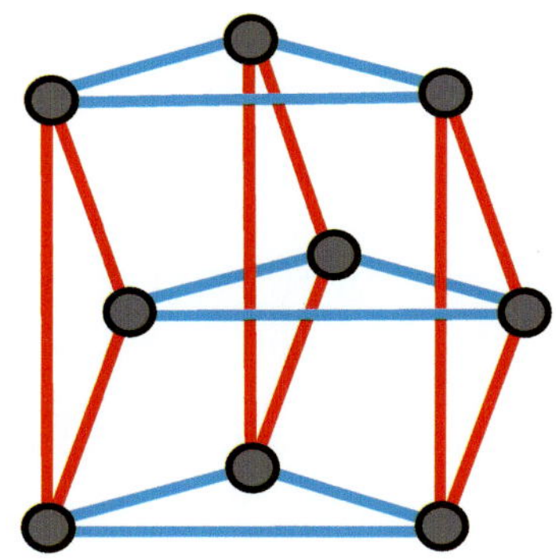

FIGURE 2.1: The lower bound construction for $r_2^3(K_4, K_4, K_k)$ given in Theorem 2.3.

blue/green subgraph uses at most $j-1 \leq k-1$ vertices if a red edge is avoided. It follows that

$$r_2^3(K_i, K_j, K_k) > (i-1)(j-1).$$

□

Theorem 2.4 ([63]) *For all* $i \leq j \leq k$,

$$r_2^3(K_i, K_j, K_k) \leq \left\lceil \frac{1}{2}\left(r_2^3(K_{i-1}, K_j, K_k) + r_2^3(K_i, K_{j-1}, K_k) + r_2^3(K_i, K_j, K_{k-1})\right) \right\rceil.$$

Proof If we let

$$\ell = r_2^3(K_{i-1}, K_j, K_k) + r_2^3(K_i, K_{j-1}, K_k) + r_2^3(K_i, K_j, K_{k-1}),$$

then by Lemma 2.1, every critical coloring for $r_2^3(K_i, K_j, K_k)$ satisfies

$$2(\deg_r(x) + \deg_b(x) + \deg_g(x)) \leq \ell - 3.$$

Thus, there will be a red/blue K_i, a red/green K_j, or a blue/green K_k if

$$\deg_r(x) + \deg_b(x) + \deg_g(x) \geq \left\lceil \frac{\ell - 2}{2} \right\rceil.$$

It follows that

$$r_2^3(K_i, K_j, K_k) \leq \left\lceil \frac{\ell - 2}{2} \right\rceil + 1 = \left\lceil \frac{\ell}{2} \right\rceil.$$

□

In the case of t colors, the following theorem was proved in [149].

Theorem 2.5 ([149]) *If* $t \geq 3$ *and* $G_1, G_2, \ldots, G_t$ *are graphs, then*

$$r_{t-1}^t(G_1, G_2, \ldots, G_t) \leq r_{t-2}^{t-1}(G_1, G_2, \ldots, G_{t-1}).$$

Moreover, if $|V(G_t)| \geq r_{t-2}^{t-1}(G_1, G_2, \ldots, G_{t-1})$, *then*

$$r_{t-1}^t(G_1, G_2, \ldots, G_t) = r_{t-2}^{t-1}(G_1, G_2, \ldots, G_{t-1}).$$

Proof Let $p = r^{t-1}_{t-2}(G_1, G_2, \ldots, G_{t-1})$ and consider a t-coloring

$$f : E(K_p) \longrightarrow \{1, 2, \ldots, t\}.$$

Define the $(t-1)$-coloring f' of K_p as follows:

$$f'(e) = \begin{cases} f(e) & \text{if } f(e) \in \{1, 2, \ldots, t-2\} \\ t-1 & \text{if } f(e) = t-1 \text{ or } f(e) = t, \end{cases}$$

where $e \in E(K_n)$. So, the coloring f' of K_p contains a copy of G_i spanned by edges in colors $1, 2, \ldots, i-1, i+1, i+2 \ldots, t-1$. It follows that

$$r^t_{t-1}(G_1, G_2, \ldots, G_t) \le r^{t-1}_{t-2}(G_1, G_2, \ldots, G_{t-1}).$$

Now assume that $|V(G_t)| \ge r^{t-1}_{t-2}(G_1, G_2, \ldots, G_{t-1})$. There exists a $(t-1)$-coloring of K_{p-1} that avoids a copy of G_i spanned by edges in colors $1, 2, \ldots, i-1, i+1, i+2 \ldots, t-1$, for every $i \in \{1, 2, \ldots, t-1\}$. Since this coloring does not have enough vertices to form a copy of G_t in colors $1, 2, \ldots, t-1$, it follows that

$$r^t_{t-1}(G_1, G_2, \ldots, G_t) \ge r^{t-1}_{t-2}(G_1, G_2, \ldots, G_{t-1}).$$

□

Theorem 2.5 generalizes a result from [62], where the $t = 3$ case was proved. As this is the case we will most commonly use, we note that this case states that

$$r^3_2(G_1, G_2, G_3) \le r(G_1, G_2),$$

with equality holding whenever $|V(G_3)| \ge r(G_1, G_2)$.

Now, we review some bounds for the more general class of $(s_1, s_2, \ldots, s_n)$-chromatic Ramsey numbers. The next two theorems offer a variation of Theorem 2.5 where $t = 3$, $s_1 = 1$, and $s_2 = 2$.

Theorem 2.6 ([164]) *For all* $i, j, k, m, n, p \in \mathbb{N}$,

$$r^3_{1,2}(K_i, K_j, K_k, K_m, K_n, K_p) \le r(K_i, K_j, K_k).$$

Furthermore, if $m \ge r(K_i, K_j)$, $n \ge r(K_i, K_k)$, *and* $p \ge r(K_j, K_k)$, *then*

$$r^3_{1,2}(K_i, K_j, K_k, K_m, K_n, K_p) = r(K_i, K_j, K_k).$$

Proof The inequality

$$r^3_{1,2}(K_i, K_j, K_k, K_m, K_n, K_p) \le r(K_i, K_j, K_k)$$

is trivial. To prove the second statement of the theorem, let $\ell = r(K_i, K_j, K_k)$ and consider a 3-coloring of $K_{\ell-1}$ that avoids red K_i, a blue K_j, and a green K_k. If $m \ge r(K_i, K_j)$, then a red/blue K_m is avoided. Similarly, if $n \ge r(K_i, K_k)$, and $p \ge r(K_j, K_k)$, then a red/green K_n and a blue/green K_p are also avoided. Hence,

$$r^3_{1,2}(K_i, K_j, K_k, K_m, K_n, K_p) \ge \ell.$$

□

Theorem 2.7 ([164]) *For all $i, j, k, m, n, p \in \mathbb{N}$,*

$$r^3_{1,2}(K_i, K_j, K_k, K_m, K_n, K_p) \le r^3_2(K_m, K_n, K_p).$$

Furthermore, if $i \ge \min\{m, n\}$, $j \ge \min\{m, p\}$, and $k \ge \min\{n, p\}$, then

$$r^3_{1,2}(K_i, K_j, K_k, K_m, K_n, K_p) = r^3_2(K_m, K_n, K_p).$$

Proof The inequality

$$r^3_{1,2}(K_i, K_j, K_k, K_m, K_n, K_p) \le r^3_2(K_m, K_n, K_p)$$

is trivial. To prove the second statement of the theorem, let let $\ell = r^3_2(K_m, K_n, K_p)$ and consider a 3-coloring of $K_{\ell-1}$ that avoids red/blue K_m, a red/green K_n, and a blue/green K_p. If $i \ge \min\{m, n\}$, then a red K_i is avoided. Similarly, if $j \ge \min\{m, p\}$ and $k \ge \min\{n, p\}$, then a blue K_j and a green K_k are avoided. Hence,

$$r^3_{1,2}(K_i, K_j, K_k, K_m, K_n, K_p) \ge \ell.$$

□

Besides Theorems 2.6 and 2.7, the following trivial upper bounds follow from the definitons of the respective Ramsey numbers:

$$r^3_{1,2}(K_i, K_j, K_k, K_m, K_n, K_p) \le r(K_i, K_p), \tag{2.2}$$
$$r^3_{1,2}(K_i, K_j, K_k, K_m, K_n, K_p) \le r(K_j, K_n), \quad \text{and}$$
$$r^3_{1,2}(K_i, K_j, K_k, K_m, K_n, K_p) \le r(K_k, K_m). \tag{2.3}$$

Theorem 2.8 provides one additional upper bound for $r^3_{1,2}(K_i, K_j, K_k, K_m, K_n, K_p)$.

Theorem 2.8 ([164]) *For $i, j, k, m, n, p \in \mathbb{N}$, let*

$$t_1 := \min\{r(K_{i-1}, K_p) + r(K_{j-1}, K_n) - 2, r^3_{1,2}(K_i, K_j, K_k, K_{m-1}, K_n, K_p) - 1\},$$
$$t_2 := \min\{r(K_{i-1}, K_p) + r(K_{k-1}, K_m) - 2, r^3_{1,2}(K_i, K_j, K_k, K_m, K_{n-1}, K_p) - 1\}, \quad \textit{and}$$
$$t_3 := \min\{r(K_{j-1}, K_n) + r(K_{k-1}, K_m) - 2, r^3_{1,2}(K_i, K_j, K_k, K_m, K_n, K_{p-1}) - 1\}.$$

Then

$$r^3_{1,2}(K_i, K_j, K_k, K_m, K_n, K_p) \le \left\lceil \frac{1}{2}(t_1 + t_2 + t_3 + 1) \right\rceil + 1.$$

Proof Consider a 3-coloring of K_ℓ that avoids a red K_i, a blue K_j, a green K_k, a red/blue K_m, a red/green K_n, and a blue/green K_p. Select a vertex x and note that $\deg_r(x) \le r(K_{i-1}, K_p) - 1$, otherwise, there would exist a red K_{i-1} (which, along with x, would form a red K_i) or a blue/green K_p. Similarly, $\deg_b(x) \le r(K_{j-1}, K_n) - 1$ and $\deg_g(x) \le r(K_{k-1}, K_m) - 1$. We also note that if

$$\deg_r(x) + \deg_b(x) \ge r^3_{1,2}(K_i, K_j, K_k, K_m, K_n, K_p),$$

then the subgraph induced by $N_r(x) \cup N_b(x)$ contains a red K_i, a blue K_j, a green K_k, a red/blue K_{m-1} (which, along with x, forms a red/blue K_i), a red/green K_n, or a blue/green K_p. Hence, we obtain the inequality

$$\deg_r(x) + \deg_b(x) \le r^3_{1,2}(K_i, K_j, K_k, K_{m-1}, K_n, K_p) - 1.$$

Similarly, we have

$$\deg_r(x) + \deg_g(x) \leq r^3_{1,2}(K_i, K_j, K_k, K_m, K_{n-1}, K_p) - 1$$

and

$$\deg_b(x) + \deg_g(x) \leq r^3_{1,2}(K_i, K_j, K_k, K_m, K_n, K_{p-1}) - 1.$$

Therefore, the following inequalities must be simultaneously satisfied:

$$\deg_r(x) + \deg_b(x) \leq t_1, \quad \deg_r(x) + \deg_g(x) \leq t_2, \quad \text{and} \quad \deg_b(x) + \deg_g(x) \leq t_3.$$

Adding these three inequalities together, we obtain

$$2(\deg_r(x) + \deg_b(x) + \deg_g(x)) \leq t_1 + t_2 + t_3.$$

So, a 3-coloring of K_ℓ will contain a red K_i, a blue K_j, a green K_k, a red/blue K_m, a red/green K_n, or a blue/green K_p if

$$2(\deg_r(x) + \deg_b(x) + \deg_g(x)) \geq t_1 + t_2 + t_3 + 1,$$

which is equivalent to

$$\ell - 1 = \deg_r(x) + \deg_b(x) + \deg_g(x) \geq \left\lceil \frac{1}{2}(t_1 + t_2 + t_3 + 1) \right\rceil.$$

The inequality

$$r^3_{1,2}(K_i, K_j, K_k, K_m, K_n, K_p) \leq \left\lceil \frac{1}{2}(t_1 + t_2 + t_3 + 1) \right\rceil + 1$$

then follows. □

In 1982, Jacobson [133] focused on the case $t = 4$, $s_1 = 1$, and $s_2 = 2$. He proved the following theorem, giving an upper bound for the chromatic Ramsey number

$$r^4_{1,2}(G_1, G_2, G_3, G_4, G_{12}, G_{34}, G_{13}, G_{24}, G_{14}, G_{23}).$$

Theorem 2.9 ([133]) *If* $s = \min\{r(G_{12}, G_{34}), r(G_{13}, G_{24}), r(G_{14}, G_{23})\}$, *then*

$$r^4_{1,2}(G_1, G_2, G_3, G_4, G_{12}, G_{34}, G_{13}, G_{24}, G_{14}, G_{23}) \leq s.$$

Proof Without loss of generality, suppose that $s = r(G_{12}, G_{34})$. Consider a 4-coloring of K_s. Viewing colors 1 and 2 as if they are the same color and colors 3 and 4 as if they are the same color, there exists a G_{12} in colors 1 and 2 or a G_{34} in colors 3 and 4. The inequality then follows. □

2.2 Chromatic Ramsey Numbers for Complete Graphs

In this section, we consider chromatic Ramsey numbers in which all of the arguments are complete graphs. The primary work on these numbers is due to Chung and Liu ([63] and [164]) and Jacobson ([133] and [132]).

2.2.1 3-Color 2-Chromatic Ramsey Numbers

We have already noted that the trivial 3-color 2-chromatic Ramsey numbers satisfy

$$r_2^3(K_1, K_j, K_k) = 1 \quad \text{and} \quad r_2^3(K_2, K_j, K_k) = j, \tag{2.4}$$

for all $2 \leq j \leq k$. Now we consider some nontrivial cases.

Theorem 2.10 ([63]) *For $k \geq 3$,*

$$r_2^3(K_3, K_3, K_k) = \begin{cases} 5 & \text{if } k \leq 5 \\ 6 & \text{if } k \geq 6. \end{cases}$$

Proof By Theorem 2.3, we see that

$$r_2^3(K_3, K_3, K_k) \geq 5,$$

for all $k \geq 3$. Now consider a 3-coloring of K_5. By Theorem 1.14, if a red/blue K_3 and a green K_3 are avoided, then there exists a green C_5. Similarly, if a red/green K_3 and a blue K_3 are avoided, then there exists a blue C_5. Together, the blue and green C_5-subgraphs of K_5 necessarily form a blue/green K_5. So, $r_2^3(K_3, K_3, K_5) \leq 5$, from which it follows that

$$r_2^3(K_3, K_3, K_3) \leq r_2^3(K_3, K_3, K_4) \leq r_2^3(K_3, K_3, K_5) = 5.$$

Now, suppose that $k \geq 6$ and consider a blue/green K_5 in which the subgraphs spanned by edges in each color form a C_5 (i.e., the unique critical coloring for $r^2(K_3)$ in blue and green that is described in Theorem 1.14). No red edge exists, so a red/blue K_3 and a red/green K_3 are avoided. Also, there are not enough vertices to form a blue/green K_k when $k \geq 6$. Hence, $r_2^3(K_3, K_3, K_k) \geq 6$. The reverse inequality follows from $r^2(K_3) = 6$ (Theorem 1.14) since avoiding a blue K_3 in a 3-coloring of K_6 results in a red/green K_3. □

Theorem 2.11 ([63]) *For $k \geq 4$,*

$$r_2^3(K_3, K_4, K_k) = \begin{cases} 7 & \text{if } k = 4 \\ 8 & \text{if } k \in \{5, 6\} \\ 9 & \text{if } k \geq 7. \end{cases}$$

Proof We separate the proof into cases, based on the value of k.

Case 1: Assume that $k = 4$. Consider a C_6 in which the edges alternate red and blue. From this C_6, form a K_6 by adding in the remaining edges and coloring all of them green. The result is the 3-coloring of K_6 given in the first image in Figure 2.2. No red/blue K_3 exists and the largest red/green or blue/green complete subgraphs are K_3's. Hence, it follows that $r_2^3(K_3, K_4, K_4) \geq 7$. The reverse inequality follows from Theorems 2.4 and 2.10 and the second equation in (2.4):

$$\begin{aligned} r_2^3(K_3, K_4, K_4) &\leq \left\lceil \frac{1}{2} \left(r_2^3(K_2, K_4, K_4) + r_2^3(K_3, K_3, K_4) + r_2^3(K_3, K_4, K_3) \right) \right\rceil \\ &= \left\lceil \frac{1}{2} (4 + 5 + 5) \right\rceil = 7. \end{aligned}$$

Case 2: Assume that $k \in \{5, 6\}$. Consider the 3-coloring of K_7 given in the second image in Figure 2.2. Since it does not contain a red/blue K_3, a red/green K_4, or a blue/green K_5. It follows that

$$r_2^3(K_3, K_4, K_6) \geq r_2^3(K_3, K_4, K_5) \geq 8.$$

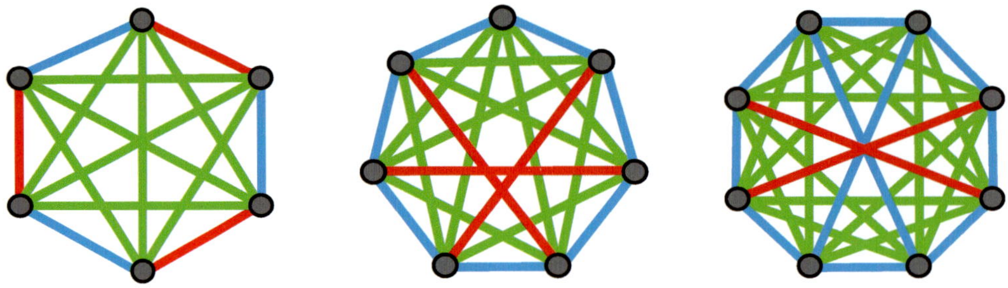

FIGURE 2.2: Critical colorings for $r_2^3(K_3, K_4, K_k)$, when $k \geq 4$.

To prove that the upper bound for both of these chromatic Ramsey numbers is also 8, consider a 3-coloring of K_8 and apply Lemma 2.1 and the second equation in (2.4) to see that every vertex x satisfies

$$\deg_r(x) + \deg_b(x) \leq r_2^3(K_2, K_4, K_k) - 1 = 3.$$

The total number of red and blue edges is at most 12. If G_r has size at most two, then deleting a single vertex from each of the red edges results in a blue/green K_6. So, we may assume that there are at least three red edges and at most nine blue edges. If G_b is bipartite, then one of its partite sets contains at least four vertices, which form a red/green K_4. So, G_b must contain a cycle of odd length. Let C be such a cycle, chosen to have minimal (odd) length.

If $|V(C)| = 3$, then there exists a red/blue K_3. If $|V(C)| = 5$, then no two nonadjacent vertices in C are joined with a blue edge as this would result in a blue C_3, contradicting the minimality of the length of C in G_b. Suppose that C is given by $v_1v_2 \cdots v_5v_1$, and denote the other vertices by v_6, v_7, and v_8. Since the subgraph induced by $\{v_6, v_7, v_8\}$ cannot be a blue K_3, we assume, without loss of generality, that edge v_6v_7 is red or green (say, red). See Figure 2.3. Note that no vertex x satisfies $\deg_b(x) \geq 4$ since if any blue edge exists among

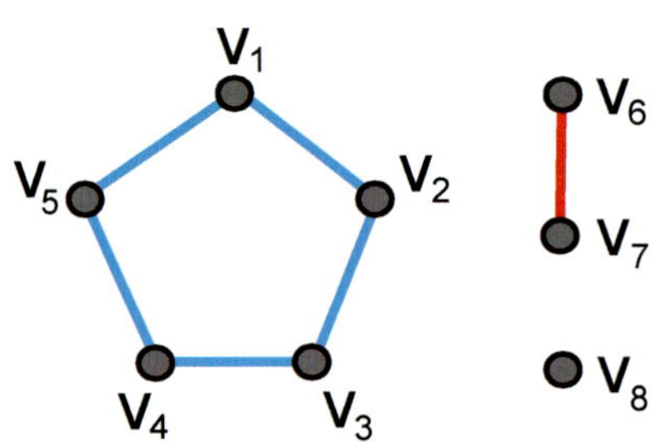

FIGURE 2.3: A blue cycle of length 5 in a 3-coloring of K_8.

the neighbors of x in G_b, a blue C_3 is formed. This would then force $N_b(x)$ to contain a red/green K_4. So, all vertices have blue degree at most 3. By Lemma 2.1 and Theorem 2.10, every vertex x satisfies

$$\deg_r(x) + \deg_g(x) \leq r_2^3(K_3, K_3, K_k) - 1 \leq 5.$$

It follows that every vertex has blue degree at least 2.

Since $2 \leq \deg_b(x) \leq 3$ for every vertex x and $|E(G_b)| \leq 9$, the only possible degree sequences in G_b are

$$2,2,2,2,2,2,2,2 \quad \text{and} \quad 3,3,2,2,2,2,2,2.$$

It follows that at least three of the vertices in C must join to both v_6 and v_7 with red or green edges. Two such vertices are not adjacent in C and these two vertices, along with v_6 and v_7, form a red/green K_4.

If $|V(C)| = 7$, then no two nonadjacent vertices in C are joined with a blue edge as this would result in a blue C_3 or a blue C_5, contradicting the minimality of the length of C in G_b. Suppose that C is given by $v_1v_2 \cdots v_7v_1$, and denote the other vertex by v_8. Since G_b has size at most 9, v_8 joins to at most two of the vertices in C with blue edges. Without loss of generality, suppose that v_1v_8 is blue. Then either the subgraph induced by $\{v_2, v_4, v_6, v_8\}$ or the subgraph induced by $\{v_3, v_5, v_7, v_8\}$ will be a red/green K_4. It follows that

$$r_2^3(K_3, K_4, K_5) \leq r_2^3(K_3, K_4, K_6) \leq 8.$$

Case 3: Assume that $k \geq 7$. Consider the 3-coloring of K_8 given in the third image in Figure 2.2. Since it does not contain a red/blue K_3, a red/green K_4, or a blue/green K_7, it follows that

$$r_2^3(K_3, K_4, K_k) \geq 9.$$

The Inequality

$$r_2^3(K_3, K_4, K_k) \leq r(K_3, K_4) = 9$$

follows from Theorem 2.5. □

Theorem 2.12 ([63]) *For $k \geq 5$,*

$$r_2^3(K_3, K_5, K_k) = \begin{cases} 9 & \text{if } k = 5 \\ 11 & \text{if } k = 6 \\ 12 & \text{if } k \in \{7, 8\} \\ 13 & \text{if } k \in \{9, 10, 11, 12, 13\} \\ 14 & \text{if } k \geq 14. \end{cases}$$

Proof We separate the proof into cases, based on the value of k.

Case 1: Assume that $k = 5$. Theorem 2.3 implies that $r_2^3(K_3, K_5, K_5) \geq 9$. To prove the reverse inequality, consider a 3-coloring of K_9. If G_r is bipartite, then by the pigeonhole principle, one of the partite sets contains at least five vertices. The subgraph induced by five such vertices forms a blue/green K_5. So, G_r must contain a cycle of odd length. Let C be such a cycle in G_r, chosen to have minimal (odd) length. If $|V(C)| = 3$, then it forms a red K_3. If $|V(C)| = 5$, then the nonadjacent vertices in C cannot be joined with a red edge or a blue edge without producing a red/blue K_3. Hence, all of the nonadjacent vertices in C join via green edges and the vertices in C induce a red/green K_5.

If $|V(C)| = 7$, then the nonadjacent vertices in C cannot be joined with a red edge without producing a red C_3 or C_5. Also, no vertex in C can be joined to two or more other vertices in C with blue edges without producing a red/blue K_3. It follows that G_b cannot have a subgraph that is isomorphic to C_7 or C_9. If G_b is bipartite, then one of the partite sets must contain at least five vertices and the subgraph induced by five such vertices forms a red/green K_5. If G_b contains a blue C_3, then it forms a blue K_3. If G_b contains a blue C_5, then the nonadjacent vertices in the C_5 cannot be joined with red or blue edges without producing a red/blue K_3. Hence, the vertices in the C_5 induce a blue/green K_5.

If $|V(C)| = 9$, then assume that C is given by $v_0v_1 \cdots v_8v_0$. No two nonadjacent vertices in C can be joined by a red edge as this would result in a red cycle with odd length less than 9, contradicting the minimality of C. If a red/blue K_3 is avoided, then edges

$$v_0v_2,\ v_1v_3,\ v_2v_4,\ v_3v_5,\ v_4v_6,\ v_5v_7,\ v_6v_8,\ v_7v_0, \text{ and } v_8v_1$$

must all be green. Following the same arguments given when $|V(C)| = 7$, the subgraph G_b is not bipartite and does not contain a blue C_3, C_5 or C_7. So, G_b must contain a blue C_9. The only possible blue edges are those of the forms v_iv_{i+3} and v_iv_{i+4}, where the indices are reduced modulo 9. We refer to such edges as having step sizes 3 and 4, respectively. We will show that the existence of the red and blue Hamiltonian cycles force the presence of a red/blue K_3. Note that no two consecutive edges in the blue C_9 can have step size 4 and no three consecutive edges in the blue C_9 can have step size 3 as either of these situations would result in a red/blue K_3. Thus, some vertex must be incident with edges in both steps sizes in the blue C_9. Without loss of generality, suppose that v_0v_3 and v_0v_5 are both blue edges. The othe edge in the blue C_9 that joins to v_5 must be v_5v_2 as v_5v_8 would result in $\{v_0, v_5, v_8\}$ forming a red/blue K_3 (see the first image in Figure 2.4).

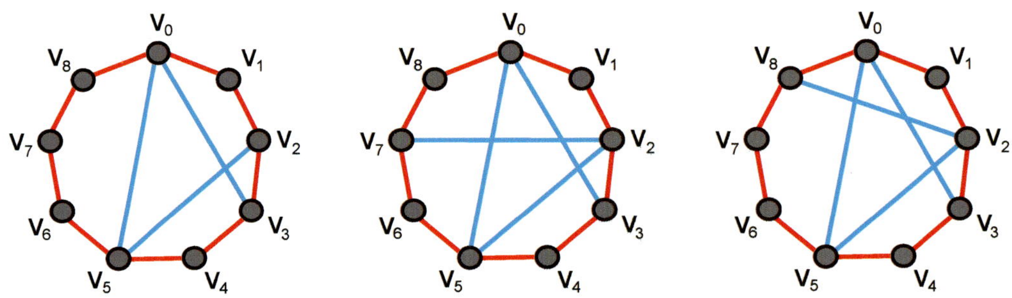

FIGURE 2.4: A 3-coloring of K_9 with red and blue Hamiltonian cycles.

The other edge in the blue C_9 that is incident with v_2 is either v_2v_7 or v_2v_8. If v_2v_7 is blue (see the second image in Figure 2.4), then either v_7v_1 or v_7v_4 will be the next blue edge. If v_7v_1 is blue, v_1v_4 is blue, v_4v_8 is blue, and there is no way to reach v_6 without forming a red/blue K_3. If v_7v_4 is the next blue edge, then v_4v_1 must be blue, v_2v_6 must be blue, and there is no way to reach v_8 without forming a red/blue K_3. If v_2v_8 is blue (see the third image in Figure 2.4), then v_8v_4 must be blue, v_4v_1 must be blue, and either v_1v_6 or v_1v_7 must be blue. In both cases, there is no way to complete the blue C_9 without forming a red/blue K_3. It follows that $r_2^3(K_3, K_5, K_5) \leq 9$.

Case 2: Assume that $k = 6$. The 3-coloring of K_{10} given in the first image in Figure 2.5 implies that $r_2^3(K_3, K_5, K_6) \geq 11$. The reverse inequality follows from Theorem 2.4, the second equation in (2.4), Theorem 2.11, and Case 1 of this proof:

$$\begin{aligned} r_2^3(K_3, K_5, K_6) &\leq \left\lceil \frac{1}{2}\left(r_2^3(K_2, K_5, K_6) + r_2^3(K_3, K_4, K_6) + r_2^3(K_3, K_5, K_5)\right)\right\rceil \\ &= \left\lceil \frac{1}{2}(5 + 8 + 9)\right\rceil = 11. \end{aligned}$$

Case 3: Assume that $k \in \{7, 8\}$. The 3-coloring of K_{11} given in the second image in Figure 2.5 implies that

$$r_2^3(K_3, K_5, K_8) \geq r_2^3(K_3, K_5, K_7) \geq 12.$$

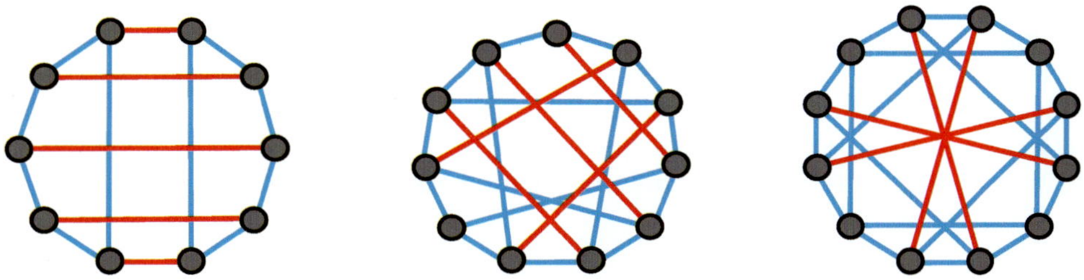

FIGURE 2.5: Critical colorings for $r_2^3(K_3, K_5, K_6)$, $r_2^3(K_3, K_5, K_7) = r_2^3(K_3, K_5, K_8)$, and $r_2^3(K_3, K_5, K_k)$, when $k \in \{9, 10, 11, 12, 13\}$. Edges not drawn are assumed to be green.

To prove that the upper bound for both of these chromatic Ramsey numbers is also 12, consider a 3-coloring of K_{12}. If a red/blue K_3, a red/green K_5, and a blue/green K_8 are avoided, then by Lemma 2.1, the second equation in (2.4), and Theorem 2.11,

$$\deg_r(x) + \deg_b(x) \leq r_2^3(K_2, K_5, K_8) - 1 = 4$$

and

$$\deg_r(x) + \deg_g(x) \leq r_2^3(K_3, K_4, K_8) - 1 = 8,$$

for any vertex x. Adding these two inequalities together, we obtain

$$2\deg_r(x) + \deg_b(x) + \deg_g(x) \leq 12.$$

This inequality, along with

$$\deg_r(x) + \deg_b(x) + \deg_g(x) = 11,$$

implies that $\deg_r(x) \leq 1$. It follows that $\deg_b(x) \in \{3, 4\}$ for every vertex x. Also, note that if a blue/green K_8 is avoided, then Theorem 1.7 implies that the subgraph $G_{b,g}$ spanned by the blue and green edges satisfies

$$|E(G_{b,g})| \leq \left(1 - \frac{1}{7}\right)\frac{12^2}{2} = 61.7143.$$

Since $|E(G_{b,g})| \in \mathbb{N}$, it follows that $|E(G_{b,g})| \leq 61$, from which we see that

$$|E(G_r)| \geq |E(K_{12})| - |E(G_{b,g})| = 66 - 61 = 5.$$

So, the edges in G_r span a matching of size 5 or 6. It follows that at most two vertices have blue degree 4 (when G_r is not a perfect matching).

We now argue that G_b does not contain a C_4-subgraph. Suppose false, and that such a cycle is given by $v_1v_2v_3v_4v_1$. If two non-adjacent vertices in the C_4 (say, v_1 and v_3) are only blue-adjacent to one of the other eight vertices, then there are at least six vertices (say, v_5, v_6, v_7, v_8, v_9, and v_{10}) that do not join to v_1 or v_3 with blue edges. If we group red and blue edges together, then $r^2(K_3) = 6$ (Theorem 1.14) implies that the subgraph induced by $\{v_5, v_6, \ldots, v_{10}\}$ contains a red/blue K_3 or a green K_3. Suppose the latter case is true and assume that the green G_3 has vertex set $\{v_5, v_6, v_7\}$. If a red/blue K_3 is avoided, then note that v_1v_3 must be green and the subgraph induced by $\{v_1, v_3, v_5, v_6, v_7\}$ is a red/green K_5. Thus, at least one of v_1 and v_3 must join to at least two of the eight vertices not in the C_4

with blue edges. The same is true for v_2 and v_4. Without loss of generality, suppose that v_1 and v_2 join to at least two of the eight vertices not in the C_4 with blue edges.

At this point, note that v_1 and v_3 cannot join to the same vertex (other than v_2 and v_4) with blue edges. This is because if v_1v_5 and v_3v_5 are both blue, then avoiding a red/blue K_3 would force v_2v_5, v_2v_4, and v_4v_5 to be green, and the subgraph induced by $\{v_1, v_2, v_3, v_4, v_5\}$ would be a blue/green K_5. Hence, we obtain the coloring shown in Figure 2.6. Now con-

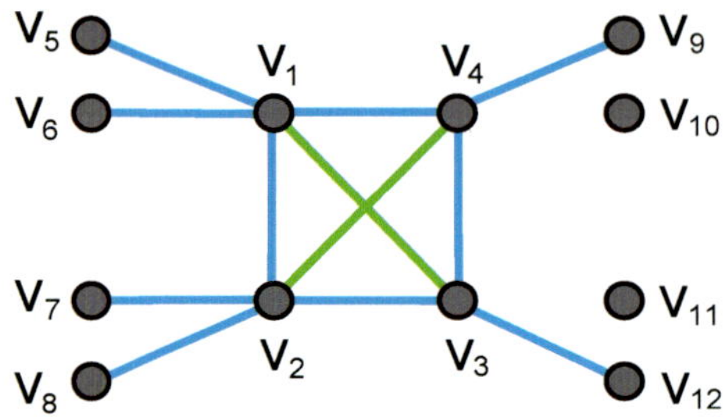

FIGURE 2.6: A 3-coloring of K_{12} that avoids a red/blue K_3 and a blue/green C_5, when a blue C_4 exists.

sider the subgraph induced by $X = \{v_5, v_6, v_{10}, v_{11}, v_{12}\}$. If it contains a red/green K_3, then that K_3, along with vertices v_2 and v_4, form a red/green K_5. It also avoids a blue K_3 since a red/blue K_3 is avoided. By Theorem 1.14, the subgraph induced by X must include a blue C_5. A similar argument can be used to show that the subgraph induced by $\{v_7, v_8, v_9, v_{10}, v_{11}\}$ contains a blue C_5. Noting that edges v_5v_6 and v_7v_8 are not blue, and the vertices v_3, v_5, v_6, v_7, and v_8 all have blue degree 3, it follows that the subgraph induced by $\{v_3, v_5, v_6, v_7, v_8\}$ is a red/green K_5. Hence, we have shown that it is not possible for a blue C_4 to exist.

Now, suppose that there exists a vertex v such that $\deg_b(v) = 4$. Since every vertex has blue degree at least 3, the vertices in $N_b(v)$ are each incident with two other edges to the remaining seven vertices. It follows that at least two of the vertices in $N_b(v)$ must join to the same vertex (other than v) with blue edges, forming a blue C_4. As we have already argued that a blue C_4 does not exist, it follows that every vertex has blue degree 3 and $|E(G_b)| = 18$.

Next, suppose that v_1v_2, v_1v_3, and v_1v_4 are blue. Then each of v_2, v_3, and v_4 must join to two vertices in $\{v_5, v_6, \ldots, v_{12}\}$ via blue edges. Avoiding a blue C_4 results in these blue edges going to six distinct vertices in $\{v_5, v_6, \ldots, v_{12}\}$. Without loss of generality, assume that v_2v_5, v_2v_6, v_3v_7, v_3v_8, v_4v_9, and v_4v_{10} are blue. Now v_{11} must have blue degree 3, so assume that v_5v_{11} is blue. Then v_6v_{11} cannot be blue without forming a blue C_4. Even if $v_{11}v_{12}$ is blue, one more blue edge must join v_{11} to $\{v_7, v_8, v_9, v_{10}\}$. Without loss of generality, suppose that v_7v_{11} is blue. Then $v_1v_2v_5v_{11}v_7v_3v_1$ is a blue C_6. Hence, we have shown that G_b must contain a blue C_6.

Finally, assume that $v_1v_2v_3v_4v_5v_6v_1$ is a blue C_6. Since $r^2(K_3) = 6$ (Theorem 1.14), there exists a green K_3, which we assume has vertex set $\{v_1, v_3, v_5\}$. Each of v_1, v_3, and v_5 must join to a unique vertex in $\{v_7, v_8, \ldots, v_{12}\}$ with blue edges. Suppose that v_1v_7, v_3v_8, and v_5v_9 are blue. Then all edges joining $\{v_1, v_3, v_5\}$ to $\{v_{10}, v_{11}, v_{12}\}$ are red or green. The subgraph induced by $\{v_{10}, v_{11}, v_{12}\}$ must contain at least one green edge which, along with v_1, v_3, and v_5, form a red/green K_5. Therefore, we have shown that

$$r_2^3(K_3, K_5, K_7) \leq r_2^3(K_3, K_5, K_8) \leq 12.$$

Case 4: Assume that $k \in \{9, 10, 11, 12, 13\}$. The 3-coloring of K_{12} given in the third image in Figure 2.5, implies that $r_2^3(K_3, K_5, K_k) \geq 13$. To prove the reverse inequality, consider a 3-coloring of K_{13}. If there are no red/blue K_3, red/green K_5, or blue/green K_k-subgraphs, then by Lemma 2.1, the second equation in 2.4, and Theorem 2.11,

$$\deg_r(x) + \deg_b(x) \leq r_2^3(K_2, K_5, K_k) - 1 = 4$$

and

$$\deg_r(x) + \deg_g(x) \leq r_2^3(K_3, K_4, K_k) - 1 = 8,$$

for any vertex x. Adding these two inequalities together, we obtain

$$2\deg_r(x) + \deg_b(x) + \deg_g(x) \leq 12.$$

This inequality, along with

$$\deg_r(x) + \deg_b(x) + \deg_g(x) = 12,$$

implies that $\deg_r(x) \leq 0$. So, all edges are blue or green and there exists a blue/green K_k for all $k \leq 13$. It follows that $r_2^3(K_3, K_5, K_k) \leq 13$.

Case 5: Assume that $k \geq 14$. Since $r(K_3, K_5) = 14$ (Theorem 1.16), Theorem 2.5 implies that $r_2^3(K_3, K_5, K_k) = 14$. □

2.2.2 3-Color (1,2)-Chromatic Ramsey Numbers

In this subsection, we focus on 3-color $(1, 2)$-chromatic Ramsey numbers for complete graphs.

Theorem 2.13 ([164]) *For $k \geq 3$,*

$$r_{1,2}^3(K_3, K_3, K_3, K_3, K_3, K_k) = \begin{cases} 5 & \text{if } k \leq 5 \\ 6 & \text{if } k \geq 6. \end{cases}$$

Proof Theorems 2.7 and 2.10 imply that

$$r_{1,2}^3(K_3, K_3, K_3, K_3, K_3, K_k) = r_2^3(K_3, K_3, K_k) = \begin{cases} 5 & \text{if } k \leq 5 \\ 6 & \text{if } k \geq 6. \end{cases}$$

□

Theorem 2.14 ([164]) *For $k \geq 4$, $r_{1,2}^3(K_3, K_3, K_3, K_3, K_4, K_k) = 6$.*

Proof In Figure 2.7, a 3-coloring of K_5 is given that lacks a a monochromatic K_3, a red/blue K_3, a red/green K_4, and a blue/green K_k. It follows that

$$r_{1,2}^3(K_3, K_3, K_3, K_3, K_4, K_k) \geq 6.$$

The reverse inequality follows from Inequality (2.3) and Theorem 1.14:

$$r_{1,2}^3(K_3, K_3, K_3, K_3, K_4, K_k) \leq r^2(K_3) = 6.$$

□

Theorem 2.15 ([164]) $r_{1,2}^3(K_3, K_3, K_4, K_4, K_4, K_4) = 8$.

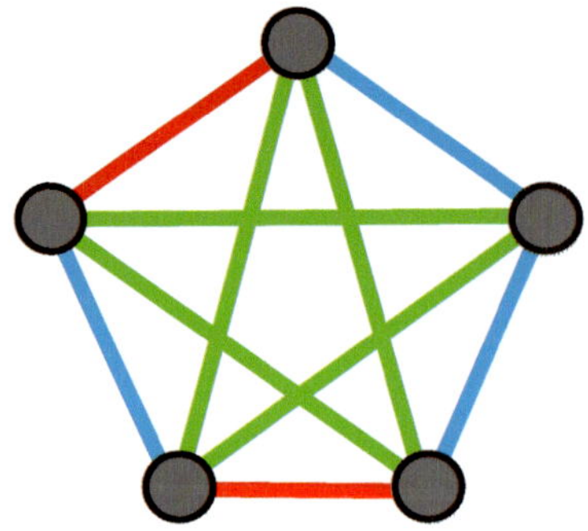

FIGURE 2.7: A 3-coloring of K_5 that avoids a monochromatic K_3, a red/blue K_3, a red/green K_4, and a blue/green K_k, when $k \geq 4$.

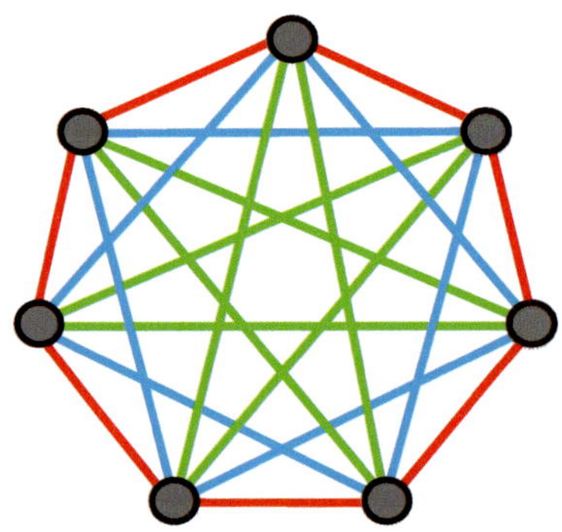

FIGURE 2.8: A 3-coloring of K_7 that avoids a monochromatic K_3 and a K_4 that uses at most two colors.

Proof The lower bound $r^3_{1,2}(K_3, K_3, K_4, K_4, K_4, K_4) \geq 8$ follows from the 3-coloring of K_7 given in Figure 2.8. To prove the reverse inequality, consider a 3-coloring of K_8 that avoids a red K_3, a blue K_3, a green K_4, and a K_4 that uses at most two colors. If $\deg_r(x) \geq 4$ for any vertex x, then $N_r(x)$ contains a red edge (which, along with x, forms a red K_3) or the subgraph induced by $N_r(x)$ is a blue/green K_4. So, $\deg_r(x) \leq 3$ for all x. Similarly, $\deg_b(x) \leq 3$ for all x, and note that these assumptions imply that $\deg_g(x) \geq 1$. If a vertex x satisfies $\deg_r(x) + \deg_g(x) \geq 6$, then by Theorem 1.14, $N_r(x) + N_g(x)$ contains a red/green K_3 (which, along with x, forms a red/green K_4) or it contains a blue K_3. So, $\deg_r(x) + \deg_g(x) \leq 5$. Similarly, $\deg_b(x) + \deg_g(x) \leq 5$. It follows that $\deg_g(x) \in \{1, 2, 3\}$ for every vertex x. We break the remainder of the proof into cases, based on the degrees of vertices in the subgraph G_g spanned by the green edges.

Case 1: Suppose that there exists a vertex x such that $\deg_g(x) = 3$. Since $\deg_r(x) + \deg_g(x) \leq 5$ and $\deg_b(x) + \deg_g(x) \leq 5$, it follows that $\deg_r(x) = \deg_b(x) = 2$. Let $N_g(x) = \{y_1, y_2, y_3\}$, $N_r(x) = \{y_4, y_5\}$, and $N_b(x) = \{y_6, y_7\}$. If the subgraph induced by $N_r(x) \cup N_g(x)$ avoids a red/green K_3 (which, along with x, would form a red/green K_4) and a blue K_3, then by Theorem 1.14, it contains a red/green C_5 and a blue C_5. We must consider two subcases.

Subcase 1.1: Assume that the subgraph induced by $N_g(x)$ contains exactly one blue edge. Without loss of generality, assume that blue C_5 in the subgraph induced by $N_r(x) \cup N_g(x)$ is given by $y_1y_3y_5y_2y_4y_1$ and the red/green C_5 is given by $y_1y_2y_3y_4y_5y_1$. In order to avoid a red K_3, y_4y_5 must be green. If both of y_1y_2 and y_2y_3 are green, then the subgraph induced by $\{x, y_1, y_2, y_3\}$ is a blue/green K_4. So, at least one of y_1y_2 and y_2y_3 must be red. Without loss of generality, assume that y_1y_2 is red (see the first image in Figure 2.9. If y_2y_3 and

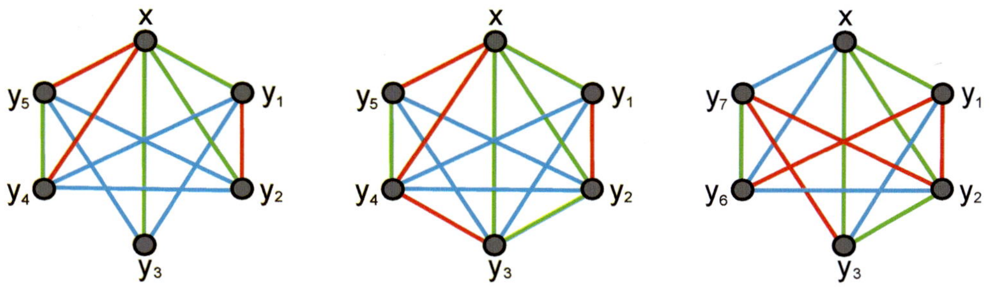

FIGURE 2.9: Subgraphs in a 3-coloring of K_8 in which the subgraph induced by $\{y_1, y_2, y_3\}$ contains a single blue edge.

y_3y_4 are both red, the subgraph induced by $\{y_1, y_2, y_3, y_4\}$ is a red/blue K_4. If y_2y_3 and y_1y_5 are both red, then the subgraph induced by $\{y_1, y_2, y_3, y_5\}$ is a red/blue K_4. If y_2y_3 is red, and y_1y_5 and y_3y_4 are both green, the the subgrpah induced by $\{y_1, y_3, y_4, y_5\}$ is a blue/green K_4. Hence, assume that y_2y_3 is green. If y_3y_4 is green, then the subgraph induced by $\{y_2, y_3, y_4, y_5\}$ is a blue/green K_4. So, assume that y_3y_4 is red (see the second image in Figure 2.9).

If the subgraph induced by $N_b(x) \cap N_g(x)$ avoids a blue/green K_3 (which, along with x, would form a blue/green K_4) and a red K_3, then by Theorem 1.14, it contains a blue/green C_5 and a red C_5. Without loss of generality, assume that the red C_5 is given by $y_1y_2y_7y_3y_6y_1$ and the blue/green C_5 is given by $y_1y_3y_2y_6y_7y_1$. If a blue K_3 is avoided, then edge y_6y_7 must be green. If the subgraph induced by $\{y_2, y_3, y_6, y_7\}$ is not a red/green K_4, then edge y_2y_6 must be blue (see the third image in Figure 2.9). At this point, we have the 3-coloring of K_8 shown in the first image in Figure 2.10.

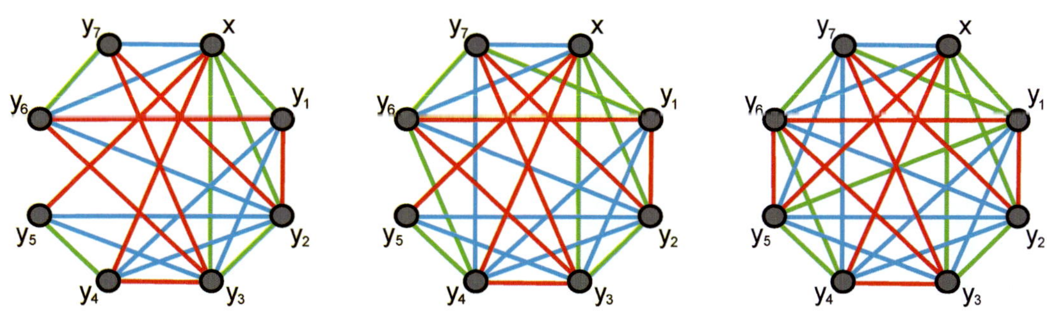

FIGURE 2.10: A 3-coloring of K_8 that avoids a red K_3, a blue K_3, a green K_4, and a K_4 that uses at most two colors.

If the subgraph induced by $\{y_1, y_2, y_4, y_6\}$ is not a red/blue K_4, then edge y_4y_6 must be green. If the subgraph induced by $\{y_3, y_4, y_6, y_7\}$ is not a red/green K_4, then edge y_4y_7 must be blue. Now y_1y_7 is part of a blue/green C_5, but in order to avoid having the subgraph induced by $\{y_1, y_4, y_7\}$ be a blue K_3, edge y_1y_7 must be green (see the second image in Figure 2.10).

Now edge y_1y_5 is part of a red/green C_5. If a blue K_3 is avoided, then edge y_5y_6 must also be red or green. In order for the subgraph induced by $\{y_1, y_5, y_5, y_7\}$ to not form a red/green K_4, edge y_5y_7 must be blue. In order for the subgraph induced by $\{y_4, y_5, y_6, y_7\}$ to not form a blue/green K_4, edge y_5y_6 must be red. For the subgraph induced by $\{y_1, y_2, y_5, y_6\}$ to not form a red/blue K_4, y_1y_5 must be green and we obtain the 3-coloring of K_8 shown in the third image in Figure 2.10. The subgraph induced by $\{y_1, y_4, y_5, y_7\}$ is a blue/green K_4, so this subcase does not occur.

Subcase 1.2: Assume that the subgraph induced by $N_g(x)$ contains two blue edges. Without loss of generality, assume that the blue C_5 in the subgraph induced by $N_r(x) \cup N_g(x)$ is given by $y_1y_2y_3y_4y_5y_1$ and the red/green C_5 is given by $y_1y_3y_5y_2y_4y_1$. If the subgraph induced by $\{x, y_1, y_2, y_3\}$ is not a blue/green K_4, then edge y_1y_3 must be red. (see the first image in Figure 2.11). In the subgraph induced by $N_b(x) \cup N_g(x)$, there exists a red C_5 and

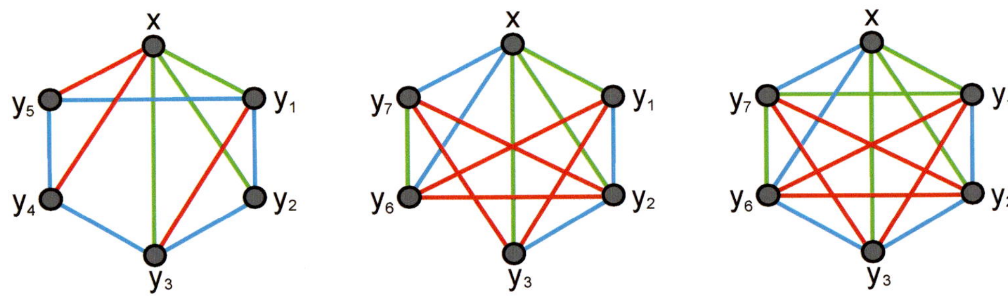

FIGURE 2.11: Subgraphs in a 3-coloring of K_8 in which the subgraph induced by $\{y_1, y_2, y_3\}$ contains two blue edges.

a blue/green C_5. Without loss of generality, assume that the red C_5 is given by $y_1y_3y_7y_2y_6y_1$ and the blue/green C_5 is given by $y_1y_2y_3y_6y_7y_1$. Since a red K_3 and a blue K_3 are avoided, edge y_6y_7 must be green, resulting in the second image in Figure 2.11.

Now edge y_1y_7 is part of a blue/green C_5, but if it is blue, then the subgraph induced by $\{y_1, y_2, y_3, y_7\}$ is a red/blue K_4. So, y_1y_7 must be green. Edge y_3y_6 is also part of a blue/green C_5, but if it is green, then the subgraph induced by $\{y_1, y_3, y_6, y_7\}$ is a red/green K_4. So, y_3y_6 must be blue (see the third image in Figure 2.11. At this point, we see that the subgraph induced by $\{y_1, y_2, y_3, y_6\}$ is a red/blue K_4, so this subcase does not occur.

Case 2: Suppose that no vertex has green degree 3 and there exists a vertex x such that $\deg_g(x) = 1$. Then $\deg_r(x) = \deg_b(x) = 3$. Corollary 1.1 implies that there are an even number of vertices that have green degree 1 (and all others have green degree 2). If there exists at least four vertices with green degree 1, then the subgraph G_{rb} spanned by the red and blue edges has size $|E(G_{rb})| \geq \frac{1}{2}(4 \cdot 6 + 4 \cdot 5) = 22$. If G_{rb} does not contain a K_4-subgraph, then Theorem 1.7 implies that

$$|E(G_{rb})| \leq \left\lfloor \left(1 - \frac{1}{3}\right) \frac{64}{2} \right\rfloor = 21,$$

which is a contradiction. Thus, exactly two vertices have green degree 1.

Denote the vertices by x, y, z_1, z_2, $\ldots$, z_6 with $\deg_g(x) = \deg_g(y) = 1$ and $\deg_g(z_i) = 2$, for all $i \in \{1, 2, \ldots, 6\}$. We obtain two subcases, based on whether or not xy is an edge in G_g (the subgraph spanned by the green edges).

Subcase 2.1: Assume that x and y are adjacent in G_g. The the subgraph of G_g induced by $\{z_1, z_2, \ldots, z_6\}$ is 2-regular, so it must be isomorphic to C_6 or $C_3 \cup C_3$. If it isomorphic to C_6 given by $z_1z_2 \cdots z_6z_1$, then the subgraph of the original 3-coloring of K_8 induced by

$\{z_1, z_3, z_5, x\}$ is a red/blue K_4. So, assume it is isomorphic to $C_3 \cup C_3$, with cycles given by $z_1z_2z_3z_1$ and $z_4z_5z_6z_4$. Not all of the edges joining x to $\{z_1, z_2, z_3\}$ are the same color (red or blue), then the subgrpah induced by $\{x, z_1, z_2, z_3\}$ is a K_4 that uses at most two colors. A similar argument applies to the edges joining x to $\{z_4, z_5, z_6\}$, leading to a coloring that isomorphic to the one shown in Figure 2.12. That is, we assume that edges xz_1, xz_2, and

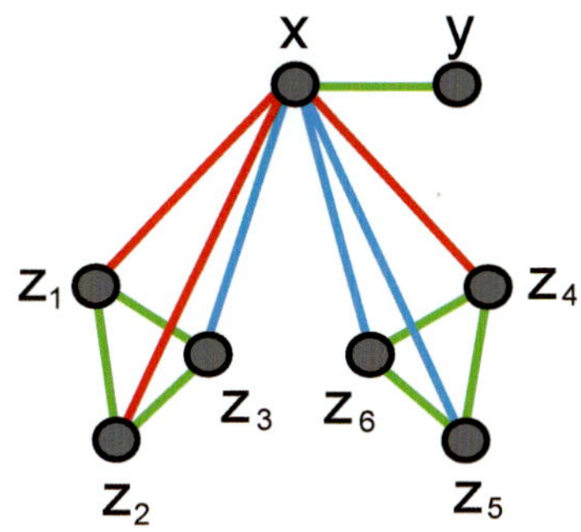

FIGURE 2.12: A 3-coloring of K_8 in which exactly two vertices have green degree 1.

xz_4 are red and edges xz_3, xz_5, and xz_6 are blue. Avoiding a red K_3 forces edges z_1z_4 and z_2z_4 to be blue. Avoiding a blue K_3 forces edges z_3z_5 and z_3z_6 to be red. If z_3z_4 is red, then the subgraph induced by $\{z_3, z_4, z_5, z_6\}$ is a red/green K_4. If z_3z_4 is blue, then the subgraph induced by $\{z_1, z_2, z_3, z_4\}$ is a blue/green K_4. Thus, this subcase does not occur.

Subcase 2.2: Assume that x and y are not adjacent in G_g. Then G_g is isomorphic to one of $P_3 \cup C_5$, $P_4 \cup C_4$, $P_5 \cup C_3$, or P_8. If it is isomorphic to $P_3 \cup C_5$, let the P_3 be given by $x_1x_2x_3$ and the C_5 be given by $y_1y_2y_3y_4y_5y_1$. Then the subgraph induced by $\{x_1, x_3, y_1, y_3\}$ is a red/blue K_4. If G_g is isomorphic to $P_4 \cup C_4$, let the P_4 be given by $x_1x_2x_3x_4$ and the C_4 be given by $y_1y_2y_3y_4y_1$. Then the subgraph induced by $\{x_1, x_3, y_1, y_3\}$ is a red/blue K_4. If G_g is isomorphic to $P_5 \cup C_3$, let the P_5 be given by $x_1x_2x_3x_4x_5$ and the C_3 be given by $y_1y_2y_3y_1$. Then the subgraph induced by $\{x_1, x_3, x_5, y_1\}$ is a red/blue K_4. If G_g is isomorphic to P_8, let it be given by $x_1x_2 \cdots x_8$. Then the subgraph induced by $\{x_1, x_3, x_5, x_7\}$ is a red/blue K_4. It follows that this subcase does not occur.

Case 3: Suppose that every vertex has green degree 2. Since G_g is 2-regular, it is isomorphic to one of C_8, $C_4 \cup C_4$, or $C_3 \cup C_5$. If it is isomorphic to C_8, let the cycle be given by $x_1x_2 \cdots x_8x_1$. Then the subgraph induced by $\{x_1, x_3, x_5, x_7\}$ is a red/blue K_4. If it is isomorphic to $C_4 \cup C_4$, let the two cycles be given by $x_1x_2x_3x_4x_1$ and $x_5x_6x_7x_8x_5$. Then the subgraph induced by $\{x_1, x_3, x_5, x_7\}$ is a red/blue K_4. If it is isomorphic to $C_3 \cup C_5$, let the cycles be given by $x_1x_2x_3x_1$ and $x_4x_5x_6x_7x_8x_4$. Each of x_1, x_2, and x_3 join to $\{x_4, x_5, x_6, x_7, x_8\}$ with either three red and two blue edges or three blue and two red edges. By the pigeonhole principle, at least two of x_1, x_2, and x_3 have the same degree sequence. Without loss of generality, suppose that x_1 and x_2 both join to $\{x_4, x_5, x_6, x_7, x_8\}$ with three red and two blue edges. Consider two subcases.

Subcase 3.1: Assume that three red edges joining x_1 to $\{x_4, x_5, x_6, x_7, x_8\}$ join to sequential vertices in the cycle. Without loss of generality, suppose that x_1x_4, x_1x_5, and x_1x_6 are red and x_1x_7 and x_1x_8 are blue. Since a red K_3 is avoided, x_4x_6 must be blue. By Theorem 1.14, if the subgraph induced by $N_r(x_1) \cup N_g(x_1)$ lacks a red/green K_3 (which, along with vertex x_1, would form a red/green K_4) and a blue K_3, then it must consist of a red/green C_5 and a blue C_5. Without loss of generality, suppose that the red/green C_5 is given by $x_2x_3x_4x_5x_6x_2$ and the blue C_5 is given by $x_2x_4x_6x_3x_5x_2$ (see the first image in Figure 2.13). If x_2x_6 is blue, then the subgraph induced by $\{x_2, x_3, x_5, x_6\}$ is a blue/green

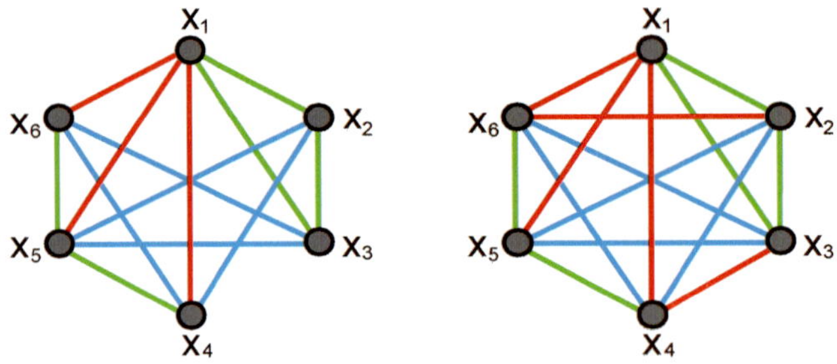

FIGURE 2.13: Subgraphs in a 3-coloring of K_8 in which x_1 joins with red edges to three sequential vertices in a green C_5.

K_4. So, x_2x_6 must be red. If x_3x_4 is blue, then the subgraph induced by $\{x_2, x_3, x_4, x_5\}$ is a blue/green K_4. So, x_3x_4 must be red (see the second image in Figure 2.13).

At this point, we have the 3-coloring of K_8 shown in the first image in Figure 2.15. Since

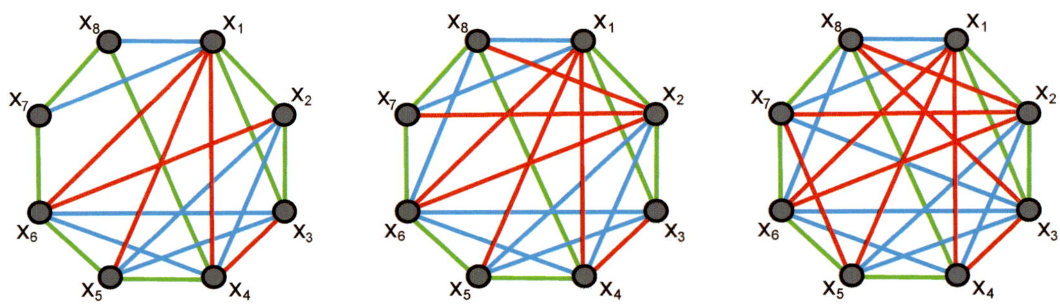

FIGURE 2.14: A 3-coloring of K_8 in which x_1 joins with red edges to three sequential vertices in a green C_5.

x_2 is joins to $\{x_4, x_5, x_6, x_7, x_8\}$ with three red edges and two blue edges, both x_2x_7 and x_2x_8 must be red. In order to avoid a red K_3, x_6x_8 must be blue (see the second image in Figure 2.13). If x_3x_7 and x_3x_8 are both red, then the subgraph induced by $\{x_2, x_3, x_7, x_8\}$ is a red/green K_4. So, x_3 must join to at least one of x_7 and x_8 with a blue edge. We find that x_3 joins to exactly one of x_7 and x_8 with a blue edge, otherwise, x_3 would have blue degree 4. If x_3x_8 is blue, then the subgraph induced by $\{x_3, x_6, x_8\}$ is a blue K_3. So, x_3x_7 is blue and x_3x_8 is red. Avoiding a blue K_3 forces edge x_5x_7 to be red (see the third image in Figure 2.13).

The only edges that have not yet been colored are x_4x_7 and x_5x_8, which cannot both be red as the subgraph induced by $\{x_4, x_5, x_7, x_8\}$ woud then form a red/green K_4. If x_5x_8 is blue, then the subgraph induced by $\{x_4, x_5, x_6, x_8\}$ is a blue/green K_4. If x_4x_7 is blue, then the subgraph induced by $\{x_4, x_5, x_6, x_7\}$ is blue/green K_4. Hence, this subcase does not occur.

Subcase 3.2: Assume that three red edges joining x_1 to $\{x_4, x_5, x_6, x_7, x_8\}$ join to vertices that are not all sequential in the cycle. Without loss of generality, suppose that x_1x_4, x_1x_5, and x_1x_7 are red and x_1x_6 and x_1x_8 are blue. Avoiding a red K_3 forces edges x_4x_7 and x_5x_7 to be blue and avoiding a blue K_3 forces edge x_6x_8 to be red (see the first image in Figure 2.15). Since the subgraph induced by $N_r(x_1) \cup N_g(x_1)$ avoids a red/green K_3 (which, along with x_1, would form a red/green K_4) and a blue K_3, Theorem 1.14 implies that it must consist of a red/green C_5 and a blue C_5. However, since edges x_4x_7 and x_5x_7 are both blue,

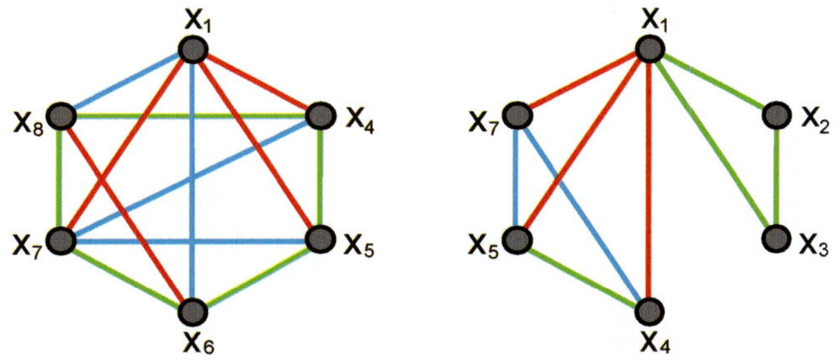

FIGURE 2.15: Subgraphs in a 3-coloring of K_8 in which x_1 joins with red edges to three vertices in a green C_5 that are not sequential.

there is no way to form the blue C_5 without edge x_2x_3 being blue (see the second image in Figure 2.15). As this edge is green, we find that this subcasse does not occur.

In all cases, we find that a 3-coloring of K_8 contains a red K_3, a blue K_3, a green K_4, or a K_4 that uses at most two colors. It follows that

$$r^3_{1,2}(K_3, K_3, K_4, K_4, K_4, K_4) \leq 8,$$

completing the proof of the theorem. □

Theorem 2.16 ([164]) *For* $j, k \geq 4$,

$$r^3_{1,2}(K_3, K_3, K_3, K_4, K_j, K_k) = \begin{cases} 8 & \text{if } j = k = 4 \\ 9 & \text{if } j \geq 4 \text{ and } k \geq 5. \end{cases}$$

Proof We consider the two cases separately.

Case 1: Suppose that $j = k = 4$. Figure 2.8 shows a 3-coloring of K_7 that lacks a monochromatic K_3 and a K_4 spanned by edges that use at most two colors. It follows that

$$r^3_{1,2}(K_3, K_3, K_3, K_4, K_4, K_4) \geq 8.$$

The upper bound follows from Theorem 2.15:

$$r^3_{1,2}(K_3, K_3, K_3, K_4, K_4, K_4) \leq r^3_{1,2}(K_3, K_3, K_4, K_4, K_4, K_4) \leq 8.$$

Case 2: Suppose that $j \geq 4$ and $k \geq 5$. The 3-coloring of K_8 given in Figure 2.16 lacks a monochromatic K_4, a red/blue K_4, and red/green K_4, and a blue/green K_5. It follows that

$$r^3_{1,2}(K_3, K_3, K_3, K_4, K_4, K_k) \geq 9.$$

This case now follows from Inequality (2.3) and Theorem 1.15:

$$9 \leq r^3_{1,2}(K_3, K_3, K_3, K_4, K_4, K_k) \leq r^3_{1,2}(K_3, K_3, K_3, K_4, K_j, K_k) \leq r(K_3, K_4) = 9.$$

□

Theorem 2.17 ([164]) $r^3_{1,2}(K_3, K_4, K_4, K_4, K_4, K_4) = 9$.

Proof In Figure 2.17, a 3-coloring of K_8 is given that avoids a red K_3 and K_4-subgraphs that use at most two edges (and hence, are not monochromatic). It follows that

$$r^3_{1,2}(K_3, K_4, K_4, K_4, K_4, K_4) \geq 9.$$

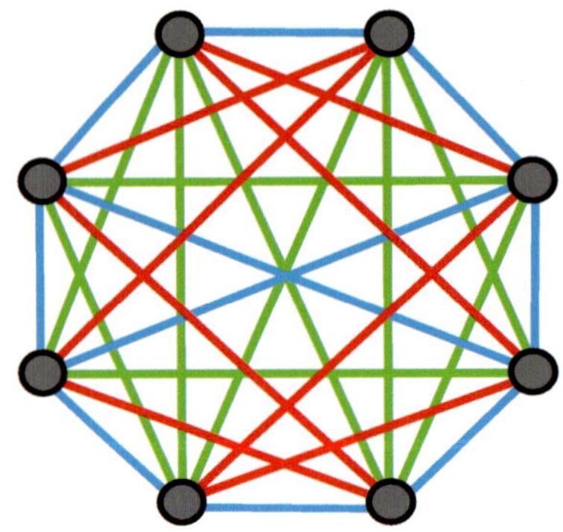

FIGURE 2.16: A critical coloring for $r^3_{1,2}(K_3, K_3, K_3, K_4, K_4, K_5)$.

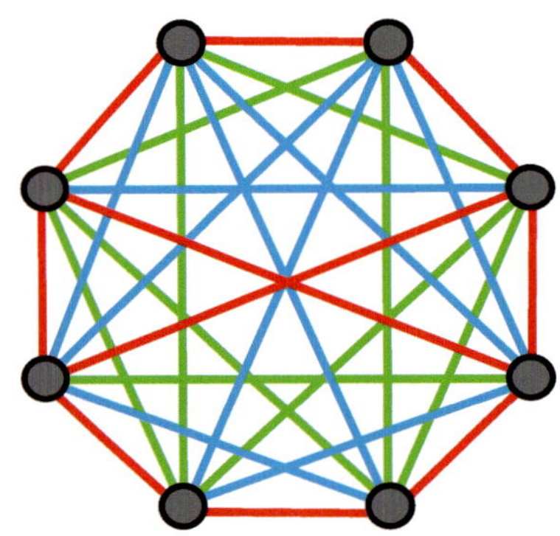

FIGURE 2.17: A critical coloring for $r^3_{1,2}(K_3, K_4, K_4, K_4, K_4, K_4)$.

Inequality (2.2) and Theorem 1.15 imply the upper bound

$$r^3_{1,2}(K_3, K_4, K_4, K_4, K_4, K_4) \leq r(K_3, K_4) = 9.$$

□

Theorem 2.18 ([164]) *For* $j, k \geq 5$,

$$r^3_{1,2}(K_3, K_3, K_3, K_5, K_j, K_k) = 14.$$

Proof Consider the 3-coloring of K_{13} given in Figure 2.18. Since a monochromatic K_3 and a K_5 that uses at most two colors are avoided, it follows that

$$r^3_{1,2}(K_3, K_3, K_3, K_5, K_5, K_5) \geq 14.$$

Inequality (2.3) and Theorem 1.16 imply that

$$14 \leq r^3_{1,2}(K_3, K_3, K_3, K_5, K_5, K_5) \leq r(K_3, K_3, K_3, K_5, K_j, K_k) \leq r(K_3, K_5) = 14.$$

□

Theorem 2.19 ([164]) *For* $i, j, k \geq 6$,

$$r^3_{1,2}(K_3, K_3, K_3, K_i, K_j, K_k) = 17.$$

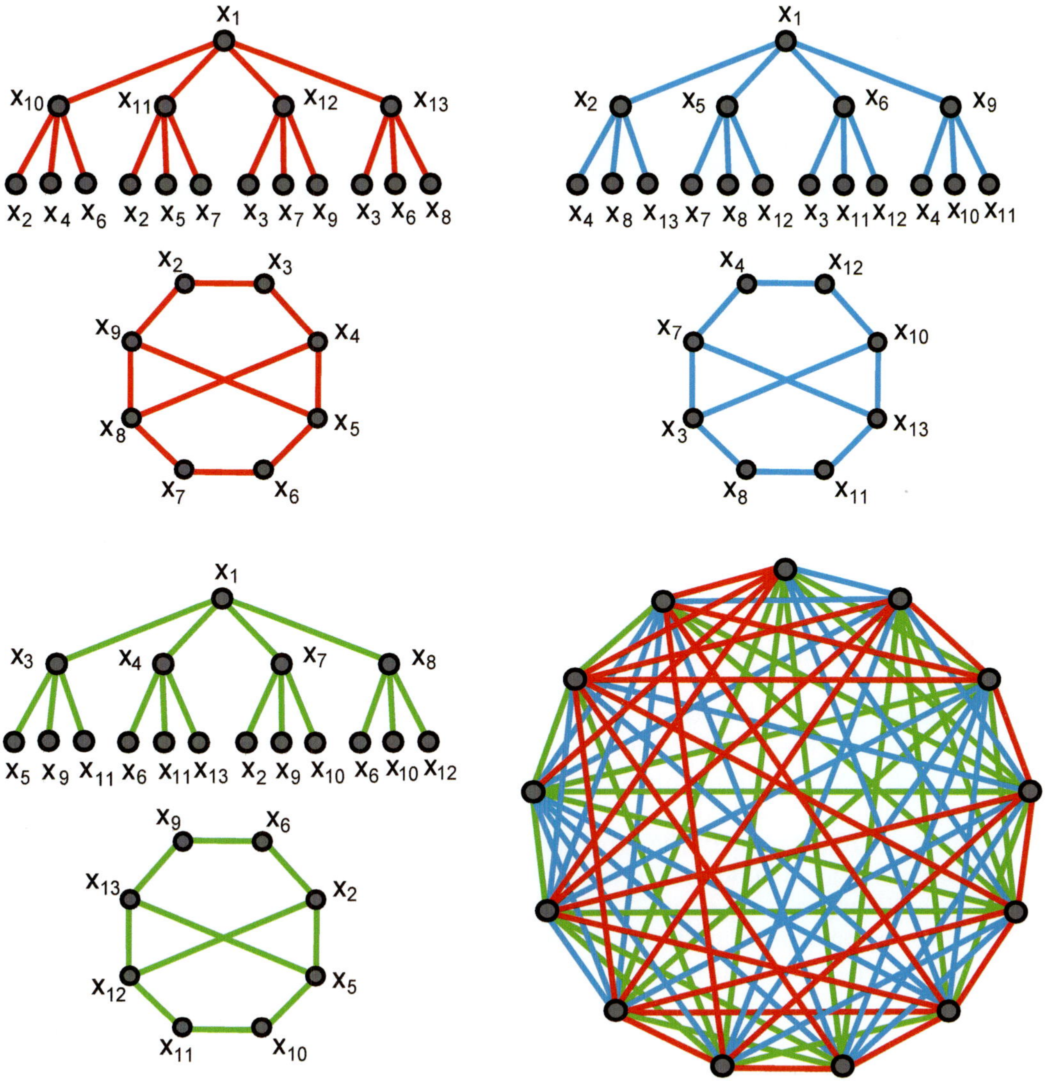

FIGURE 2.18: A critical coloring for $r^3_{1,2}(K_3, K_3, K_3, K_5, K_5, K_5)$.

Proof This theorem is an immediate consequence of Theorems 1.18 and 2.6, with the observation that $i, j, k \geq r^2(K_3) = 6$ (Theorem 1.14):

$$r^3_{1,2}(K_3, K_3, K_3, K_i, K_j, K_k) = r^3(K_3) = 17.$$

□

2.2.3 4-Color (1,2)-Chromatic Ramsey Numbers

We now turn our attention to the evaluation of 4-color $(1, 2)$-chromatic Ramsey numbers. These numbers were first considered by Jacobson [132, 133] in 1980.

Theorem 2.20 ([133]) *For all $n \geq 3$,*

$$r^4_{1,2}(K_3, K_3, K_3, K_3, K_3, K_3, K_3, K_3, K_3, K_n) = \begin{cases} 5 & \text{if } n = 3 \\ 6 & \text{if } n \geq 4. \end{cases}$$

Proof We handle the cases $n = 3$ and $n \geq 4$ separately.

Case 1: Suppose that $n = 3$. The first image in Figure 2.19 shows a 4-coloring of K_4 in which every K_3-subgraph is rainbow. It follow that $r^4_{1,2}(K_3) \geq 5$. To prove the reverse

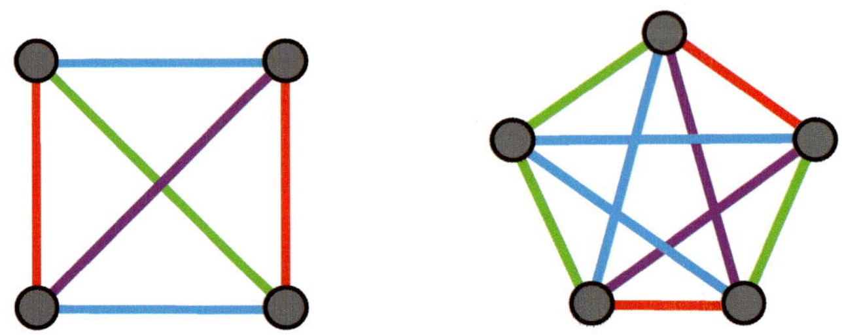

FIGURE 2.19: A critical coloring for $r^4_{1,2}(K_3, K_3, K_3, K_3, K_3, K_3, K_3, K_3, K_3, K_n)$ and a 4-coloring of K_5 that avoids a monochromatic K_3, a red/blue K_3, a green/purple K_3, a red/green K_3, a blue/purple K_3, a red/purple K_3, and a blue/green K_4. .

inequality, consider a 4-coloring of K_5. If a vertex is incident with two edges of the same color, then regardless of the color of the edge formed by the other endpoints of the two edges, a K_3 exists that is either monochromatic or that uses only two colors. The only other possibility is that every vertex is incident with edges that have distinct colors. In a 4-coloring of K_5, this forces the subgraph spanned by edges in any one color to have an odd number of vertices of odd degree, which is impossible. Hence, $r^4_{1,2}(K_3) \leq 5$.

Case 2: Suppose that $n \geq 4$. The second image in Figure 2.19 shows a 4-coloring of K_5 that avoids a monochromatic K_3, a red/blue K_3, a green/purple K_3, a red/green K_3, a blue/purple K_3, a red/purple K_3, and a blue/green K_4. It follows that

$$r^4_{1,2}(K_3, K_3, K_3, K_3, K_3, K_3, K_3, K_3, K_3, K_n) \geq 6.$$

The reverse inequality follows from Theorems 1.14 and 2.9:

$$r^4_{1,2}(K_3, K_3, K_3, K_3, K_3, K_3, K_3, K_3, K_3, K_n) \leq r(K_3, K_3) = 6.$$

□

Theorem 2.21 ([133]) *If $n_1 \geq 3$ and n_2, n_3, and n_4 are all greater than 3, then*

$$r^4_{1,2}(K_3, K_3, K_3, K_3, K_3, K_3, K_{n_1}, K_{n_2}, K_{n_3}, K_{n_4}) = 9.$$

Proof The lower bound follows from the 4-coloring of K_8 given in Figure 2.20 that avoids a monochromatic K_3, a red/blue K_3, a green/purple K_3, a red/green K_4, a blue/purple K_4, a red/purple K_4, and a blue/green K_4. It follows that

$$r^4_{1,2}(K_3, K_3, K_3, K_3, K_3, K_4, K_{n_1}, K_{n_2}, K_{n_3}, K_{n_4}) \geq 9.$$

The upper bound follows from Theorems 1.15 and 2.9:

$$r^4_{1,2}(K_3, K_3, K_3, K_3, K_3, K_4, K_{n_1}, K_{n_2}, K_{n_3}, K_{n_4}) \leq r(K_3, K_4) = 9.$$

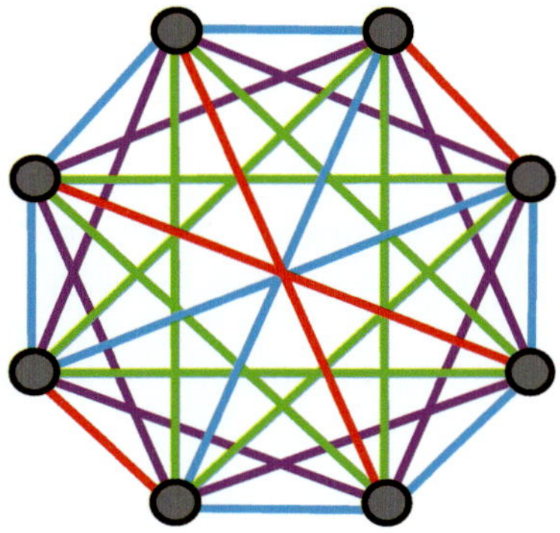

FIGURE 2.20: A critical coloring for $r^4_{1,2}(K_3, K_3, K_3, K_3, K_3, K_3, K_3, K_3, K_3, K_n)$.

□

Next, we will consider the evaluation of

$$r^4_{1,2}(K_3, K_3, K_3, K_3, K_3, K_4, K_3, K_4, K_3, K_4).$$

This $(1, 2)$-chromatic Ramsey number was proved in Jacobson's dissertation [132]. It requires that we first prove a couple of preliminary results concerning 4-colorings of K_8.

Lemma 2.2 ([132]) *Consider a 4-coloring of K_8 that avoids a red/blue K_3, a red/green K_3, a red/purple K_3, a green/purple K_4, a blue/purple K_4, and a blue/green K_4. In each of the subgraphs G_{rb}, G_{rg}, and G_{rp}, two distinct cycles of length 4 cannot share more than one edge.*

Proof Without loss of generality, suppose that the subgraph G_{rb} has two distinct cycles of length 4 that have at least two edges in common. Note that if they have three or four edges in common, then they cannot be distinct. So, suppose that they have exactly two edges in common. If the two common edges do not have a vertex in common, then a red/blue K_3 must exist. For example, if wx and yz are the edges common to two distinct cycles of length 4, then one cycle must be given by $wxyzw$ and the other must be given by $wxzyw$, from which it follows that the subgraph induced by $\{w, x, y, z\}$ is a red/blue K_4.

Now assume that the two common edges have a vertex in common. Suppose that they are given by wx and xy, z_1 and z_2 are the other vertices that make up the two cycles, and z_3, z_4, and z_5 are the other vertices. That is, the cycles are given by $wxyz_1w$ and $wxyz_2w$. Note that at least two of the vertices in $\{z_3, z_4, z_5\}$ must be independent, otherwise the subgraph induced by $\{z_3, z_4, z_5\}$ is a red/blue K_3. Without loss of generality, assume that z_3 and z_4 are not adjacent in G_{rb}. Also, note that $wy \notin E(G_{rb})$ since a red/blue K_3 is avoided. If w or y were to be incident with at least four edges that are red or blue, then the red/blue neighborhood would be an independent set of order 4 as the presence of a red or blue edge in the neighborhood would result in a red/blue K_3. As w and y already have degree 3 in G_{rb}, they cannot join to z_3 or z_4 with red or blue edges. Therefore, the subgraph induced by $\{w, y, z_3, z_4\}$ is a green/purple K_4 in the original 4-coloring of K_8. □

Theorem 2.22 ([132]) *If a 4-coloring of K_8 exists that avoids a red/blue K_3, a red/green K_3, a red/purple K_3, a green/purple K_4, a blue/purple K_4, and a blue/greeen K_4, then it contains exactly one red edge.*

Proof Consider a 4-coloring of K_8 and suppose that it lacks a red/blue K_3, a red/green K_3, a red/purple K_3, a green/purple K_4, a blue/purple K_4, and a blue/green K_4 and select an arbitrary vertex x. If $\deg_r(x)+\deg_b(x) \geq 4$, then the subgraph induced by $N_r(x) \cup N_b(x)$ either contains a red or blue edge (which, along with vertex x, forms a red/blue K_3) or all of its edges are green and purple (forming a green/purple K_4). It follows that $\deg_r(x) + \deg_b(x) \leq 3$. Similarly, we also see that $\deg_r(x) + \deg_g(x) \leq 3$ and $\deg_r(x) + \deg_p(x) \leq 3$. If $\deg_r(x) \in \{2,3\}$, these inequalities imply that $\deg_b(x), \deg_g(x), \deg_p(x) \in \{0,1\}$, which is impossible since

$$\deg_r(x) + \deg_b(x) + \deg_g(x) + \deg_p(x) = 7.$$

So,

$$\deg_r(x) \in \{0,1\} \quad \text{and} \quad \deg_b(x), \deg_g(x), \deg_p(x) \in \{2,3\}, \tag{2.5}$$

for any vertex x. In particular, note that if a vertex is incident with a red edge, then it is incident with exactly two blue, two green, and two purple edges.

By Theorem 2.16, this 4-coloring of K_8 must contain at least one red edge. Let xy be a red edge and denote the other vertices by $z_1, z_2, \ldots, z_6$. From (2.5), all edges joining $\{x,y\}$ to $\{z_1, z_2, \ldots, z_6\}$ are blue, green, and purple. In order for the degree condtions given in (2.5) to be satisfied, each of x and y must join to $\{z_1, z_2, \ldots, z_6\}$ with two blue edges, two green edges, and two purple edges. Without loss of generality, assume that xz_1 and xz_2 are blue. If xz_i and yz_i are the same color for any $i \in \{1,2,\ldots,6\}$, then the subgraph induced by $\{x,y,z_i\}$ is a K_3 that uses at most two colors, one of which is red. So, xz_i and yz_i are different colors for all $i \in \{1,2,\ldots,6\}$. Hence, edges yz_1 and yz_2 cannot be blue. Consider two cases.

Case 1: Assume that yz_1 and yz_2 are the same color (which we assume is green). Since x and y must each be incident with two purple edges and $N_p(x) \cap N_p(y) = \emptyset$, without loss of generality, assume that xz_3, xz_4, yz_5, and yz_6 are purple. Then xz_5 and xz_6 must be green and yz_3 and yz_4 must be blue. Avoiding a monochromatic K_3 (which follows from avoiding a K_3 that uses at most two colors, one of which is red) forces z_1z_2 to be purple, z_3z_4 to be green, and z_5z_6 to be blue (see Figure 2.21).

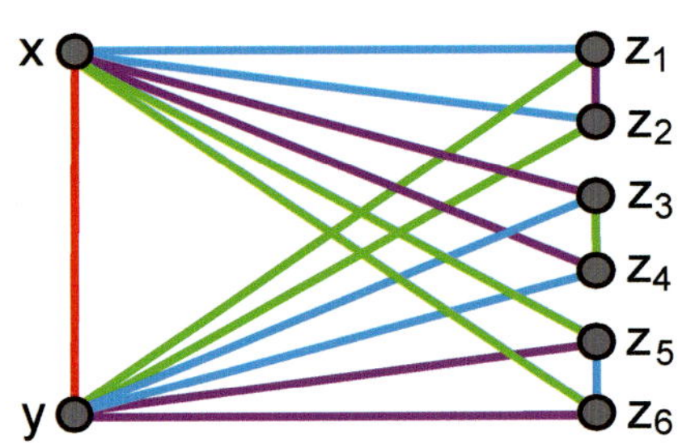

FIGURE 2.21: A 4-coloring of K_8 that avoids a red/blue K_3, a red/green K_3, a red/purple K_3, a green/purple K_4, a blue/purple K_4, and a blue/green K_4. This coloring corresponds with Case 1 in the proof of Theorem 2.22.

If there exists a second red edge, then without loss of generality, we assume that it is edge z_2z_3. Now z_1z_3 cannot be red or purple since either possiblity would make the subgraph induced by $\{z_1, z_2, z_3\}$ be a red/purple K_3. Also, z_1z_3 cannot be blue since this would result in xz_1z_3yx and xz_2z_3yx being two red/blue cycles of length 4 that have two edges in common, contradicting Lemma 2.2. Hence, z_1z_3 must be green.

Now edge z_2z_4 cannot be red or green as the subgraph induced by $\{z_2, z_3, z_4\}$ would then be a red/green K_3. It also cannot be blue as this would make both of the red/blue cycles xz_2z_4yx and $z_2z_4yz_3z_2$ have two eges in common, contradicting Lemma 2.2. So, edge z_2z_4 must be purple. If edge z_1z_4 is red or green, then the subgraph induced by $\{z_1, z_3, z_4\}$ is a red/green K_3. If z_1z_4 is blue or purple, then the subgraph induced by $\{x, z_1, z_2, z_3\}$ is a blue/purple K_4. Thus, this case does not occur.

Case 2: Assume that yz_1 and yz_2 are different colors (assume yz_1 is green and yz_2 is purple). Since x must be incident with two green and two purple edges, we assume that xz_3 and xz_4 are green and xz_5 and xz_6 are purple. The vertex y must also be incident with two blue, two green, and two purple edges. Assume that yz_3 is purple, yz_5 is green, and yz_4 and yz_6 are blue (see Figure 2.22).

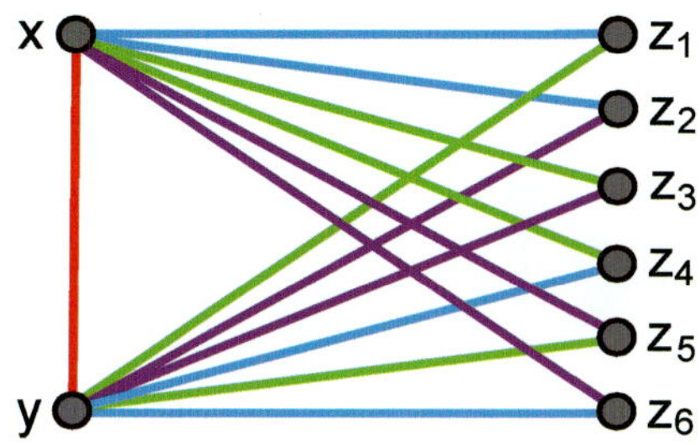

FIGURE 2.22: A 4-coloring of K_8 that avoids a red/blue K_3, a red/green K_3, a red/purple K_3, a green/purple K_4, a blue/purple K_4, and a blue/green K_4. This coloring corresponds with Case 2 in the proof of Theorem 2.22.

Since a K_3 that uses at most two colors, one of which is red, is avoided, the cycle $z_1z_2z_3z_4z_6z_5z_1$ does not contain any red edges. Relative to such a cycle, if there exists a red edge other than xy, it must either join two vertices that have a single common neighbor in the cycle $z_1z_2z_3z_4z_6z_5z_1$ or two vertices that do not have a common neighbor in the cycle $z_1z_2z_3z_4z_6z_5z_1$. Without loss of generality, we consider the possibilities where z_1z_3 or z_1z_4 are red, giving us two subcases.

Subcase 2.1: Suppose that z_1z_3 is red. Then z_1 and z_3 must each be incident with exactly two blue, two green, and two purple edges. If edge z_1z_4 is red or green, then the red/green cycles $z_1z_3xyz_1$ and $z_1z_3xz_4z_1$ have two edges in common, contradicting Lemma 2.2. So, z_1z_4 must be blue or green. Also, z_1z_2 cannot be red or blue as this would result in the subgraph induced by $\{x, z_1, z_2\}$ being a red/blue K_3. Hence, z_1z_2 is green or purple. We complete the remainder of this subcase with subsubcases based on the possible colors of z_1z_4 and z_1z_2.

Subsubcase 2.1.1: Assume that z_1z_4 is blue and z_1z_2 is green (see the first image in Figure 2.23). Edge z_2z_3 cannot be red or purple as this would force the subgraph induced

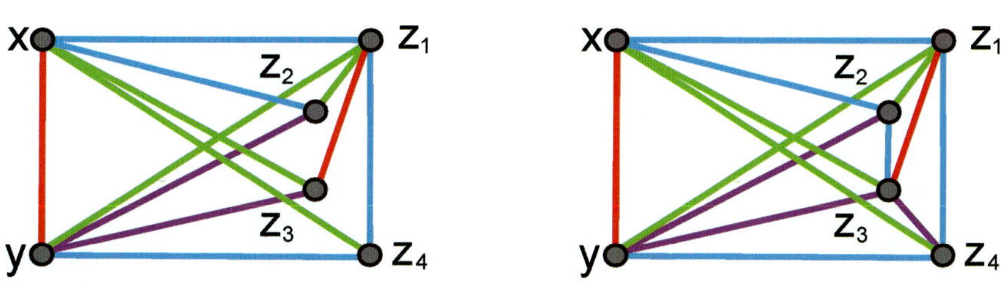

FIGURE 2.23: A subgraph of a 4-coloring of K_8, corresponding to Subsubcase 2.1.1 in the proof of Theorem 2.22.

by $\{y, z_2, z_3\}$ to be a red/purple K_3, and it cannot be green as this would force the subgraph induced by $\{z_1, z_2, z_3\}$ to be a red/green K_3. So, z_2z_3 must be blue. Edge z_3z_4 cannot be red or blue as this would force the subgraph induced by $\{z_1, z_3, z_4\}$ to be a red/blue K_3, and it cannot be green as this would force the subgraph induced by $\{x, z_3, z_4\}$ to be a green K_3. Hence, z_3z_4 must be purple (see the second image in Figure 2.23). Now edge z_2z_4 cannot be blue or purple as this would lead to the subgraph induced by $\{y, z_2, z_3, z_4\}$ being a blue/purple K_4. It also cannot be green since the subgraph induced by $\{x, z_1, z_2, z_4\}$ would be a blue/green K_4. Finally, if z_2z_4 is red, then xz_2z_4yx and xz_1z_4yx are two red/blue cycles of length four that have two edges in common, contradicting Lemma 2.2. Thus, this subsubcase does not occur.

Subsubcase 2.1.2: Assume that z_1z_4 is blue and z_1z_2 is purple (see the first image in Figure 2.24). Edge z_3z_4 cannot be red or blue as this would force the subgraph induced by

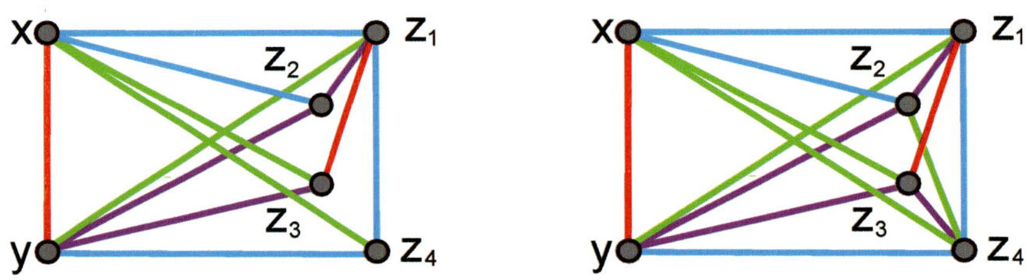

FIGURE 2.24: A subgraph of a 4-coloring of K_8, corresponding to Subsubcase 2.1.2 in the proof of Theorem 2.22.

$\{z_1, z_3, z_4\}$ to be a red/blue K_3, and it cannot be green as this would force the subgraph induced by $\{x, z_3, z_4\}$ to be a green K_3. So, z_3z_4 must be purple. Edge z_2z_4 cannot be red or blue as this would result in the red/blue cycles $xz_1z_4z_2x$ and xz_1z_4yx having two edges in common, contradicting Lemma 2.2. It also cannot be purple as this would result in the red/purple cycles $yz_2z_1z_3y$ and $yz_2z_4z_3y$ having two edges in common, contradicting Lemma 2.2. So, z_2z_4 must be green (see the second image in Figure 2.24). Edge z_2z_3 cannot be red or purple as this would result in the subgraph induced by $\{z_1, z_2, z_3\}$ being a red/purple K_3. Hence, z_2z_3 must be blue or green.

Edge z_1z_5 cannot be red or green as this would force the subgraph induced by $\{y, z_1, z_5\}$ to be a red/green K_3, and it cannot be blue since z_1 is already incident with two blue edges. Hence, z_1z_5 must be purple. Edge z_3z_5 cannot be red or green as this would result in the red/green cycles xz_3z_5yx and xz_3z_1yx having two edges in common, contradicting Lemma 2.2. It also cannot be purple since z_3 is already incident with two purple edges. Hence, z_3z_5 must be blue, resulting in the image shown in Figure 2.25.

Finally, consider the subgraph induced by $\{z_2, z_3, z_5, z_6\}$. Since z_3 is incident with one red edge and two purple edges, the edges z_2z_3 and z_3z_6 must be blue or green. Edge z_2z_5 cannot be red or purple as this would force the subgraph induced by $\{z_1, z_2, z_5\}$ to be a red/purple K_3. Edge z_5z_6 cannot be red or purple as this would force the subgraph induced by $\{x, z_5, z_6\}$ to be a red/purple K_3. Finally, edge z_2z_6 must be blue or green since z_1 is already incident with one red edge and two purple edges. It follows that the subgraph induced by $\{z_2, z_3, z_5, z_6\}$ is a blue/green K_4. Hence, this subsubcase does not occur.

Subsubcase 2.1.3: Assume that z_1z_4 is green and z_1z_2 is green or purple (see Figure 2.26). Consider the red/green cycles $xz_3z_1z_4x$ and xz_4z_1yx, which have two edges in common, contradicting Lemma 2.2. Hence, this subsubcase does not occur.

Subcase 2.2: Suppose that z_1z_4 is red. Then z_1 and z_4 must each be incidddent with exactly two blue, two green, and two purple edges. If edge z_2z_4 is red or purple, then the

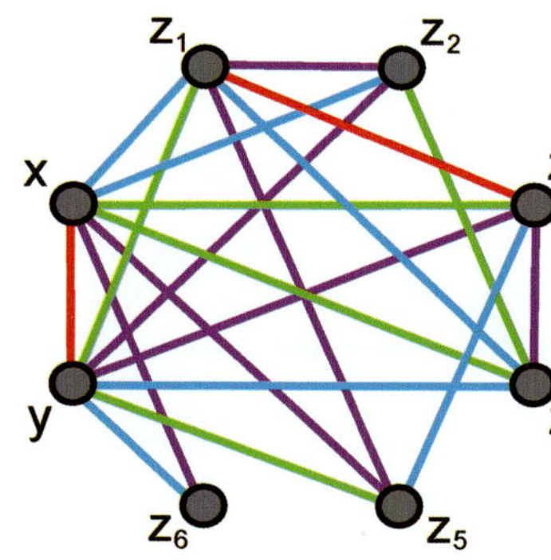

FIGURE 2.25: A subgraph of a 4-coloring of K_8, corresponding to Subsubcase 2.1.2 in the proof of Theorem 2.22.

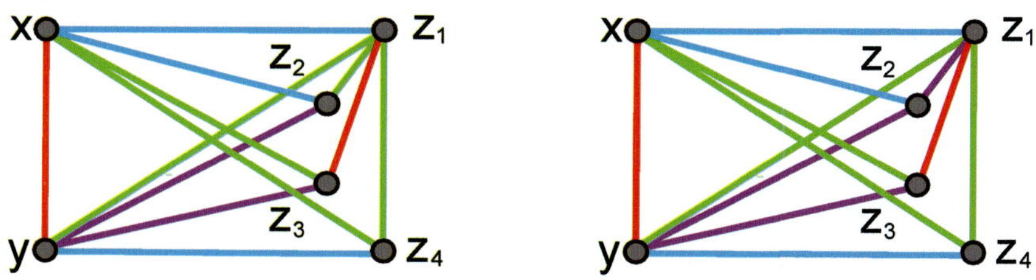

FIGURE 2.26: Possible subgraphs of a 4-coloring of K_8, corresponding to Subsubcase 2.1.3 in the proof of Theorem 2.22.

subgraph induced by $\{y, z_2, z_3\}$ is a red/purple K_3. So, z_2z_4 must be blue or green. Edge z_3z_4 cannot be red or green as this would force the subgraph induced by $\{x, z_3, z_4\}$ to be a red/green K_3. Hence, z_3z_4 is blue or purple. We complete the remainder of this subcase with subsubcases based on the possible colors of z_2z_4 and z_3z_4.

Subsubcase 2.2.1: Assume that z_2z_4 is blue and z_3z_4 is blue or purple (see Figure 2.27). Then the red/blue cycles xz_1z_4yx and xz_2z_4yx have two edges in common, contradicting

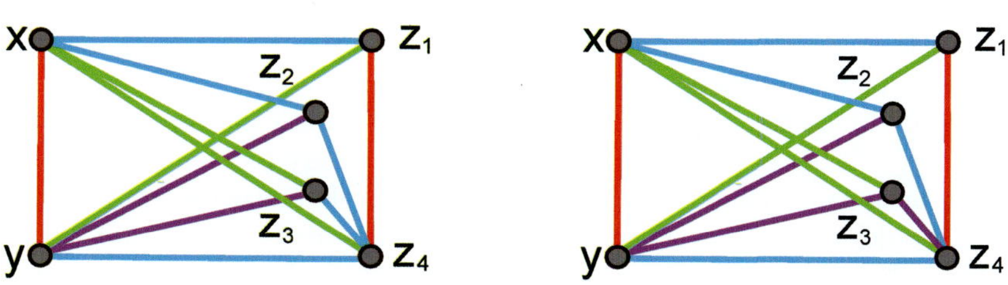

FIGURE 2.27: A subgraph of a 4-coloring of K_8, corresponding to Subsubcase 2.2.1 in the proof of Theorem 2.22.

Lemma 2.2. So, this subsubcase does not occur.

Subsubcase 2.2.2: Assume that z_2z_4 is green and z_3z_4 is blue (see Figure 2.28). Edge z_2z_3 cannot be red or purple as this would result in the subgraph induced by $\{y, z_2, z_3\}$

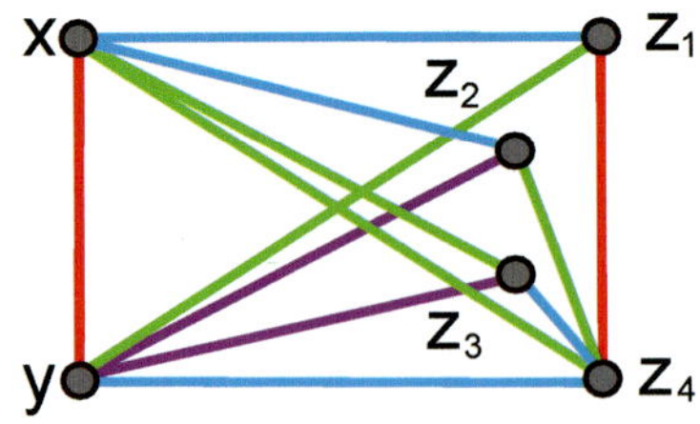

FIGURE 2.28: A subgraph of a 4-coloring of K_8, corresponding to Subsubcase 2.2.2 in the proof of Theorem 2.22.

being a red/purple K_3. Hence, z_2z_3 is blue or green. Either way, the subgraph induced by $\{x, z_2, z_3, z_4\}$ is a blue/green K_4. So, this subsubcase does not occur.

Subsubcase 2.2.3: Assume that z_2z_4 is green and z_3z_4 is purple (see the first image in Figure 2.29). Edge z_1z_2 is not red or green as this would result in the subgraph induced by

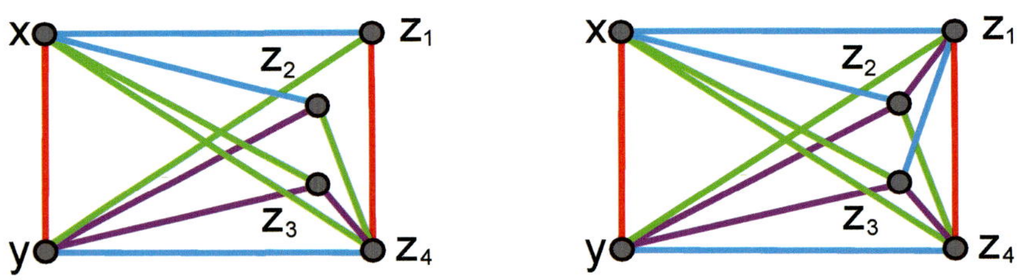

FIGURE 2.29: A subgraph of a 4-coloring of K_8, corresponding to Subsubcase 2.2.3 in the proof of Theorem 2.22.

$\{z_1, z_2, z_4\}$ being a red/green K_3. It is also not blue since this would result in the subgraph induced by $\{x, z_1, z_2\}$ being a blue K_3. So, z_1z_2 must be purple. Edge z_1z_3 cannot be red or purple since this would result in the subgraph induced by $\{z_1, z_3, z_4\}$ being a red/purple K_3. It also cannot be green since this would result in the red/green cycles xz_4z_1yx and xz_3z_1yx having two edges in common, contradicting Lemma 2.2. So, z_1z_3 must be blue (see the second image in Figure 2.29).

Edge z_1z_5 cannot be red or green, otherwise the subgraph induced by $\{y, z_1, z_5\}$ would be a red/green K_3. It also cannot be blue since z_1 is already incident with two blue edges. So, z_1z_5 must be purple. Edge z_2z_5 is not blue or purple since the subgraph induced by $\{x, z_1, z_2, z_5\}$ would then form a blue/purple K_4. It is also not red as this would force the subgraph induced by $\{z_1, z_2, z_5\}$ to be a red/purple K_3. Hence, z_2z_5 is green, and the subgraph induced by $\{y, z_1, z_2, z_5\}$ is a green/purple K_4 (see Figure 2.30). So, this subsubcase does not occur.

We have shown in all cases that a 4-coloring of K_8 that lacks a red/blue K_3, a red/green K_3, a red/purple K_3, a green/purple K_4, a blue/purple K_4, and a blue/green K_4 cannot contain more than one red edge. □

Theorem 2.23 ([132]) $r^4_{1,2}(K_3, K_3, K_3, K_3, K_3, K_4, K_3, K_4, K_3, K_4) = 8.$

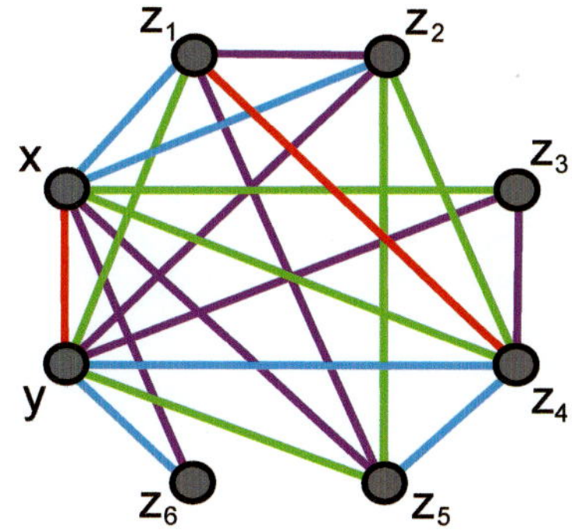

FIGURE 2.30: A subgraph of a 4-coloring of K_8, corresponding to Subsubcase 2.2.3 in the proof of Theorem 2.22.

Proof Since every 3-coloring of K_8 using blue, green, and purple is a 4-coloring with color 1 (red) omitted, Theorem 2.16 implies the lower bound

$$r^4_{1,2}(K_3, K_3, K_3, K_3, K_3, K_4, K_3, K_4, K_3, K_4) \geq r^3_{1,2}(K_3, K_3, K_3, K_4, K_4, K_4) = 8.$$

To prove the reverse inequality, consider a 4-coloring of K_8 and suppose that it lacks a red/blue K_3, a red/green K_3, a red/purple K_3, a green/purple K_4, a blue/purple K_4, and a blue/green K_4. Theorem 2.22 implies that there exists exactly one red edge.

Let w be a vertex that is not incident with the red edge. Without loss of generality, suppose that wx_1, wx_2, and wx_3 are blue, wy_1 and wy_2 are green, and wz_1 and wz_2 are purple. There exists a pair of points (either y_1 and y_2 or z_1 and z_2) such that the edges joining the pair of vertices to $\{x_1, x_2, x_3\}$ are not red. Without loss of generality, assume that the edges that join $\{x_1, x_2, x_3\}$ to $\{y_1, y_2\}$ are not red. If a K_3 is avoided that uses at most two colors, one of which is red, then the subgraph induced by $\{x_1, x_2, x_3\}$ does not contain any red or blue edges and the edge y_1y_2 is not red or green. The subgraph induced by $\{x_1, x_2, x_3, y_1, y_2\}$ does not contain a blue/green K_3 (since a blue/green K_4 would be formed by including vertex w) or a purple K_3. By Theorem 1.14, this subgraph consists of a blue/green C_5 and a purple C_5. This leads to two possibilities: the subgraph induced by $\{x_1, x_2, x_3\}$ contains either one or two purple edges, resulting in two cases.

Case 1: Suppose that the subgraph induced by $\{x_1, x_2, x_3\}$ contains a single purple edge. Without loss of generality, suppose that x_1x_3 is purple and assume that $x_1x_3y_2x_2y_1x_1$ is the purple C_5. Then $x_1x_2x_3y_1y_2x_1$ is a blue/green C_5. In order to avoid a monochromatic K_3 edges x_1x_2 and x_2x_3 must be green and edge y_1y_2 must be blue (see the first image in Figure 2.31). If edge x_1y_2 is green, then the subgraph induced by $\{x_1, x_2, x_3, y_2\}$ is a green/purple K_4. If edge x_3y_1 is green, then the subgraph induced by $\{x_1, x_2, x_3, y_1\}$ is a green/purple K_4. So, both x_1y_2 and x_3y_1 are blue, and the subgraph induced by $x_1, x_3, y_1, y_2\}$ is a blue/purple K_4.

Case 2: Suppose that the subgraph induced by $\{x_1, x_2, x_3\}$ contains two purple edges. Without loss of generality, suppose that x_1x_2 and x_2x_3 are purple and assume that $x_1x_3x_xy_1y_2x_1$ is the purple C_5. Then $x_1x_3y_2x_2y_1x_1$ is a blue/green C_5 (see the second image in Figure 2.31). In order to avoid a monochromatic K_3, x_1x_3 must be green. Avoiding a blue/purple K_3 (which, along with w, would form a blue/purple K_4) and a red/green K_3, Theorem 1.14 implies that the subgraph induced by $\{x_1, x_2, x_3, z_1, z_2\}$ consists of a blue/purple C_5 and a red/green C_5. Without loss of generality, suppose that $x_1x_2x_3z_1z_2x_1$ is a blue/purple C_5 and $x_1x_3z_2x_2z_1x_1$ is a red/green C_5. Avoiding a purple K_3 forces edge

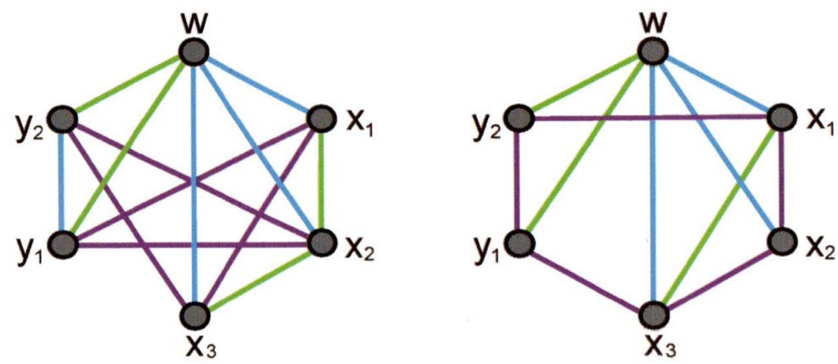

FIGURE 2.31: Subgraphs of a 4-coloring of K_8.

z_1z_2 to be blue. Exactly one of the edges x_1z_1, x_2z_1, x_2z_2, and x_3z_2 must be red and the other edges in the cycle $x_1x_3z_2x_2z_1x_1$ are green (see Figure 2.32). Note that the subgraphs

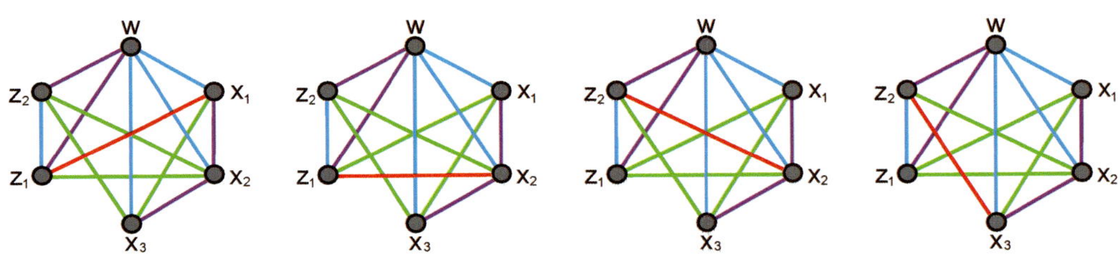

FIGURE 2.32: Possible location of the red edge in subgraphs of a 4-coloring of K_8.

with x_1z_1 and x_3z_2 red are isomorphic (the first and fourth images in Figure 2.32), as are the subgraphs with x_2z_1 and x_2z_2 red (the second and third images in Figure 2.32), leading to two subcases.

Subcase 2.1: Assume that x_1z_1 is red (equivalently, x_3z_2 is red). If edge x_1z_2 is purple, then the subgraph induced by $\{x_1, x_2, x_3, z_2\}$ is a green/purple K_4. If x_1z_2 is blue, then the subgraph induced by $\{w, x_1, z_2\}$ is a red/blue K_3.

Subcase 2.2: Assume that x_2z_1 is red (equivalently, x_2z_2 is red). If edge x_3z_1 is purple, then the subgraph induced by $\{x_2, x_3, z_1\}$ is a red/purple K_3. So, x_3z_1 must be blue. If edge x_1z_2 is blue, then the subgraph induced by $\{x_1, x_3, z_1, z_2\}$ is a blue/green K_4. So, x_1z_2 must be purple, and the subgraph induced by $\{x_1, x_2, x_3, z_2\}$ is a green/purple K_4.

In all cases, we see that such a 4-coloring of K_8 does not exist. Hence,

$$r^4_{1,2}(K_3, K_3, K_3, K_3, K_3, K_4, K_3, K_4, K_3, K_4) \leq 8.$$

□

Theorem 2.24 ([133]) *If* $n_1, n_3 \geq 3$ *and* $n_2, n_4 \geq 5$, *or if* $n_1, n_2, n_3, n_4 \geq 4$, *then*

$$r^4_{1,2}(K_3, K_3, K_3, K_3, K_3, K_5, K_{n_1}, K_{n_2}, K_{n_3}, K_{n_4}) = 14.$$

Proof Theorems 2.9 and 1.16 imply the upper bound

$$r^4_{1,2}(K_3, K_3, K_3, K_3, K_3, K_5, K_{n_1}, K_{n_2}, K_{n_3}, K_{n_4}) \leq r(K_3, K_5) = 14.$$

To show that 14 is also a lower bounds, we consider the two cases separately.

Case 1: Assume that $n_1, n_3 \geq 3$ and $n_2, n_4 \geq 5$. Since every 3-coloring of K_8 using blue, green, and purple is a 4-coloring with color 1 (red) omitted, Theorem 2.18 implies the lower bound

$$r_{1,2}^4(K_3, K_3, K_3, K_3, K_3, K_5, K_{n_1}, K_{n_2}, K_{n_3}, K_{n_4}) \geq r(K_3, K_3, K_3, K_5, K_5, K_5) = 14.$$

Case 2: Assume that $n_1, n_2, n_3, n_4 \geq 4$. Let $S_r := \{\pm 2\}$, $S_b := \{\pm 3\}$, $S_g := \{\pm 1, \pm 4\}$, and $S_p := \{\pm 5, \pm 6\}$, and factor K_{13} into the Cayley graphs $\mathrm{Cay}(\mathbb{Z}_{13}; S_r)$, $\mathrm{Cay}(\mathbb{Z}_{13}; S_b)$, $\mathrm{Cay}(\mathbb{Z}_{13}; S_g)$, and $\mathrm{Cay}(\mathbb{Z}_{13}; S_p)$, colored red, blue, green, and purple, respectively (see Figure 2.33). From this 4-coloring of K_{13}, it follows that

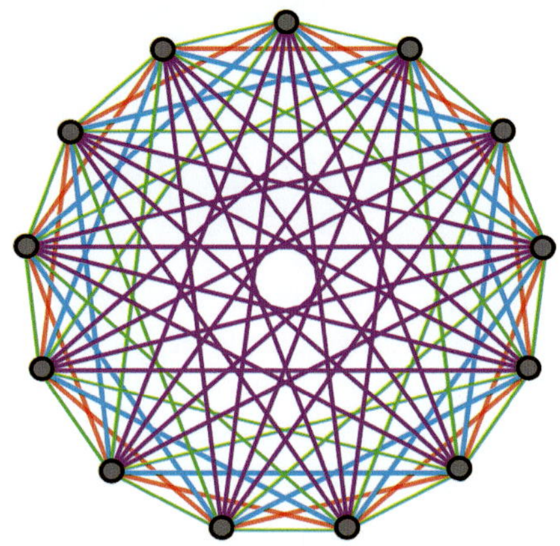

FIGURE 2.33: A 4-coloring of K_{13} that avoids a monochromatic K_3, a red/blue K_3, a green/purple K_5, a red/green K_4, a blue/purple K_4, a red/pruple K_4, and a blue/green K_4.

$$r_{1,2}^4(K_3, K_3, K_3, K_3, K_3, K_5, K_{n_1}, K_{n_2}, K_{n_3}, K_{n_4}) \geq 14.$$

□

Theorem 2.25 ([133]) *If* $n_1, n_3 \geq 3$ *and* $n_2, n_4 \geq 6$, *then*

$$r_{1,2}^4(K_3, K_3, K_3, K_3, K_3, K_6, K_{n_1}, K_{n_2}, K_{n_3}, K_{n_4}) = 17.$$

Proof Since every 3-coloring of K_8 using blue, green, and purple is a 4-coloring with color 1 (red) omitted, Theorem 2.19 implies the lower bound

$$r_{1,2}^4(K_3, K_3, K_3, K_3, K_3, K_6, K_{n_1}, K_{n_2}, K_{n_3}, K_{n_4}) \geq r(K_3, K_3, K_3, K_6, K_6, K_6) = 17.$$

To prove the reverse inequality, consider a 4-coloring of K_{17}. If we view red and blue as if they are the same color, Theorem 1.18 can be applied to the resulting 3-coloring of K_{17} to see that there is a red/blue K_3, a green K_3, or a purple K_3. It follows that

$$r_{1,2}^4(K_3, K_3, K_3, K_3, K_3, K_6, K_{n_1}, K_{n_2}, K_{n_3}, K_{n_4}) \leq r^3(K_3) = 17.$$

□

Theorem 2.26 ([133]) *If* $n_1, n_2, n_3, n_4 \geq 5$, *then*

$$r_{1,2}^4(K_3, K_3, K_3, K_3, K_4, K_4, K_{n_1}, K_{n_2}, K_{n_3}, K_{n_4}) = 18.$$

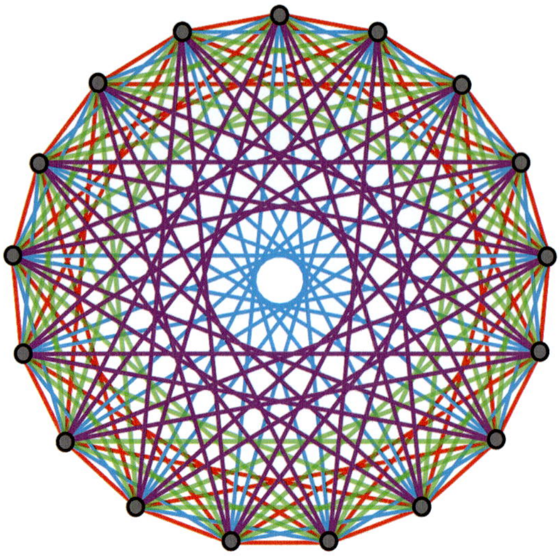

FIGURE 2.34: A 4-coloring of K_{13} that avoids a monochromatic K_3, a red/blue K_4, a green/purple K_4, a red/green K_5, a blue/purple K_5, a red/pruple K_5, and a blue/green K_5.

Proof For the lower bound, define $S_r := \{\pm 1, \pm 4\}$, $S_b := \{\pm 2, \pm 8\}$, $S_g := \{\pm 3, \pm 5\}$, and $S_p := \{\pm 6, \pm 7\}$. Then K_{17} can be factored into the Cayley graphs $\mathrm{Cay}(\mathbb{Z}_{17}; S_r)$, $\mathrm{Cay}(\mathbb{Z}_{17}; S_b)$, $\mathrm{Cay}(\mathbb{Z}_{17}; S_g)$, and $\mathrm{Cay}(\mathbb{Z}_{17}; S_p)$ in colors red, blue, green, and purple, respectively (see Figure 2.34). From this 4-coloring of K_{17}, it follows that

$$r^4_{1,2}(K_3, K_3, K_3, K_3, K_4, K_4, K_{n_1}, K_{n_2}, K_{n_3}, K_{n_4}) \geq 18$$

Theorems 1.17 and 2.9 imply the upper bound

$$r^4_{1,2}(K_3, K_3, K_3, K_3, K_4, K_4, K_{n_1}, K_{n_2}, K_{n_3}, K_{n_4}) \leq r^2(K_4) = 18.$$

□

2.2.4 4-Color (1,3)-Chromatic Ramsey Numbers

In this subsection, we prove two results concerning 4-color $(1,3)$-chromatic Ramsey numbers for complete graphs.

Theorem 2.27 ([164]) *For $k \geq 4$,*

$$r^4_{1,3}(K_3, K_3, K_3, K_3, K_4, K_4, K_4, K_k) = \begin{cases} 7 & \text{if } k = 4 \\ 8 & \text{if } k \geq 5. \end{cases}$$

Proof Consider the cases separately.

Case 1: Suppose that $k = 4$. The 4-coloring of K_6 shown in the first image in Figure 2.35 avoids a monochromatic K_3 and every K_4-subgraph uses all four colors. It follows that

$$r^4_{1,3}(K_3, K_3, K_3, K_3, K_4, K_4, K_4, K_4) \geq 7.$$

To prove the reverse inequality, consider a 4-coloring of K_7 and denote by G_r, G_b, G_g, and G_p the subgraphs spanned by the red, blue, green, and purple edges, respectively.

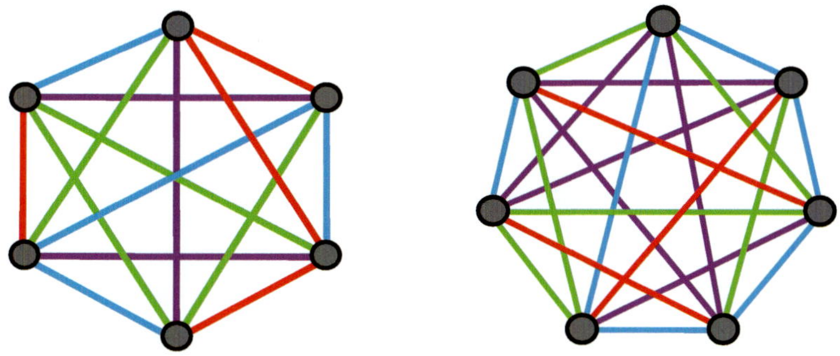

FIGURE 2.35: Critical colorings for $r^4_{1,3}(K_3, K_3, K_3, K_3, K_4, K_4, K_4, K_k)$, where $k \geq 4$.

Since K_7 has size 21, at least one of G_r, G_b, G_g, and G_g has at most five edges. Without loss of generality, suppose that G_r has at most five edges and lacks a K_3-subgraph. If G_r only contains cycles of even length, then Theorem 1.4 implies that it is bipartite. By the pigeonhole principle, one of its partite sets contains at least four vertices, which form an independent set in G_r. In the original K_7, these four vertices form a blue/green/purple K_4. If G_r contains a cycle of length five, then that cycle must use all five edges. Without loss of generality, suppose that $x_1x_2x_3x_4x_5x_1$ is the C_5-subgraph and denote the other two vertices by x_6 and x_7. Then $\{x_1, x_3, x_6, x_7\}$ is an independent set of vertices in G_r, so they form a blue/green/purple K_4 in the original K_7. Since G_r has at most five edges, it is not possible for it to contain a C_7-subgraph. In all cases, we find that there exists a blue/green/purple K_4, and it follows that

$$r^4_{1,3}(K_3, K_3, K_3, K_3, K_4, K_4, K_4, K_4) \leq 7.$$

Case 2: Suppose that $k \geq 5$. The 4-coloring of K_7 shown in the second image in Figure 2.35 avoids a monochromatic K_3, a red/blue/green K_4, a red/blue/purple K_4, a red/green/purple K_4, and a blue/green/purple K_5. It follows that

$$r^4_{1,3}(K_3, K_3, K_3, K_3, K_4, K_4, K_4, K_k) \geq 8,$$

whenever $k \geq 5$.

To prove the reverse inequality, consider a 4-coloring of K_8 and denote by G_r, G_b, G_g, and G_p the subgraphs spanned by the red, blue, green, and purple edges, respectively. At least one of G_b, G_g, and G_p contains at most nine edges. Otherwise, each of these subgraphs would each contain at least ten edges, contradicting the fact the K_8 has size 28. Without loss of generality, suppose that G_b has at most nine edges and avoids a monochromatic K_3. Consider several subcases.

Subcase 2.1: If G_b only contains cycles of even length, then Theorem 1.4 implies that it is bipartite. By the pigeonhole principle, one of its partite sets contains at least four vertices, which form an independent set in G_b. In the original K_8, these four vertices form a red/green/purple K_4.

Subcase 2.2: If G_b contains a cycle of length 5, assume it is given by $x_1x_2x_3x_4x_5x_1$. Since it avoids a monochromatic K_3, no other edges exist in the subgraph induced by $\{x_1, x_2, x_3, x_4, x_5\}$. Denote the other vertices by x_6, x_7, and x_8, and consider three subsubcases, based on the number of edges in the subgraph induced by $\{x_6, x_7, x_8\}$.

Subcase 2.2.1: Suppose that the subgraph induced by $\{x_6, x_7, x_8\}$ is empty (see the first image in Figure 2.36). Since there are at most four edges joining $\{x_1, x_2, x_3, x_4, x_5\}$

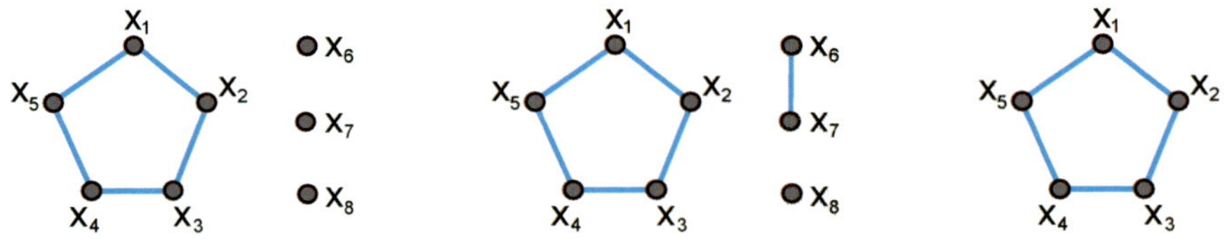

FIGURE 2.36: Possible blue subgraphs when G_b contains a C_5.

to $\{x_6, x_7, x_8\}$, one of x_1, x_2, x_3, x_4, and x_5 is not adjacent to any of x_6, x_7, or x_8 in G_b (suppose it is vertex x_1). Then $\{x_1, x_6, x_7, x_8\}$ is an independent set of vertices in G_b, which corresponds with a red/green/purple K_4 in the original K_8.

Subcase 2.2.2: Suppose that the subgraph induced by $\{x_6, x_7, x_8\}$ contains a single edge. Without loss of generality, suppose that x_6x_7 is an edge in G_b (see the second image in Figure 2.36). Each of x_6, x_7, and x_8 are adjacent to at most two of the vertices in $\{x_1, x_2, x_3, x_4, x_5\}$, otherwise, a K_3-subgraph is formed. If x_8 is not adjacent to any of the vertices in $\{x_1, x_2, x_3, x_4, x_5\}$, then one of x_6 and x_7 is adjacent to at most one of the vertices in $\{x_1, x_2, x_3, x_4, x_5\}$. Without loss of generality, suppose that x_6 is not adjacent to x_1, x_2, x_3, and x_4. Then $\{x_1, x_3, x_6, x_8\}$ is an independent set of vertices in G_b, which corresponds with a red/green/purple K_4 in the original K_8.

If x_8 is adjacent to exactly one of the vertices in $\{x_1, x_2, x_3, x_4, x_5\}$, say x_1, then one of x_6 and x_7 is adjacent to at most one of the vertices in $\{x_1, x_2, x_3, x_4, x_5\}$. So, it is possible to select two two nonadjacent vertices from $\{x_1, x_2, x_3, x_4, x_5\}$ that are not adjacent to x_6 or x_8. These two vertices, along with x_6 and x_8, form a red/green/purple K_4 in the original K_8.

If x_8 is adjacent to exactly two of the vertices in $\{x_1, x_2, x_3, x_4, x_5\}$, then it must be nonadjacent with two nonadjacent vertices in $\{x_1, x_2, x_3, x_4, x_5\}$. Without loss of generality, suppose that x_8 is not adjacent to x_1 or x_3. At most one edge joins $\{x_6, x_7\}$ to $\{x_1, x_2, x_3, x_4, x_5\}$, so suppose that x_6 does not join to $\{x_1, x_2, x_3, x_4, x_5\}$. Then $\{x_1, x_3, x_6, x_8\}$ induces a red/green/purple K_4 in the original K_8.

Subcase 2.2.3: Suppose that the subgraph induced by $\{x_6, x_7, x_8\}$ contains two edges. Without loss of generality, suppose that x_6x_7 and x_7x_8 are edges in G_b (see the third image in Figure 2.36). Then at most two edges join $\{x_6, x_8\}$ to $\{x_1, x_2, x_3, x_4, x_5\}$. Hence, it is possible to select two vertices from $\{x_1, x_2, x_3, x_4, x_5\}$ that are not adjacent and do not join to $\{x_6, x_8\}$. These two vertices, along with x_6 and x_8, form a red/green/purple K_4 in the original K_8.

Subcase 2.3: If G_b contains a cycle of length 7 and does not contain a cycle of length 5. Assume that the cycle of length 7 is given by $x_1x_2x_3x_4x_5x_6x_7x_1$ and that the other vertex is x_8. Besides this cycle, the subgraph induced by $\{x_1, x_2, \ldots, x_7\}$ does not contain any additional edges as such an edge would result in a K_3 or K_5-subgraph. Hence, there exists at most two edges joining x_8 to $\{x_1, x_2, \ldots, x_7\}$. If the vertices x_8 joins to are adjacent, say x_1x_8 and x_2x_8 are edges, then $\{x_3, x_5, x_7, x_8\}$ forms a red/green/purple K_4 in the original K_8. If the vertices x_8 joins to are a distance of two edges apart on the cycle, say x_1x_8 and x_3x_8 are edges, then $\{x_2, x_4, x_6, x_8\}$ forms a red/green/purple K_4 in the original K_8. If the vertices x_8 joins to are a distance of three edges apart on the cycle, say x_1x_8 and x_4x_8 are edges, then $\{x_2, x_5, x_7, x_8\}$ forms a red/green/purple K_4 in the original K_8.

In all cases, we have shown that there exists a red/green/purple K_4. It follows that

$$r^4_{1,3}(K_3, K_3, K_3, K_3, K_4, K_4, K_4, K_k) \le 8,$$

whenever $k \ge 5$. □

Theorem 2.28 ([164]) *For $i \geq 4$, $j \geq 5$, and $k \geq 5$,*

$$r^4_{1,3}(K_3, K_3, K_3, K_3, K_4, K_i, K_j, K_k) = 9.$$

Proof The 4-coloring of K_8 shown in Figure 2.37 avoids a monochromatic K_3, a red/blue/green K_4, a red/blue/purple K_4, a red/green/purple K_5, and a blue/green/purple K_5. It follows that

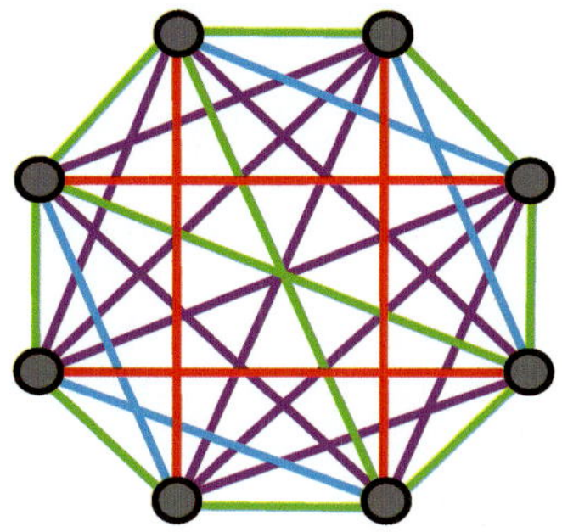

FIGURE 2.37: A critical coloring for $r^4_{1,3}(K_3, K_3, K_3, K_3, K_4, K_i, K_j, K_k)$, where $i \geq 4$, $j \geq 5$, and $k \geq 5$.

$$r^4_{1,3}(K_3, K_3, K_3, K_3, K_4, K_4, K_5, K_6) \geq 9.$$

The upper bound follows from Theorem 1.15, from which it follows that every 3-coloring of K_9 contains a purple K_3 or a red/blue/green K_4:

$$r^4_{1,3}(K_3, K_3, K_3, K_3, K_4, K_4, K_5, K_6) \leq r(K_3, K_4) = 9.$$

□

2.3 Chromatic Ramsey Numbers for Other Graphs

Now we turn our attention to chromatic Ramsey numbers for graphs that are not complete. The main cases that have been studied include stars and paths.

2.3.1 Chromatic Ramsey Numbers Involving Stars

In 1972, Harary [121] proved that

$$r(K_{1,i}, K_{1,j}) = \begin{cases} i + j - 1 & \text{if } i \text{ and } j \text{ are both even} \\ i + j & \text{otherwise.} \end{cases} \tag{2.6}$$

Burr and Roberts [55] extended this result to the multicolor setting in 1973. To state their result, suppose that $n_i \geq 2$ for all $i \in \{1, 2, \ldots, t\}$, and define

$$N_t := \sum_{i=1}^{t} (n_i - 1).$$

If k is the number of elements in $\{n_1, n_2, \ldots, n_t\}$ that are even, then

$$r(K_{1,n_1}, K_{1,n_2}, \ldots, K_{1,n_t}) = \begin{cases} N_t + 1 & \text{if } k \geq 2 \text{ is even} \\ N_t + 2 & \text{otherwise.} \end{cases}$$

Next, we will determine the value of the chromatic Ramsey number $r_2^3(K_{1,i}, K_{1,j}, K_{1,k})$. The following lemma offers an upper bound for $r_2^3(K_{1,i}, K_{1,j}, K_{1,k})$ that will be useful in our evaluation.

Lemma 2.3 ([62]) *For all $i \leq j \leq k$,*

$$r_2^3(K_{1,i}, K_{1,j}, K_{1,k}) \leq \min\left\{i + j, \lceil (i + j + k)/2 \rceil\right\}.$$

Proof Let x be a vertex in a 3-coloring of K_p. If a red/blue $K_{1,i}$ and a red/green $K_{1,j}$ are avoided, then

$$\deg_r(x) + \deg_b(x) \leq i - 1 \quad \text{and} \quad \deg_r(x) + \deg_g(x) \leq j - 1.$$

It follows that

$$p = \deg_r(x) + \deg_b(x) + \deg_g(x) + 1 \leq i + j - 1.$$

So, if $p \geq i + j$, there exists a red/blue $K_{1,i}$ or a red/green $K_{1,j}$. Hence,

$$r_2^3(K_{1,i}, K_{1,j}, K_{1,k}) \leq i + j. \tag{2.7}$$

If a blue/green $K_{1,k}$ is also avoided, then

$$\deg_b(x) + \deg_g(x) \leq k - 1$$

for every vertex x. It follows that

$$2(\deg_r(x) + \deg_b(x) + \deg_g(x)) \leq i + j + k - 3,$$

from which we see that

$$p \leq \lfloor (i + j + k - 3)/2 \rfloor + 1.$$

So, if

$$p \geq \lfloor (i + j + k - 3)/2 \rfloor + 2 = \lceil (i + j + k)/2 \rceil,$$

then there exists a red/blue $K_{1,i}$, a red/green $K_{1,j}$, or a blue/green $K_{1,k}$. It follows that

$$r_2^3(K_{1,i}, K_{1,j}, K_{1,k}) \leq \lceil (i + j + k)/2 \rceil. \tag{2.8}$$

Together, Inequalities (2.7) and (2.8) imply the inequality given in the statement of the lemma. □

The following theorem appeared in [62] in 1977 (see also [64]), but the proof contained a few mistakes and was incomplete. These issues are addressed in the proof given below.

Theorem 2.29 ([62]) *If $i \leq j \leq k$, then*

$$r_2^3(K_{1,i}, K_{1,j}, K_{1,k}) = \begin{cases} i + j - 1 & \text{if } k \geq i + j \text{ and } i \text{ and } j \text{ are both even} \\ i + j & \text{if } k \geq i + j \text{ and } i \text{ is odd or } j \text{ is odd} \\ \lceil (i + j + k)/2 \rceil - 1 & \text{if } k < i + j \text{ and } (i, j, k) \in S \\ \lceil (i + j + k)/2 \rceil & \text{if } k < i + j \text{ and } (i, j, k) \notin S, \end{cases}$$

where $S = \{(i, j, k) \mid i + j + k \equiv 3 \pmod 4$ and not all of i, j, and k are odd$\}$.

Proof When $k \geq i+j$, $k+1 \geq i+j+1 \geq r(K_{1,i}, K_{1,j})$, so Equation (2.6) and Theorem 2.5 imply that

$$r_2^3(K_{1,i}, K_{1,j}, K_{1,k}) = \begin{cases} i+j-1 & \text{if } k \geq i+j \text{ and } i \text{ and } j \text{ are both even} \\ i+j & \text{if } k \geq i+j \text{ and } i \text{ is odd or } j \text{ is odd.} \end{cases}$$

When $k < i+j$, we separate the remainder of the proof into cases, based on the value of $i+j+k$ modulo 4. The inequality

$$r(K_{1,i}, K_{1,j}, K_{1,k}) \leq \lceil (i+j+k)/2 \rceil$$

follows from Lemma 2.3, providing the upper bound for all cases in which $(i, j, k) \notin S$.

Case 1: Assume that $i+j+k \equiv 0 \pmod 4$. We will provide a 3-coloring of $K_{(i+j+k)/2-1}$ that avoids a red/blue $K_{1,i}$, a red/green $K_{1,j}$, and a blue/ green $K_{1,k}$. There are two subcases to consider.

Subcase 1.1: Assume that exactly one of i, j, and k is even. Without loss of generality, suppose that i is even and j and k are both odd. By Theorem 1.6, $K_{(i+j+k)/2-1}$ can be factored into $((i+j+k)/2-2)/2$ spanning cycles. Color $((i+j-k)/2-1)/2$ of the spanning cycles red, $((i-j+k)/2-1)/2$ of the spanning cycles blue, and $(-i+j+k)/4$ of the spanning cycles green. Then the largest red/blue star has size

$$(i+j-k)/2 - 1 + (i-j+k)/2 - 1 = i-2,$$

the largest red/green star has size

$$(i+j-k)/2 - 1 + (-i+j+k)/2 = j-1,$$

and the largest blue/green star has size

$$(i-j+k)/2 - 1 + (-i+j+k)/2 = k-1.$$

It follows that

$$r(K_{1,i}, K_{1,j}, K_{1,k}) \geq (i+j+k)/2 = \lceil (i+j+k)/2 \rceil.$$

Subcase 1.2: Assume that all of i, j, and k are even. First, we color the edges of $K_{(i+j+k)/2-2}$, which has even order. By Theorem 1.5, it can be factored into $(i+j+k)/2-3$ perfect matchings. Color $(i+j-k)/2-1$ of the perfect matchings red, $(i-j+k)/2-1$ of the perfect matchings blue, and $(-i+j+k)/2-1$ of the perfect matchings green. The largest red/blue star has size

$$(i+j-k)/2 - 1 + (i-j+k)/2 - 1 = i-2,$$

the largest red/green star has size

$$(i+j-k)/2 - 1 + (-i+j+k)/2 - 1 = j-2,$$

and the largest blue/green star has size

$$(i-j+k)/2 - 1 + (-i+j+k)/2 - 1 = k-2.$$

Introduce a new vertex x and join it to the $K_{(i+j+k)/2-2}$ with $(i+j-k)/2$ red edges, $(i-j+k)/2-1$ blue edges, and $(-i+j+k)/2-1$ green edges. While the vertices in the $K_{(i+j+k)/2-2}$ still avoid a red/blue $K_{1,i}$, a red/green $K_{1,j}$, and a blue/green $K_{1,k}$, vertex

x is the center of a red/blue star with size $i-1$, a red/green star with size $j-1$, and a blue/green star with size $k-2$. It follows that

$$r(K_{1,i}, K_{1,j}, K_{1,k}) \geq (i+j+k)/2 = \lceil (i+j+k)/2 \rceil.$$

Case 2: Assume that $i+j+k \equiv 1 \pmod 4$. In this case, we see that $\lceil (i+j+k)/2 \rceil = (i+j+k+1)/2$ and $K_{(i+j+k-1)/2}$ has even order. By Theorem 1.5, $K_{(i+j+k-1)/2}$ can be factored into $(i+j+k-3)/2$ perfect matchings. Color $(i+j-k-1)/2$ of the perfect matchings red, $(i-j+k-1)/2$ of the perfect matchings blue, and $(-i+j+k-1)/2$ of the perfect matchings green. Then the largest red/blue star has size

$$(i+j-k-1)/2 + (i-j+k-1)/2 = i-1,$$

the largest red/green star has size

$$(i+j-k-1)/2(-i+j+k-1)/2 = j-1,$$

and the largest blue/green star has size

$$(i-j+k-1)/2 + (-i+j+k-1)/2 = k-1.$$

It follows that

$$r(K_{1,i}, K_{1,j}, K_{1,k}) \geq \lceil (i+j+k)/2 \rceil.$$

Case 3: Assume that $i+j+k \equiv 2 \pmod 4$. Then $\lceil (i+j+k)/2 \rceil = (i+j+k)/2$ and $K_{(i+j+k)/2-1}$ has even order. By Theorem 1.5, $K_{(i+j+k)/2-1}$ factors into $(i+j+k)/2-2$ perfect matchings. Color $(i+j-k)/2$ of the perfect matchings red, $(i-j+k)/2-1$ of the perfect matchings blue, and $(-i+j+k)/2-1$ of the perfect matchings green. Then the largest red/blue star has size

$$(i+j-k)/2 + (i-j+k)/2 - 1 = i-1,$$

the largest red/green star has size

$$(i+j-k)/2 + (-i+j+k)/2 - 1 = j-1,$$

and the largest blue/green star has size

$$(i-j+k)/2 - 1 + (-i+j+k)/2 - 1 = k-2.$$

It follows that

$$r(K_{1,i}, K_{1,j}, K_{1,k}) \geq \lceil (i+j+k)/2 \rceil.$$

Case 4: Assume that $i+j+k \equiv 3 \pmod 4$. Then $\lceil (i+j+k)/2 \rceil = (i+j+k+1)/2$, which is even. We divide this case into subcases based upon whether or not (i,j,k) is in S.

Subcase 4.1: Assume that $(i,j,k) \notin S$. That is, suppose that i, j, and k are all odd. As with Cases 1, 2, and 3, the upper bound

$$r(K_{1,i}, K_{1,j}, K_{1,k}) \leq \lceil (i+j+k)/2 \rceil$$

follows from Lemma 2.3. For the lower bound, consider $K_{(i+j+k+1)/2-1}$, which has odd order. By Theorem 1.6, $K_{(i+j+k+1)/2-1}$ can be factored into $(i+j+k-3)/4$ spanning cycles. Color $(i+j-k-1)/4$ of the spanning cycles red, $(i-j+k-1)/4$ of the spanning cycles blue, and $(-i+j+k-1)/4$ of the spanning cycles green. Then the largest red/blue star has size

$$(i+j-k-1)/2 + (i-j+k-1)/2 = i = 1,$$

the largest red/green star has size

$$(i+j-k-1)/2+(-i+j+k-1)/2=j-1,$$

and the largest blue/green star has size

$$(i-j+k-1)/2+(-i+j+k-1)/2=k-1.$$

It follows that

$$r(K_{1,i},K_{1,j},K_{1,k}) \geq \lceil (i+j+k)/2 \rceil.$$

Subcase 4.2: Assume that $(i,j,k) \in S$. That is, suppose that not all of i, j, and k are odd. We start by showing that every Gallai 3-coloring of $K_{(i+j+k-1)/2}$ contains a red/blue $K_{1,i}$, a red/green $K_{1,j}$, or a blue green $K_{1,k}$. Suppose false, then such a coloring would satisfy

$$\deg_r(x)+\deg_b(x)+\deg_g(x)=(i+j+k-3)/2,$$

for every vertex x, as well as

$$\deg_r(x)+\deg_b(x) \leq i-1, \ \deg_r(x)+\deg_g(x) \leq j-1, \text{ and } \deg_b(x)+\deg_g(x) \leq k-1.$$

It follows that

$$\deg_r(x) \geq (i+j-k-1)/2, \ \deg_b(x) \geq (i-j+k-1)/2, \text{ and } \deg_g(x) \geq (-i+j+k-1)/2.$$

As these degrees sum to $(i+j+k-3)/2$, every vertex must have red degree $(i+j-k-1)/2$, blue degree $(i-j+k-1)/2$, and green degree $(-i+j+k-1)/2$. Without loss of generality, suppose that k is even. Then

$$(i+j+k-1)/2 \quad \text{and} \quad (i+j-k-1)/2=(i+j+k-1)/2-k$$

are both odd. Since G_r would then be a graph of odd order in which every vertex has odd degree, no such Gallai 3-coloring of $K_{(i+j+k-1)/2}$ exists. It follows that

$$r(K_{1,i},K_{1,j},K_{1,k}) \leq \lceil (i+j+k)/2 \rceil - 1.$$

To prove the reverse inequality, note that $\lceil (i+j+k)/2 \rceil - 2 = (i+j+k-3)/2$, which is even. By Theorem 1.5, $K_{(i+j+k-3)/2}$ can be factored into $(i+j+k-5)/2$ perfect matchings. Color $(i+j-k-1)/2$ of the perfect matchings red, $(i-j+k-1)/2$ of the perfect matchings blue, and $(-i+j+k-3)/2$ of the perfect matchings green. Then the largest red/blue star has size

$$(i+j-k-1)/2+(i-j+k-1)/2=i-1,$$

the largest red/green star has size

$$(i+j-k-1)/2+(-i+j+k-3)/2=j-2,$$

and the largest blue/green star has size

$$(i-j+k-1)/2+(-i+j+k-3)/2=k-2.$$

It follows that

$$r(K_{1,i},K_{1,j},K_{1,k}) \geq \lceil (i+j+k)/2 \rceil - 1,$$

completing the proof in all cases. □

In addition to the previous theorem, several cases of the chromatic Ramsey number $r_2^3(K_{1,i},K_j,K_k)$ were determined in [64] (also, see [62]). In 2017, the problem of determining the $(t-1)$-chromatic Ramsey number for stars was completed by Khamseh and Omidi in [149]. Although we state their main result below without proof, we will prove a few special diagonal cases (i.e., weakened Ramsey numbers for stars) in Section 3.2.

Theorem 2.30 ([149]) *Suppose that $t \geq 3$ and $n_i \geq 2$ for all $i \in \{1, 2, \ldots, t\}$ such that $n_t \leq r_{t-2}^{t-1}(K_{1,n_1}, K_{1,n_2}, \ldots, K_{1,n_{t-1}})$. Let*

$$N_t := \sum_{i=1}^{t}(n_i - 1) \quad \text{and} \quad M_t := \left\lfloor \frac{N_t + 1 - 1}{t-1} \right\rfloor.$$

If $N_t = q(t-1) + k$, where $0 \leq k < t - 1$, then

$$r_{t-1}^{t}(K_{1,n_1}, K_{1,n_2}, \ldots, K_{1,n_t}) = \begin{cases} M_t & \text{if } M_t \text{ is odd, } k = 0, \text{ and some } n_i \text{ is even} \\ M_t + 1 & \text{otherwise.} \end{cases}$$

Khamseh and Omidi [149] went a step further by considering the case where one of the arguments is a matching rather than a star.

Theorem 2.31 ([149]) *Suppose that $t \geq 3$, $n_i \geq 2$ for all $i \in \{1, 2, \ldots, t-1\}$, and $N_{t-1} := \sum_{i=1}^{t-1}(n_i - 1)$. Then we have the following:*

1. *If $2s \geq r_{t-2}^{t-1}(K_{1,n_1}, K_{1,n_2}, \ldots, K_{1,n_{t-1}})$, then*
$$r_{t-1}^{t}(K_{1,n_1}, K_{1,n_2}, \ldots, K_{1,n_{t-1}}, sK_2) = r_{t-2}^{t-1}(K_{1,n_1}, K_{1,n_2}, \ldots, K_{1,n_{t-1}}).$$

2. *If $2s < r_{t-2}^{t-1}(K_{1,n_1}, K_{1,n_2}, \ldots, K_{1,n_{t-1}})$ and $N_{t-1} < (2t-3)s - t + 2$, then*
$$r_{t-1}^{t}(K_{1,n_1}, K_{1,n_2}, \ldots, K_{1,n_{t-1}}, sK_2) = 2s.$$

3. *If $2s < r_{t-2}^{t-1}(K_{1,n_1}, K_{1,n_2}, \ldots, K_{1,n_{t-1}})$ and $N_{t-1} \geq (2t-3)s - t + 2$, then*
$$r_{t-1}^{t}(K_{1,n_1}, K_{1,n_2}, \ldots, K_{1,n_{t-1}}, sK_2) = \left\lceil \frac{N_{t-1} + s}{t-1} \right\rceil + 1.$$

2.3.2 3-Color (1,2)-Chromatic Ramsey Numbers Involving Stars

In this subsection, we consider chromatic Ramsey numbers of the form

$$r_{1,2}^{3}(K_3, K_3, K_3, K_{1,i}, K_{1,j}, K_{1,k}),$$

where $2 \leq i \leq j \leq k$. These numbers were first considered in Jacobson's dissertation [132] in 1980. As this dissertation is not readily available, and the proofs regarding these chromatic Ramsey numbers are incomplete (and in a few cases, incorrect), we provide detailed proofs of the theorems we cover.

Before we consider some specific cases of i, j, and k, note that Theorem 1.18 implies the following upper bound:

$$r_{1,2}^{3}(K_3, K_3, K_3, K_{1,i}, K_{1,j}, K_{1,k}) \leq r^3(K_3) = 17.$$

Also, Theorem 2.29 can be used to obtain improved upper bounds for certain values of i, j, and k:

$$r_{1,2}^{3}(K_3, K_3, K_3, K_{1,i}, K_{1,j}, K_{1,k}) \leq r_2^3(K_{1,i}, K_{1,j}, K_{1,k}).$$

Theorem 2.32 ([132]) *For $2 \leq j \leq k$,*

$$r_{1,2}^{3}(K_3, K_3, K_3, K_{1,2}, K_{1,j}, K_{1,k}) = \begin{cases} 3 & \text{if } j = 2 \\ 4 & \text{if } j = k = 3 \\ 5 & \text{if } j \geq 3 \text{ and } k \geq 4. \end{cases}$$

Proof Consider the following cases.

Case 1: Suppose that $j = 2$. Since all of the arguments have order at least 3, the lower bound

$$r^3_{1,2}(K_3, K_3, K_3, K_{1,2}, K_{1,2}, K_{1,k}) \geq 3$$

is trivial. On the other hand, in any 3-coloring of K_3, either there exists a red edge (which along with any other edge forms a red/blue or red/green $K_{1,2}$), or all edges are blue and green. In the latter case, at least two edges must be the same color, forming a blue $K_{1,2}$ or a green $K_{1,2}$. It follows that

$$r^3_{1,2}(K_3, K_3, K_3, K_{1,2}, K_{1,2}, K_{1,k}) \leq 3.$$

Case 2: Suppose that $j = k = 3$. The first image in Figure 2.38 avoids a monochromatic K_3, a red/blue $K_{1,2}$, a red/green $K_{1,3}$, and a blue/green $K_{1,3}$. It follows that

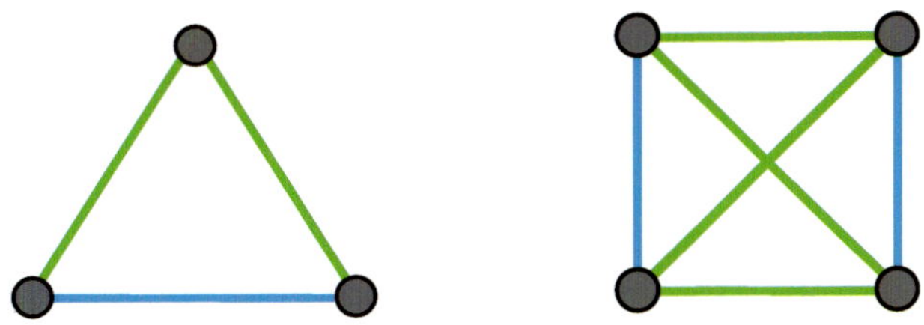

FIGURE 2.38: Critical colorings for $r^3_{1,2}(K_3, K_3, K_3, K_{1,2}, K_{1,j}, K_{1,k})$ when $j \geq 3$.

$$r^3_{1,2}(K_3, K_3, K_3, K_{1,2}, K_{1,3}, K_{1,3}) \geq 4.$$

To prove the reverse inequality, consider a 3-coloring of K_4 and select a vertex x. If $\deg_r(x)+\deg_b(x) \geq 2$, then x is the center vertex of a red/blue $K_{1,2}$. Otherwise, $\deg_g(x) \geq 2$ and x is the center vertex for either a red/green $K_{1,3}$ or a blue/green $K_{1,3}$. It follows that

$$r^3_{1,2}(K_3, K_3, K_3, K_{1,2}, K_{1,3}, K_{1,3}) \leq 4.$$

Case 3: Suppose that $j \geq 3$ and $k \geq 4$. The second image in Figure 2.38 avoids a monochromatic K_3, a red/blue $K_{1,2}$, a red/green $K_{1,j}$, and a blue/green $K_{1,k}$. It follows that

$$r^3_{1,2}(K_3, K_3, K_3, K_{1,2}, K_{1,j}, K_{1,k}) \geq 5.$$

To prove the reverse inequality, consider a 3-coloring of K_5 and select a vertex x. If $\deg_r(x)+\deg_b(x) \geq 2$, then x is the center vertex of a red/blue $K_{1,2}$. Otherwise, $\deg_g(x) \geq 3$. If $N_g(x)$ contains a green edge, then that edge, along with vertex x, forms a green K_3. So, all of the edges in $N_g(x)$ must be red and blue, from which it follows that $N_g(x)$ contains a red/blue K_3. Since $K_{1,2}$ is a subgraph of K_3, it follows that

$$r^3_{1,2}(K_3, K_3, K_3, K_{1,2}, K_{1,j}, K_{1,k}) \leq 5.$$

□

Theorem 2.33 ([132]) *For* $3 \leq j \leq k$,

$$r^3_{1,2}(K_3, K_3, K_3, K_{1,3}, K_{1,j}, K_{1,k}) = \begin{cases} 5 & \text{if } 3 \leq j \leq k \leq 4 \\ 6 & \text{if } (j,k) \in \{(3,k),(4,k),(5,5)\}, \text{ where } k \geq 5 \\ 7 & \text{if } j \geq 5 \text{ and } k \geq 6. \end{cases}$$

Proof Consider the following cases.

Case 1: Suppose that $3 \le j \le k \le 4$. The first image in Figure 2.39 shows aa 3-coloring of K_4 that avoids a monochromatic K_3 and a $K_{1,3}$ that uses at most two colors. It follows

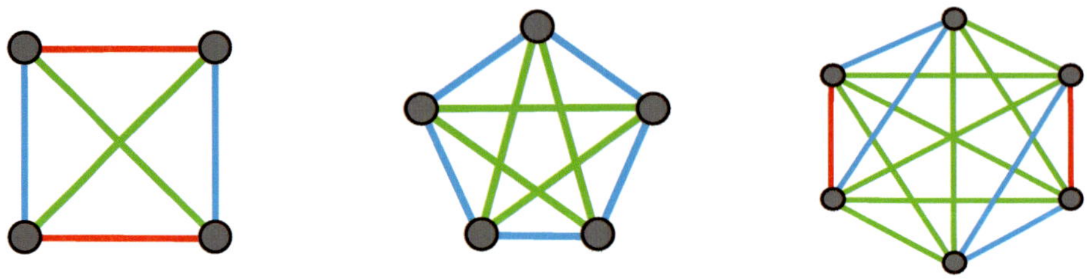

FIGURE 2.39: Critical colorings for $r^3_{1,2}(K_3, K_3, K_3, K_{1,3}, K_{1,j}, K_{1,k})$.

that

$$r^3_{1,2}(K_3, K_3, K_3, K_{1,3}, K_{1,3}, K_{1,3}) \ge 5.$$

To prove the upper bound, consider a 3-coloring of K_5 that avoids a monochromatic K_3, a red/blue $K_{1,3}$, a red/green $K_{1,4}$, and a blue/green $K_{1,4}$. Every vertex must have red degree 1, blue degree 1, and green degree 2. By Corollary 1.1, this is not possible as the subgraph spanned by the red edges would have an odd number of vertices with odd degree. It follows that

$$r^3_{1,2}(K_3, K_3, K_3, K_{1,3}, K_{1,4}, K_{1,4}) \le 5.$$

Hence,

$$r^3_{1,2}(K_3, K_3, K_3, K_{1,3}, K_{1,j}, K_{1,k}) = 5,$$

for all $3 \le j \le k \le 4$.

Case 2: Suppose that $j = 3$ and $k \ge 5$. The second image in Figure 2.39 shows a 3-coloring of K_5 that avoids a monochromatic K_3, a red/blue $K_{1,3}$, a red/green $K_{1,3}$, and a blue/green $K_{1,5}$. It follows that

$$r^3_{1,2}(K_3, K_3, K_3, K_{1,3}, K_{1,3}, K_{1,5}) \ge 6.$$

To prove the upper bound, consider a 3-coloring of K_6 and select a vertex x. If a red/blue $K_{1,3}$ is avoided, then $\deg_r + \deg_b(x) \le 2$. So, $\deg_g(x) \ge 3$, and we see that a red/green $K_{1,3}$ exists. Hence,

$$r^3_{1,2}(K_3, K_3, K_3, K_{1,3}, K_{1,3}, K_{1,k}) \le 6,$$

for all $k \ge 5$.

Case 3: Suppose that $j = 4$ and $k \ge 5$. As with the previous case, the second image in Figure 2.39 implies that

$$r^3_{1,2}(K_3, K_3, K_3, K_{1,3}, K_{1,4}, K_{1,5}) \ge 6.$$

To prove the upper bound, consder a 3-coloring of K_6 and select a vertex x. Avoiding a a red/blue $K_{1,3}$ forces $\deg_g(x) \ge 3$. If x is incident with a red edge, then it is the center vertex for a red/green $K_{1,4}$. Otherwise, all edges incident with x are blue and green and x is the center vertex for a blue/green $K_{1,5}$. It follows that

$$r^3_{1,2}(K_3, K_3, K_3, K_{1,3}, K_{1,4}, K_{1,k}) \le 6,$$

for all $k \ge 5$.

Case 4: Suppose that $j = k = 5$. The second image in Figure 2.39 once again gives the lower bound

$$r^3_{1,2}(K_3, K_3, K_3, K_{1,3}, K_{1,5}, K_{1,5}) \geq 6.$$

To prove the upper bound, consider a 3-coloring of K_6 and select a vertex x. Avoiding a red/blue $K_{1,3}$ forces $\deg_g(x) \geq 3$. If no red/green or blue/green $K_{1,5}$ exists, then x must be incident with exactly one red edge, one blue edge, and three green edges. Of course, x was chosen arbitrarily, so this property is true of every vertex. So the subgraph G_{rb} spanned by the red and blue edges is 2-regular. The only 2-regular graphs of order 6 are C_6 and $2C_3$. If $G_{rb} \cong C_6$ with cycle $x_1x_2x_3x_4x_5x_6x_1$, then the subgraph induced by $\{x_1, x_3, x_5\}$ is a green K_3. If $G_{rb} \cong 2C_3$, then at least two edges in a C_3-subgraph must be the same color (from among red and blue), contradicting the observation that every vertex is incident with exactly one red and one blue edge. Hence,

$$r^3_{1,2}(K_3, K_3, K_3, K_{1,3}, K_{1,5}, K_{1,5}) \leq 6.$$

Case 5: Suppose that $j \geq 5$ and $k \geq 6$. The third image in Figure 2.39 shows a 3-coloring of K_6 that avoids a monochromatic K_3, a red/blue $K_{1,3}$, a red/green $K_{1,5}$, and a blue/green $K_{1,6}$. It follows that

$$r^3_{1,2}(K_3, K_3, K_3, K_{1,3}, K_{1,5}, K_{1,6}) \geq 7.$$

To prove the upper bound, consider a 3-coloring of K_7. If a red/blue $K_{1,3}$ is avoided, then every vertex x satisfies $\deg_r(x) + \deg_b(x) \leq 2$, from which it follows that $\deg_g(x) \geq 4$. By Theorem 1.2, the subgraph G_g spanned by the green edges has at least 14 edges. Theorem 1.7 implies that the Turán number satisfies

$$ex(7, K_3) \leq \left\lfloor \left(1 - \frac{1}{2}\right) \frac{49}{2} \right\rfloor = 12,$$

from which it follows that there exists a green K_3. Hence,

$$r^3_{1,2}(K_3, K_3, K_3, K_{1,3}, K_{1,j}, K_{1,k}) \leq 7,$$

for all $j \geq 5$ and $k \geq 6$. □

Theorem 2.34 ([132]) *For $4 \leq j \leq k$,*

$$r^3_{1,2}(K_3, K_3, K_3, K_{1,4}, K_{1,j}, K_{1,k}) = \begin{cases} 6 & \text{if } j = k = 4 \\ 7 & \text{if } (j,k) \in \{(4,k), (5,k), (6,6)\}, \text{ where } k \geq 5 \\ 9 & \text{if } j \geq 6 \text{ and } k \geq 7. \end{cases}$$

Proof Consider the following cases.

Case 1: Suppose that $j = k = 4$. The first image in Figure 2.40 shows a 3-coloring of K_5 that avoids a monochromatic K_3 and in which every vertex is incident with edges in all three colors. It follows that

$$r^3_{1,2}(K_3, K_3, K_3, K_{1,4}, K_{1,4}, K_{1,4}) \geq 6.$$

To prove the reverse inequality, consider a 3-coloring of K_6 and select a vertex x. Among the five edges incident with x, some color appears on at most one of the edges. Without loss of generality, suppose that at most one of the edges incident with x is red. Then x is the center vertex for a blue/green $K_{1,4}$. Hence,

$$r^3_{1,2}(K_3, K_3, K_3, K_{1,4}, K_{1,4}, K_{1,4}) \leq 6.$$

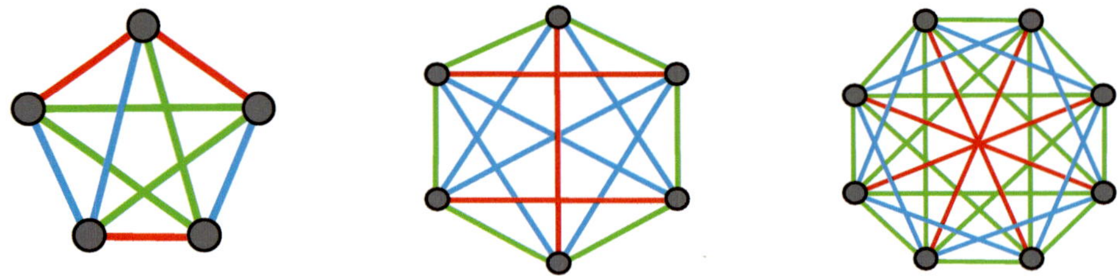

FIGURE 2.40: Critical colorings for $r^3_{1,2}(K_3, K_3, K_3, K_{1,4}, K_{1,j}, K_{1,k})$.

Case 2: Suppose that $j \in \{4, 5\}$ and $k \geq 5$. The second image in Figure 2.40 shows a 3-coloring of K_6 that avoids a monochromatic K_3 and in which every vertex is incident with one red edge, two blue edges, and two green edges. It follows that

$$r^3_{1,2}(K_3, K_3, K_3, K_{1,4}, K_{1,4}, K_{1,5}) \geq 7.$$

To prove the upper bound, consider a 3-coloring of K_7 that avoids a red/blue $K_{1,4}$ and a red/green $K_{1,5}$. For any vertex x, $\deg_r(x) + \deg_b(x) \leq 3$, from which it follows that $\deg_g(x) \geq 3$. Also, $\deg_r(x) + \deg_g(x) \leq 4$ implies that $\deg_b(x) \geq 2$. It is not possible for every vertex to be incident with exactly three green edges as this would contradict Corollary 1.1 since the subgraph G_g spanned by the green edges would have an odd number of vertices with odd degree. Hence, there exists some vertex y such that $\deg_g(y) \geq 4$. Such a vertex is required to have red degree 0, blue degree 2 and green degree 4. Vertices that are incident with a red edge have red degree 1, blue degree 2, and green degree 3. Hence, the subgraph G_b spanned by the blue edges is 2-regular. The only 2-regular graphs of order 7 are C_7 and $C_4 \cup C_3$. The second case is not possible since a blue K_3 is avoided. Let $x_1x_2x_3x_4x_5x_6x_7x_1$ be the blue C_7 and assume that x_1 satisfies $\deg_g(x_1) = 4$. In order to avoid a monochromatic K_3 edges x_3x_5 and x_4x_6 must be red (see Figure 2.43). Now edge x_3x_6 cannot be green as

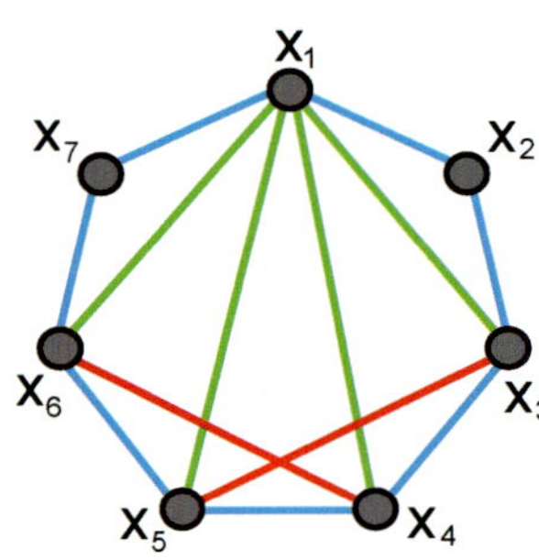

FIGURE 2.41: A 3-coloring of K_7 that avoids a monochromatic K_4, a red/blue $K_{1,4}$, and a red/green $K_{1,5}$.

this would make the subgraph induced by $\{x_1, x_3, x_6\}$ be a green K_3. It also cannot be red or blue without forming a red/blue $K_{1,4}$. It follows that

$$r^3_{1,2}(K_3, K_3, K_3, K_{1,4}, K_{1,5}, K_{1,k}) \leq 7,$$

for all $k \geq 5$. Therefore,

$$r^3_{1,2}(K_3, K_3, K_3, K_{1,4}, K_{1,j}, K_{1,k}) = 7,$$

whenever $j \in \{4, 5\}$ and $k \geq 5$.

Case 3: Suppose that $j = k = 6$. As with the previous case, the second image in Figure 2.40 provides the lower bound

$$r^3_{1,2}(K_3, K_3, K_3, K_{1,4}, K_{1,6}, K_{1,6}) \geq 7.$$

To prove the reverse inequality, consider a 3-coloring of K_7 that avoids a monochromatic K_3, a red/blue $K_{1,4}$, a red/green $K_{1,6}$, and a blue/green $K_{1,6}$. Avoiding a red/blue $K_{1,4}$ forces $\deg_r(x) + \deg_b(x) \leq 3$, which implies that $\deg_g(x) \geq 3$. Avoiding a red/green $K_{1,6}$ forces $\deg_r(x) + \deg_g(x) \leq 5$, which implies that $\deg_b(x) \geq 1$. Avoiding a blue/green $K_{1,6}$ forces $\deg_b(x) + \deg_g(x) \leq 5$, which implies that $\deg_r(x) \geq 1$. Not all of the vertices can have green degree 3 as this would contradict Corollary 1.1. So, some vertex must have green degree 4, red degree 1, and blue degree 1. Without loss of generality, suppose that x_1 is one such vertex with x_1x_2 red, x_1x_7 blue, and x_1x_3, x_1x_4, x_1x_5, and x_1x_6 green. In order to avoid a green K_3, the subgraph induced by $\{x_3, x_4, x_5, x_6\}$ is a red/blue K_4. Avoiding a red/blue $K_{1,4}$ then forces edges x_2x_3, x_2x_4, x_2x_5, x_2x_6, x_7x_3, x_7x_4, x_7x_5, and x_7x_6 to be green (see Figure 2.45). Now edge x_2x_7 cannot be red or green since a red/green $K_{1,6}$

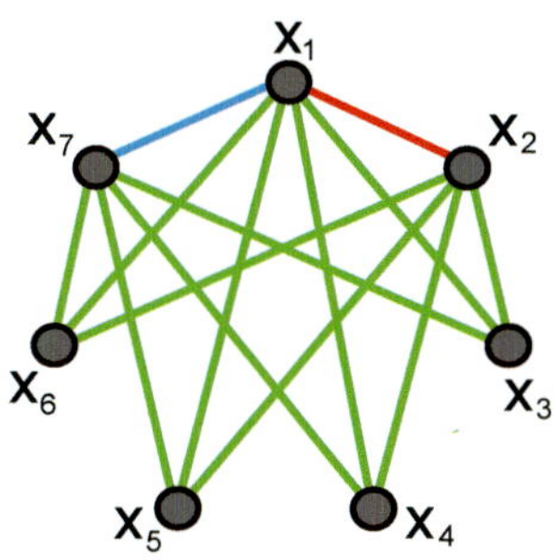

FIGURE 2.42: A 3-coloring of K_7 that avoids a monochromatic K_4, a red/blue $K_{1,4}$, a red/green $K_{1,6}$, and a blue/green $K_{1,6}$.

would then be formed with vertex x_2 as its center. It also cannot be blue since a blue/green $K_{1,6}$ would then be formed with x_7 as its center. It follows that

$$r^3_{1,2}(K_3, K_3, K_3, K_{1,4}, K_{1,6}, K_{1,6}) \leq 7.$$

Case 4: Suppose that $j \geq 6$ and $k \geq 7$. Let $S_r := \{4\}$, $S_b := \{\pm 2\}$, and $S_g := \{\pm 1, \pm 3\}$ and factor K_8 into the Cayley graphs $\text{Cay}(\mathbb{Z}_8; S_r)$, $\text{Cay}(\mathbb{Z}_8; S_b)$, and $\text{Cay}(\mathbb{Z}_8; S_g)$, colored red, blue, and green, respectively (see the third image in Figure 2.40). This 3-coloring of K_8 avoids a monochromatic K_3, a red/blue $K_{1,4}$, a red/green $K_{1,6}$, and a blue/green $K_{1,7}$. It follows that

$$r^3_{1,2}(K_3, K_3, K_3, K_{1,4}, K_{1,j}, K_{1,k}) \geq 9,$$

for all $j \geq 6$ and $k \geq 7$.

To prove the reverse inequality, consider a 3-coloring of K_9 that lacks a monochromatic K_3 and a red/blue $K_{1,4}$. Then for any vertex x, $\deg_r(x) + \deg_b(x) \leq 3$, which implies that $\deg_g(x) \geq 5$. It is not possible that all of the vertices have green degree 5 as this would contradict Corollary 1.1. So, some vertex must have green degree at least 6. Denote such a vertex by y and note that the subgraph incduced by $N_g(y)$ cannot contain any green edges without producing a green K_3. Hence, the subgraph induced by $N_g(y)$ is a red/blue K_n for some $n \geq 6$. By Theorem 1.14, there exists a red K_3 or a blue K_3. It follows that

$$r^3_{1,2}(K_3, K_3, K_3, K_{1,4}, K_{1,j}, K_{1,k}) \leq 9,$$

for all $j \geq 6$ and $k \geq 7$. □

Theorem 2.35 ([132]) *For $5 \leq j \leq k$,*

$$r^3_{1,2}(K_3, K_3, K_3, K_{1,5}, K_{1,j}, K_{1,k}) = \begin{cases} 8 & \text{if } j = 5 \text{ and } k \in \{5,6\} \\ 9 & \text{if } (j,k) \in \{(5,7),(6,7),(7,7),(6,6)\} \\ 10 & \text{if } j = 7 \text{ and } k \geq 8 \\ 11 & \text{if } k \geq j \geq 8. \end{cases}$$

Proof Consider the following cases.

Case 1: Suppose that $j = 5$ and $k \in \{5, 6\}$. Let $S_r := \{\pm 1\}$, $S_b := \{\pm 2\}$, and $S_g := \{\pm 3\}$ and factor K_7 into the Cayley graphs Cay($\mathbb{Z}_7; S_r$), Cay($\mathbb{Z}_7; S_b$), and Cay($\mathbb{Z}_7; S_g$), colored red, blue, and green, respectively (see the first image in Figure 2.43). The resulting 3-coloring

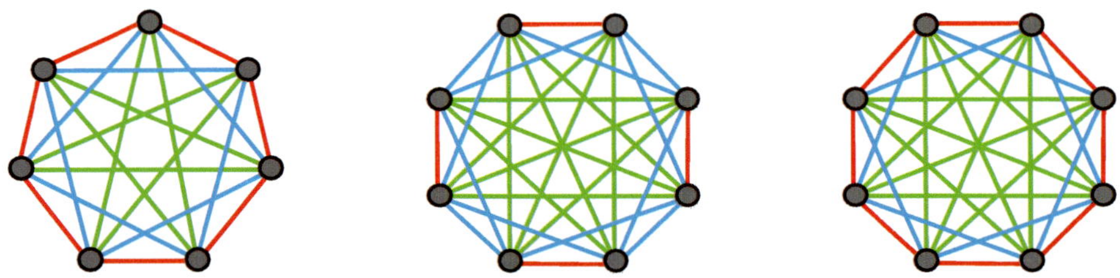

FIGURE 2.43: Critical colorings for $r^3_{1,2}(K_3, K_3, K_3, K_{1,5}, K_{1,j}, K_{1,k})$ when $5 \leq j \leq k \leq 7$.

of K_7 lacks a monochromatic K_3 and a $K_{1,5}$ that uses at most two colors. It follows that

$$r^3_{1,2}(K_3, K_3, K_3, K_{1,5}, K_{1,5}, K_{1,5}) \geq 8.$$

The upper bound follows from Theorem 2.29:

$$r^3_{1,2}(K_3, K_3, K_3, K_{1,5}, K_{1,5}, K_{1,6}) \leq r^3_2(K_{1,5}, K_{1,5}, K_{1,6}) = 8.$$

Case 2: Suppose that $j \in \{5, 6, 7\}$ and $k = 7$. The 3-coloring of K_8 shown in the second image in Figure 2.43 avoids a monochromatic K_3 and every vertex has red degree 1, blue degree 3, and green degree 3. Hence, it does not contain a red/blue $K_{1,5}$, a red/green K_5, or a blue/green $K_{1,7}$. It follows that

$$r^3_{1,2}(K_3, K_3, K_3, K_{1,5}, K_{1,5}, K_{1,7}) \geq 9.$$

To prove the upper bound, consider a 3-coloring of K_9 that avoids a monochromatic K_3. If a red/blue $K_{1,5}$ is avoided, then for every vertex x, $\deg_r(x) + \deg_b(x) \leq 4$, which implies that $\deg_g(x) \geq 4$. If a red/green $K_{1,7}$ is avoided, then $\deg_r(x) + \deg_g(x) \leq 6$, which implies that $\deg_b(x) \geq 2$. If a blue/green $K_{1,7}$ is avoided, then $\deg_b(x) + \deg_g(x) \leq 6$, which implies that $\deg_r(x) \geq 2$. Since every vertex is incident with exactly eight edges, we find that every vertex has red degree 2, blue degree 2, and green degree 4. Let xy be a green edge and note that since a green K_3 is avoided, $N_g(x) \cap N_g(y) = \emptyset$. Let $N_g(x) = \{x_1, x_2, x_3, y\}$, $N_g(y) = \{y_1, y_2, y_3, x\}$, and denote the ninth vertex by z (see the first image in Figure 2.44). The subgraphs induced by $N_g(x)$ and $N_g(y)$ do not contain any green edges. Hence, each x_i must join to exactly three of the vertices in $\{y_1, y_2, y_3, z\}$ with green edges and each y_j must join to exactly three of the vertices in $\{x_1, x_2, x_3, z\}$ with green edges. Also, exactly four of the vertices in $\{x_1, x_2, x_3, y_1, y_2, y_3\}$ join to z with green edges. Consider two subcases.

Subcase 2.1: Suppose that z joins to all three of the vertices in one of the sets $\{x_1, x_2, x_3\}$ and $\{y_1, y_2, y_3\}$ with green edges. Without loss of generality, assume that zx_1, zx_2, zx_3, and

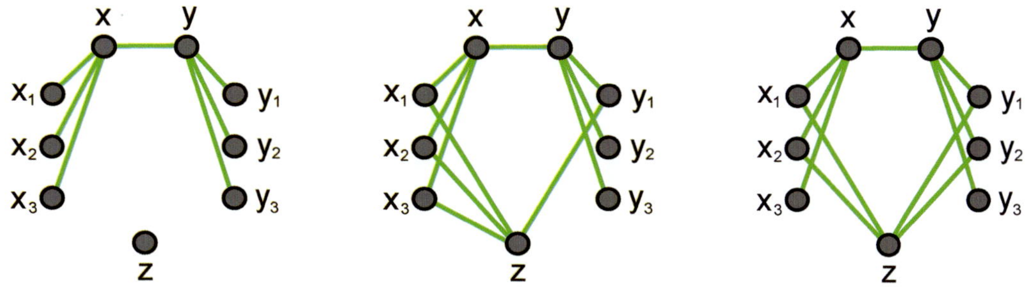

FIGURE 2.44: A 3-coloring of K_9 that avoids a monochromatic K_3, a red/blue $K_{1,5}$, a red/green $K_{1,7}$, and a blue/green $K_{1,7}$.

zy_1 are green (see the second image in Figure 2.44). Vertex y_1 must join to at least two of the vertices in $\{x_1, x_2, x_3\}$ with green edges. If x_iy_1 is green, then the subgraph induced by $\{x_i, y_1, z\}$ is a green K_3, which is a contradiction.

Subcase 2.2: Suppose that z joins to exactly two vertices in each of the sets $\{x_1, x_2, x_3\}$ and $\{y_1, y_2, y_3\}$ with green edges. Without loss of generality, suppose that zx_1, zx_2, zy_1, and zy_2 are green (see the third image in Figure 2.44). Vertex y_1 must join to at least two of the vertices in $\{x_1, x_2, x_3\}$ with green edges. In particular, even if x_3y_1 is green, y_1 must join to at least one element in $\{x_1, x_2\}$ with a green edge. Once again, if x_iy_1 is green, then the subgraph induced by $\{x_i, y_1, z\}$ is a green K_3, which is a contradiction.

In both subcases, we arrived at a contradiction. Therefore, every 3-coloring of K_9 contains a monochromatic K_3, a red/blue $K_{1,5}$, a red/green $K_{1,7}$, or a blue/green $K_{1,7}$. It follows that

$$r^3_{1,2}(K_3, K_3, K_3, K_{1,5}, K_{1,7}, K_{1,7}) \leq 9,$$

and we see that

$$r^3_{1,2}(K_3, K_3, K_3, K_{1,5}, K_{1,j}, K_{1,k}) = 9,$$

for all $j \in \{5, 6, 7\}$ and $k = 7$.

Case 3: Suppose that $j = k = 6$. The 3-coloring of K_8 shown in the third image in Figure 2.43 avoids a monochromatic K_3 and every vertex has red degree 2, blue degree 2, and green degree 3. Hence, it does not contain a red/blue $K_{1,5}$, a red/green K_6, or a blue/green $K_{1,6}$. It follows that

$$r^3_{1,2}(K_3, K_3, K_3, K_{1,5}, K_{1,6}, K_{1,6}) \geq 9.$$

The reverse inequality follows from Theorem 2.29:

$$r^3_{1,2}(K_3, K_3, K_3, K_{1,5}, K_{1,6}, K_{1,6}) \leq r^3_2(K_{1,5}, K_{1,6}, K_{1,6}) = 9.$$

Case 4: Suppose that $j = 7$ and $k \geq 8$. The 3-coloring of K_9 given in Figure 2.45 avoids a monochromatic K_3, a red/blue $K_{1,5}$, a red/green K_7, and a blue/green K_8. It follows that

$$r^3_{1,2}(K_3, K_3, K_3, K_{1,5}, K_{1,7}, K_{1,8}) \geq 10.$$

To prove the upper bound, consider a 3-coloring of K_{10} that avoids a monochromatic K_3, a red/blue $K_{1,5}$, and a red/green $K_{1,7}$. For any vertex x, $\deg_r(x) + \deg_b(x) \leq 4$, which implies that $\deg_g(x) \geq 5$. Suppose that xy is a green edge and note that $N_g(x) \cap N_g(y) = \emptyset$, which forces x and y to each have green degree exactly 5. Without loss of generality, suppose that $N_g(x) = \{x_1, x_2, x_3, x_4, y\}$ and $N_g(y) = \{y_1, y_2, y_3, y_4, x\}$ (see Figure 2.46). In order for x_i to have green degree at least 5, it must join to y_1, y_1, y_3, and y_4 with green edges.

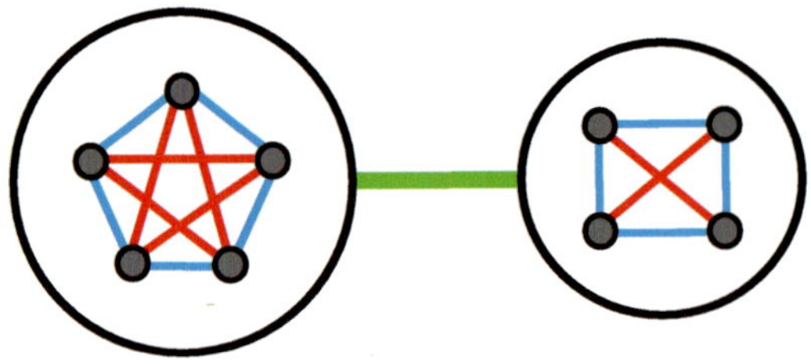

FIGURE 2.45: A critical coloring for $r^3_{1,2}(K_3, K_3, K_3, K_{1,5}, K_{1,7}, K_{1,8})$.

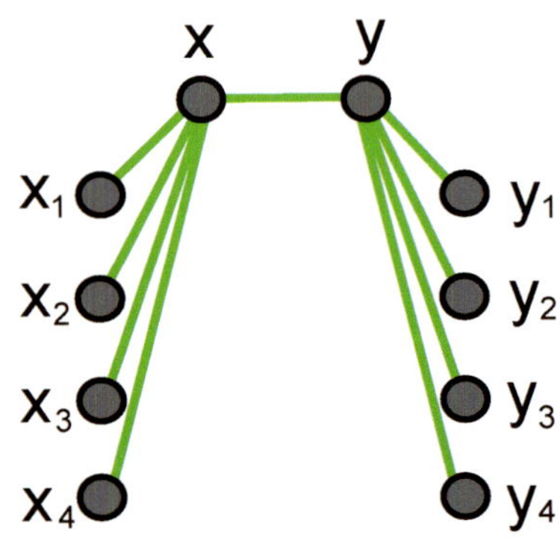

FIGURE 2.46: A 3-coloring of K_{10} that avoids a green K_3 and in which every vertex has green degree at least 5.

Likewise, in order for y_j to have green degree at least 5, it must join to x_1, x_2, x_3, and x_4 with green edges. At this point, we see that the subgraph G_b spanned by the green edges is isomorphic to $K_{5,5}$ with partite sets $N_g(x)$ and $N_g(y)$. The subgraph induced by $N_g(x)$ is a red/blue K_5 that avoids a monochromatic K_3, so by Theorem 1.14, the red and blue subgraphs of this K_5 must each be isomorphic to C_5. However, a red/green $K_{1,7}$ can now be formed with center vertex y, giving a contradiction. Thus, every 3-coloring of K_{10} contains a monochromatic K_3, a red/blue $K_{1,5}$, or a red/green $K_{1,7}$. It follws that

$$r^3_{1,2}(K_3, K_3, K_3, K_{1,5}, K_{1,7}, K_{1,k}) \leq 10,$$

for all $k \geq 8$.

Case 5: Suppose that $k \geq j \geq 8$. The 3-coloring of K_{10} given in Figure 2.47 avoids a monochromatic K_3, a red/blue $K_{1,5}$, a red/greenK_8, and a blue/green K_8. It follows that

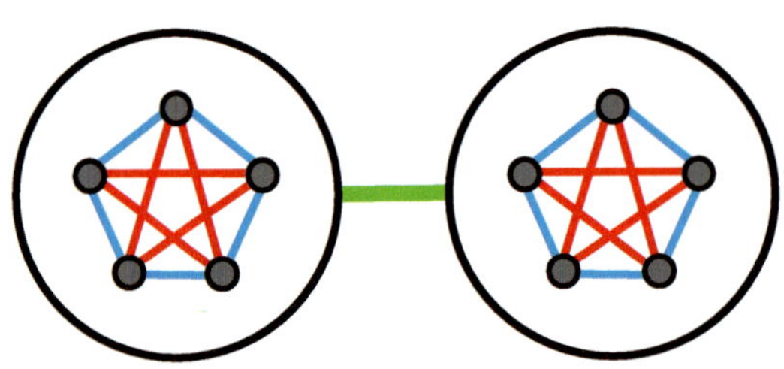

FIGURE 2.47: A critical coloring for $r^3_{1,2}(K_3, K_3, K_3, K_{1,5}, K_{1,8}, K_{1,8})$.

$$r^3_{1,2}(K_3, K_3, K_3, K_{1,5}, K_{1,8}, K_{1,8}) \geq 11.$$

The reverse inequality follows from Theorem 2.29:

$$r^3_{1,2}(K_3, K_3, K_3, K_{1,5}, K_{1,8}, K_{1,8}) \leq r^3_2(K_{1,5}, K_{1,8}, K_{1,8}) = 11.$$

□

Theorem 2.36 ([132]) *For* $k \in \{6, 7\}$,

$$r^3_{1,2}(K_3, K_3, K_3, K_{1,6}, K_{1,6}, K_{1,k}) = 9.$$

Proof Let $S_r := \{\pm 2\}$, $S_b := \{\pm 3\}$, and $S_g := \{\pm 1, 4\}$ and factor K_8 into the Cayley graphs $\text{Cay}(\mathbb{Z}_8; S_r)$, $\text{Cay}(\mathbb{Z}_8; S_b)$, and $\text{Cay}(\mathbb{Z}_8; S_g)$, colored red, blue, and green, respectively (see the first image in Figure 2.48). The resulting 3-coloring of K_8 avoids a monochromatic

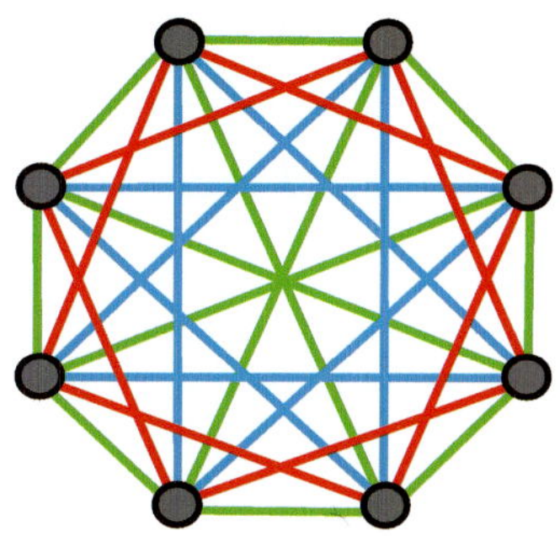

FIGURE 2.48: A critical coloring for $r^3_{1,2}(K_3, K_3, K_3, K_{1,6}, K_{1,6}, K_{1,6})$.

K_3, a red/blue $K_{1,6}$, a red/green $K_{1,6}$, and a blue/green $K_{1,6}$. It follows that

$$r^3_{1,2}(K_3, K_3, K_3, K_{1,6}, K_{1,6}, K_{1,6}) \geq 9.$$

To prove the upper bound, consider a 3-coloring of K_9 that avoids a monochromatic K_3, a red/blue $K_{1,6}$, a red/green $K_{1,6}$, and a blue/green $K_{1,7}$. If x is any vertex, then $\deg_r(x) + \deg_b(x) \leq 5$, from which it follows that $\deg_g(x) \geq 3$. Since a red/green $K_{1,6}$ is avoided, $\deg_r(x) + \deg_g(x) \leq 5$, from which it follows that $\deg_b(x) \geq 3$. Since a blue/green $K_{1,7}$ is avoided, $\deg_b(x) + \deg_g(x) \leq 6$, from which it follows that $\deg_\text{r}(\text{x}) \geq 2$. Since the red, blue, and green degrees must sum to 8, it follows that every vertex has red degree 2, blue degree 3, and green degree 3. This is not possible as the subgraph G_b spanned by the blue edges would contain an odd number of vertices of odd degree, contradicting Corollary 1.1. Thus,

$$r^3_{1,2}(K_3, K_3, K_3, K_{1,6}, K_{1,6}, K_{1,7}) \leq 9,$$

from which it follows that

$$r^3_{1,2}(K_3, K_3, K_3, K_{1,6}, K_{1,6}, K_{1,k}) = 9,$$

for all $k \in \{6, 7\}$. □

In his dissertation, Jacobson [132] claims to have determined several other chromatic Ramsey numbers of the form

$$r^3_{1,2}(K_3, K_3, K_3, K_{1,i}, K_{1,j}, K_{1,k}).$$

However, the lack of details given prevent us from continuing with their evaluation here. Thus, for larger i, j, and k, we leave the evaluations as an open problem (see Problem 3 in Section 5.1).

2.3.3 Chromatic Ramsey Numbers Involving Paths

Throughout this subsection, assume that $2 \le i \le j \le k$. In 1967, Gerencsér and Gyárfás [105] proved that

$$r(P_i, P_j) = j + \left\lfloor \frac{i}{2} \right\rfloor - 1. \tag{2.9}$$

This result, along with Theorem 2.5, implies that

$$r_2^3(P_i, P_j, P_k) = j + \left\lfloor \frac{i}{2} \right\rfloor - 1,$$

whenever $k \ge j + \lfloor \frac{i}{2} \rfloor - 1$. The case where $k < j + \lfloor \frac{i}{2} \rfloor - 1$ was handled by Meenakshi and Sundararaghavan [179] in 1986. In order to work through their evalaution, we need the following lemma.

Lemma 2.4 ([179]) *Let $k \ge 3$ and $j = k - 2\lfloor \frac{k+4}{6} \rfloor$. Then*

$$r_2^3(P_j, P_j, P_k) \le k - 1.$$

Proof Write $k = 6q + r$ for $q, r \in \mathbb{Z}$ such that $0 \le r < 6$. Then

$$j = k - 2\left\lfloor \frac{6q + r + 4}{6} \right\rfloor = k - 2\left(q + \left\lfloor \frac{r+4}{6} \right\rfloor\right).$$

Equation (2.9) and Theorem 2.5 imply that

$$\begin{aligned} r_2^3(P_j, P_j, P_k) &\le j + \left\lfloor \frac{j}{2} \right\rfloor - 1 \\ &= k - 2q - 2\left\lfloor \frac{r+4}{6} \right\rfloor + \left\lfloor \frac{6q + r - 2q + 2\lfloor \frac{r+4}{6} \rfloor}{2} \right\rfloor - 1 \\ &= k - 1 - 2\left\lfloor \frac{r+4}{6} \right\rfloor + \left\lfloor \frac{r}{2} \right\rfloor \\ &\le k - 1. \end{aligned}$$

The last inequality follows from checking that $-2\lfloor \frac{r+4}{6} \rfloor + \lfloor \frac{r}{2} \rfloor \in \{-1, 0\}$ for each value of $r \in \{0, 1, \ldots, 5\}$. □

Theorem 2.37 ([179]) *For all $i, j, k \in \mathbb{N}$ such that $2 \le i \le j \le k < j + \lfloor \frac{i}{2} \rfloor - 1$,*

$$r_2^3(P_i, P_j, P_k) = \left\lfloor \frac{i + 2j + 4k - 2}{6} \right\rfloor.$$

Proof Let $n = n(i, j, k) := \lfloor \frac{i+2j+4k-2}{6} \rfloor$ and partition the vertex set of K_{n-1} into three subsets X, Y, and Z such that

$$|X| = \left\lceil \frac{2k + j - i - 1}{3} \right\rceil, \quad |Y| = \left\lfloor \frac{2j - 2k + i - 2}{6} \right\rfloor, \quad \text{and} \quad |Z| = \left\lfloor \frac{k - j + i - 2}{3} \right\rfloor.$$

We leave it to the reader to confirm that $n - 1 = |X| + |Y| + |Z|$. Color the edges red in the subgraph induced by Y, color the edges blue in the subgraph induced by Z, and color the edges green in the subgraph induced by X. Color all edges red that join vertices in Y to vertices in $X \cup Z$, and color the edges blue that join vertices in X to vertices in Z (see Figure 2.49). The longest red/blue path has order

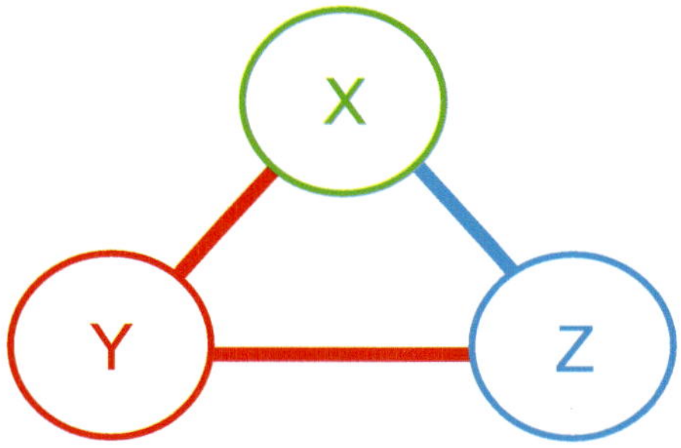

FIGURE 2.49: Constructing a critical coloring for $r_2^3(P_i, P_j, P_k)$.

$$2|Y| + 2|Z| + 1 = 2\left\lfloor \frac{2j - 2k + i - 2}{6} \right\rfloor + 2\left\lfloor \frac{k - j + i - 2}{3} \right\rfloor + 1 \leq i - 1,$$

the longest red/green path has order

$$|X| + 2|Y| = \left\lceil \frac{2k + j - i - 1}{3} \right\rceil + 2\left\lfloor \frac{2j - 2k + i - 2}{6} \right\rfloor \leq j - 1,$$

and the longest blue/green path has order

$$|X| + |Z| = \left\lceil \frac{2k + j - i - 1}{3} \right\rceil + \left\lfloor \frac{k - j + i - 2}{3} \right\rfloor \leq k - 1.$$

It follows that $r_2^3(P_i, P_j, P_k) \geq n$.

To prove the reverse inequality, first assume that $i \leq 4$. Then $k < j + \lfloor \frac{i}{2} \rfloor - 1$ implies that $i = 4$, $j = k$, and

$$n(4, j, j) = \left\lfloor \frac{4 + 2j + 4j - 2}{6} \right\rfloor = \left\lfloor j + \frac{1}{3} \right\rfloor = j.$$

We proceed by induction on $j \geq 4$ to prove that $r_2^3(P_4, P_j, P_j) \leq j$. For the base case $j = 4$, consider a 3-coloring of K_4 that avoids a red/blue P_4. Then there exists a vertex x such that $\deg_g(x) \geq 2$. Let $y, z \in N_g(x)$ and denote the fourth vertex by w. Then $yxzw$ is either a red/green P_4 or a blue/green P_4. If follows that $r_2^3(P_4, P_4, P_4) \leq 4$. Now suppose that $j > 4$ and $r_2^3(P_4, P_{j-1}, P_{j-1}) = j - 1$. Consider a 3-coloring of K_j that lacks a red/blue P_4. If there exists a vertex x such that $\deg_g(x) \geq j - 3$, then the subgraph induced by $V(K_j) \setminus \{x\}$ contains either a red/green P_{j-1} or a blue/green P_{j-1} by the inductive hypothesis. Reintroducing vertex x, either of these paths of order $j - 1$ can be extended to a P_j in the corresponding colors. If $\deg_g(x) < j - 3$, then $\deg_{rb}(x) \geq 3$ for every vertex x. Suppose that $w, z \in N_{rb}(y)$ for some vertex y. Then w must be red-adjacent or blue-adjacent to some vertex $v \notin \{y, z\}$. It follows that $vwyz$ is a red/blue P_4, which is a contradiction. Hence, $r_2^3(P_4, P_j, P_j) \leq j$.

Using the $i \leq 4$ case as the base case, we proceed by (strong) induction on $i + j + k$ to prove that $r_2^3(P_i, P_j, P_k) \leq n$ when $i \geq 5$. Suppose that

$$r_2^3(P_{i'}, P_{j'}, P_{k'}) \leq \left\lfloor \frac{i' + 2j' + 4k' - 2}{6} \right\rfloor,$$

whenever $i' + j' + k' < i + j + k$, and consider a 3-coloring of $K_{n(i,j,k)}$. We consider several cases based on the values of i, j, and k.

Case 1: Suppose that $i = j = k$. Without loss of generality, suppose that x satisfies

$$\deg_r(x) \geq \max\{\deg_b(y), \deg_g(y)\}, \tag{2.10}$$

for every $y \in V(K_n)$. Consider the subgraph induced by $V(K_n) \setminus \{x\}$. By the inductive hypothesis,

$$r_2^3(P_{i-2}, P_{i-2}, P_i) \leq i - 1 + \left\lfloor \frac{i-2}{6} \right\rfloor < i + \left\lfloor \frac{i-2}{6} \right\rfloor = n(i, i, i).$$

Hence, there exists a red/blue P_{i-2}, a red/green P_{i-2}, or a blue/green P_i. In the last case, we are done, so assume one of the first two cases occurs. Without loss of generality, suppose that $x_1 x_2 \cdots x_{i-2}$ is a red/blue P_{i-2}.

Subcase 1.1: If $xx_1, xx_{i-2} \in E(G_{rb})$, then $x_1 x_2 \cdots x_{i-2} x x_1$ is a red/blue C_{i-1}. Denote by X the set of vertices in K_n that are not in the red/blue C_{i-1}, and note that

$$|X| = i + \left\lfloor \frac{i-2}{6} \right\rfloor - (i-1) = \left\lfloor \frac{i+4}{6} \right\rfloor.$$

If any vertex in X joins to any vertex in the red/blue C_{i-1} with a red or blue edge, then together, they form a red/blue P_i. So, assume that all edges joining the vertices in X to the vertices in the red/blue C_{i-1} are green. By Lemma 2.4,

$$r_2^3(P_i, P_{i-2\lfloor (i+4)/6 \rfloor}, P_{i-2\lfloor (i+4)/6 \rfloor}) \leq i - 1.$$

It follows that the subgraph induced by the vertices in the red/blue C_{i-1} contains a red/blue P_i, a red/green $P_{i-2\lfloor (i+4)/6 \rfloor}$, or a blue/green $P_{i-2\lfloor (i+4)/6 \rfloor}$. In the first case, we are done, so suppose, without loss of generality, that there exists a red/green $P_{i-2\lfloor (i+4)/6 \rfloor}$. Let Y be the set of vertices in the red/blue C_{i-1} that are not contained in the red/green $P_{i-2\lfloor (i+4)/6 \rfloor}$ and note that

$$|Y| = i - 1 - \left(i - 2\left\lfloor \frac{i+4}{6} \right\rfloor\right) = -1 + 2\left\lfloor \frac{i+4}{6} \right\rfloor \geq \left\lfloor \frac{i+4}{6} \right\rfloor.$$

Since all of the edges joining X to Y are green, $X \cup Y$ contains a green $P_{2\lfloor (i+4).6 \rfloor}$ that shares an end vertex with the red/green $P_{i-2\lfloor (i+4)/6 \rfloor}$. Together, these two paths form a red/green P_i.

Subcase 1.2: If $xx_1 \in E(G_{rb})$ and $xx_{i-2} \in E(G_g)$, then $xx_1 x_2 \cdots x_{i-2}$ is a red/blue P_{i-1}. Let $Y = V(K_n) \setminus \{x, x_1, x_2, \ldots, x_{i-2}\}$. If there exists a vertex $y \in Y$ such that xy is red or blue, then $yxx_1x_2 \cdots x_{i-2}$ is a red/blue P_i. So, assume that all edges xy, where $y \in Y$, are green. Then $\deg_g(x) \geq n - i$. Inequality (2.10) implies that $\deg_r(x) \geq n - i$, and it follows that x is adjacent to at least $n - i$ vertices in the red/blue path $x_1 x_2 \cdots x_{i-2}$. Suppose that xx_ℓ is red for some ℓ such that $2 \leq \ell \leq i - 3$. If edge $x_{\ell-1} x_{i-2}$ is red or blue, then

$$xx_\ell x_{\ell+1} \cdots x_{i-2} x_{\ell-1} x_{\ell-2} \cdots x_1 x$$

is a red/blue C_{i-1}. By the argument given in the proof of Subcase 1.1, there must exist a red/blue P_i. So, edge $x_{\ell-1} x_{i-2}$ must be green. Of course, this is true for each vertex x_ℓ, where $2 \leq \ell \leq i - 3$, for which xx_ℓ is red or blue, and there are at least $n - i + 1$ such vertices, since $\deg_r(x) \geq \deg_g(x) \geq n - i + 1$. It follows that vertex x_{i-2} satisfies

$$\deg_g(x_{i-2}) \geq n - (i-1) + 1 + n - (i-1) = 2\left\lfloor \frac{i+4}{6} \right\rfloor + 1.$$

By Inequality (2.10),

$$\deg_r(x) \geq \left\lfloor \frac{i+4}{6} \right\rfloor + 1 \geq 2\left\lfloor \frac{i+4}{6} \right\rfloor + 1,$$

which can only occur if $\lfloor \frac{i+4}{6} \rfloor = 0$. However, this contradicts the assumption that $i \geq 5$. Thus, there exists a red/blue P_i.

Subcase 1.3: Assume that $xx_1, xx_{i-2} \in E(G_g)$. Let $Y = V(K_n) \setminus \{x, x_1, x_2, \ldots, x_{i-2}\}$. Denote by n_1 the number of vertices in $N_r(x) \cap \{x_1, x_2 \ldots, x_{i-2}\}$ and denote by n_2 the number of vertices in $N_r(x) \cap Y$. Hence, $\deg_r(x) = n_1 + n_2$. Suppose that $\ell \in \{2, 3, \ldots, i-3\}$ is such that xx_ℓ is red. If edge $x_{\ell-1}x_{i-2}$ is red or blue, then $x_\ell x_{\ell+1} \cdots x_{i-2}x_{\ell-1}x_{\ell-2} \cdots x_1$ is a red/blue P_{i-2} in which one endpoint joins to x with a red or blue edge, and the other endpoint joins to x with a green edge. This is now equivalent to the situation considered in Subcase 1.2. So, edge $x_{\ell-1}x_{i-2}$ must be green for every such ℓ. Hence, x_{i-2} is adjacent to at least n_1 vertices in $\{x_1, x_2, \ldots x_{i-3}\}$ with green edges. Now consider a vertex $y \in Y$ such that xy is red. Then yx_{i-2} must be green if a red/blue P_i is to be avoided. It follows that

$$\deg_g(x_{i-2}) \geq n_1 + n_2 + 1,$$

which contradicts Inequality (2.10) since $\deg_r(x) = n_1 + n_2$. Thus, there exists a red/blue P_i.

Case 2: Assume that $i = j < k$. If there exists a vertex x that satisfies

$$\mathrm{deg_r(x)} \geq \max\{\mathrm{deg_r(y)}, \mathrm{deg_b(y)}, \mathrm{deg_g(y)}\},$$

for all $y \in V(K_n)$, then $r_2^3(P_{i-2}, P_{i-2}, P_k) \leq n(i,j,k) - 1$, and we repeat the arguments used in Case 1 to prove that there exists a red/blue P_i or a red/green P_i. So, without loss of generality, suppose that there exists a vertex x that satisfies

$$\mathrm{deg_b(x)} \geq \max\{\mathrm{deg_r(y)}, \mathrm{deg_b(y)}, \mathrm{deg_g(y)}\},$$

for all $y \in V(K_n)$. By the inductive hypothesis, $r_2^3(P_{i-2}, P_j, P_{k-1}) \leq n(i,j,k) - 1$. So, the subgraph induced by $V(K_n) \setminus \{x\}$ contains a red/blue P_{i-2}, a red/green P_j, or a blue/green P_{k-1}. If there exists a red/blue P_{i-2}, then a proof similar to that of Case 1 can be used. If there is a red/green P_j, we are done, so assume that there is a blue/green P_{k-1}. Then a proof similar to that of Case 1 can again be used. It follows that in all cases, there exists a red/blue P_i, a red/green P_j, or a blue/green P_k.

Case 3: Assume that $i < j \leq k < j + \lfloor \frac{i}{2} \rfloor - 1$. Select a vertex x such that

$$\deg_r(x) + \deg_b(x) \leq \deg_r(y) + \deg_b(y),$$

for all $y \in V(K_n)$. If $\deg_r(x) + \deg_b(x) \geq \lceil \frac{j}{2} \rceil$ and G_{rb} is connected, then by Theorem 1.8, it contains a red/blue $P_{2\lceil \frac{j}{2} \rceil}$. Hence, it contains a red/blue P_i. Otherwise, the subgraph G_{rb} is disconnected and each of its components has order at least $\lceil \frac{j}{2} \rceil$. In this case, a green P_j can be formed using the green edges joining two distinct components.

If $\deg_r(x) + \deg_b(x) < \lceil \frac{j}{2} \rceil$, then $\deg_g(x) \geq n - 1 - \lfloor \frac{j}{2} \rfloor$. By the inductive hypothesis, $r_2^3(P_i, P_{j-1}, P_{k-1}) \leq n(i,j,k) - 1$, so there exists a red/blue P_i, a red/green P_{j-1}, or a blue/green P_{k-1} in the subgraph induced by $V(K_n) \setminus \{x\}$. In the first case, we are done, so assume that there is either a red/green P_{j-1} or a blue/green P_{k-1}. Since $\deg_g(x) \geq n - 1 - \lfloor \frac{j}{2} \rfloor$, x must be adjacent via green edges to two consecutive vertices in the path, allowing it to be extended to a red/green P_j or a blue/green P_k.

In all cases, we find that $r_2^3(P_i, P_j, P_k) \leq n$, completing the proof of the theorem. □

In 2012, Khamseh and Omidi [148] determined the value of $r_{t-1}^t(P_{i_1}, P_{i_2}, \ldots, P_{t_t})$ when $t = 4$, and in 2018, Khamseh [145] determined the value when $t = 5$. Bucić and Khamseh [30] completed the evaluation of $r^t_{t-1}(P_{i_1}, P_{i_2}, \ldots, P_{t_t})$ for all t in 2021. We state their theorem without proof.

Theorem 2.38 ([30]) *If $t \geq 3$ and $2 \leq i_1 \leq i_2 \leq \cdots \leq i_t$, then*

$$r^t_{t-1}(P_{i_1}, P_{i_2}, \ldots, P_{i_t}) = \begin{cases} \left\lfloor \frac{i_1+2i_2+\cdots+2^{t-1}i_t-2}{2^t-2} \right\rfloor & \text{if } i_t < r^t_{t-1}(P_{i_1}, P_{i_2}, \cdots, P_{i_{t-1}}) \\ r^t_{t-1}(P_{i_1}, P_{i_2}, \cdots, P_{i_{t-1}}) & \text{otherwise.} \end{cases}$$

Building off of these results involving paths, Khamseh [146, 147] considered the t-color $(t-1)$-chromatic Ramsey number for linear forests (see also [148]). A *linear forest* is the disjoint union of nontrivial paths. An (n, j) *linear forest* is a linear forest of order n in which j of the paths have odd order. Let c_i denote the number of connected components of an (n_i, j_i) linear forest L_i. The $(t-1)$-chromatic Ramsey number $r^t_{t-1}(L_1, L_2, \ldots, L_t)$ was considered for various values of $t \leq 5$ in [146], with a more general result being proved in [147]. We state this result without proof.

Theorem 2.39 ([147]) *For $t \geq 3$, let $2 \leq n_1 \leq n_2 \leq \cdots \leq n_t$, $j_1 \leq 1$, and*

$$j_2(c_2-1) = j_3(c_3-1) = \cdots = j_{t-1}(c_{t-1}-1) = 0.$$

Then

$$r^t_{t-1}(L_1, L_2, \ldots, L_t) = \begin{cases} \left\lfloor \frac{\sum_{i=0}^{t-1} 2^i n_{i+1}-2}{2(2^{t-1}-1)} \right\rfloor & \text{if } n_t < r^{t-1}_{t-2}(L_1, L_2, \ldots, L_{t-1}) \\ r^{t-1}_{t-2}(L_1, L_2, \ldots, L_{t-1}) & \text{otherwise.} \end{cases}$$

3

Weakened Ramsey Numbers

For $t \geq 3$ and $1 \leq s \leq t-1$, define the *weakened Ramsey number* $r_s^t(G)$ to be the least $n \in \mathbb{N}$ such that every t-coloring of K_n contains a subgraph isomorphic to G that is spanned by edges that use at most s colors. The name "weakened Ramsey number" first appeared in Harborth and Möller's paper [123] in 1999 and these numbers are the diagonal cases of chromatic Ramsey numbers. They are closely related to the set-coloring Ramsey numbers studied in [5], [8], [57], [74], [85], [88], and [211].

3.1 Weakened Ramsey Numbers for Complete Graphs

In Theorem 2.10, we saw that $r_2^3(K_3) = 5$. In addition to this weakened Ramsey number, Chung and Liu [63] also determined the value of $r_2^3(K_4)$. In the authors' original proof, the second case in the proof contained a mistake. The proof given below corrects this error.

Theorem 3.1 ([63]) $r_2^3(K_4) = 10$.

Proof The lower bound $r_2^3(K_4) \geq 10$ follows from Theorem 2.3 (also, see Figure 2.1). To prove the upper bound, consider a 3-coloring of K_{10} using the colors red, blue, and green. By Lemma 2.1 and Theorem 2.11 the following inequalities hold for any vertex x:

$$\deg_R(x) + \deg_B(x) \leq 6, \quad \deg_R(x) + \deg_G(x) \leq 6, \quad \text{and} \quad \deg_B(x) + \deg_G(x) \leq 6.$$

Since $\deg_R(x) + \deg_B(x) + \deg_G(x) = 9$, it follows that

$$\deg_R(x) = \deg_B(x) = \deg_G(x) = 3$$

for every vertex x. Grouping together the colors blue and green, $r(K_3, K_4) = 9$ (Theorem 1.15) implies that there exists a red K_3, or a blue/green K_4. In the latter case, we are done, so assume there is a red K_3 and label its vertices v_1, v_2, and v_3. If any two of v_1, v_2, and v_3 join to a vertex $x \notin \{v_1, v_2, v_3\}$ with red edges, then a red $K_4 - e$ is formed. Regardless of the color of the missing edge, a K_4 is formed with edges using at most two colors. So, each vertex v_1, v_2, and v_3 must have red degree 3, and we obtain a red K_3 with three pendant edges as in Figure 3.1. Label the vertices at the ends of the pendant edges by v_4, v_5, and v_6 (so that v_1v_4, v_2v_5, and v_3v_6 are red) and label the remaining vertices v_7, v_8, v_9, and v_{10}. We now consider three cases.

Case 1: Suppose that the subgraph induced by $\{v_4, v_5, v_6\}$ contains either two or three red edges. Then at most two red edges join $\{v_4, v_5, v_6\}$ to $\{v_7, v_8, v_9, v_{10}\}$. In the subgraph induced by $\{v_7, v_8, v_9, v_{10}\}$, at least two vertices have red degree 3 and the other vertices have red degree 2. It follows that this induced subgraph contains a red K_4 or a red $K_4 - e$. In the latter case, a K_4 spanned by edges using at most two colors is formed.

DOI: 10.1201/9781003498865-3

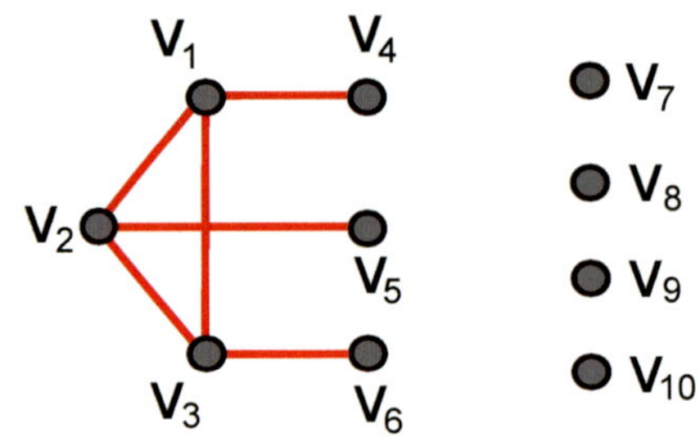

FIGURE 3.1: Red edges in a 3-coloring of K_{10}.

Case 2: Suppose that the subgraph induced by $\{v_4, v_5, v_6\}$ contains exactly one red edge. Without loss of generality, assume that v_4v_5 is red. Then v_6 joins to exactly two vertices in $\{v_7, v_8, v_9, v_{10}\}$ with red edges and v_4 joins to exactly one vertex in $\{v_7, v_8, v_9, v_{10}\}$ with a red edge. It follows that some vertex in $\{v_7, v_8, v_9, v_{10}\}$ (say, v_7) joins to both v_4 and v_6 using only blue or green edges. Then the subgraph induced by $\{v_2, v_4, v_6, v_7\}$ is a blue/green K_4.

Case 3: Suppose that the subgraph induced by $\{v_4, v_5, v_6\}$ does not contain any red edges. Then there exist six red edges joining $\{v_4, v_5, v_6\}$ to $\{v_7, v_8, v_9, v_{10}\}$ with exactly two such edges incident with each of v_4, v_5, and v_6. Some vertex in $\{v_7, v_8, v_9, v_{10}\}$ (say, v_7) is incident with at most one of these six red edges. Without loss of generality, assume that if v_7 is incident with one of these red edges, then v_4v_7 is red. Then the edges joining v_7 to $\{v_5, v_6\}$ are blue or green and the subgraph induced by $\{v_1, v_5, v_6, v_7\}$ is a blue/green K_4.

In all cases, we find that there exists a K_4 spanned by edges using at most two colors. It follows that $r_2^3(K_4) \leq 10$, completing the proof of the theorem. □

We now turn our attention to some general results involving weakened Ramsey numbers for complete graphs of order at least 4. The next few theorems were proved by Harborth and Möller [123] in 1999.

Theorem 3.2 ([123]) *For all $j \geq 1$, $n \geq 4$, and $1 \leq i \leq n/4$,*

$$r^t_{\binom{n}{2}-i}(K_n) = n + j,$$

where $\binom{n+j-1}{2} < t + i \leq \binom{n+j}{2}$.

Proof If a t-coloring of K_{n+j-1} uses all t colors, then the inequality $\binom{n+j-1}{2} < t + i$ implies that at most $i - 1$ edges have colors that are repeated. So, in every K_n-subgraph, at least $\binom{n}{2} - i + 1$ colors are used. It follows that $r^t_{\binom{n}{2}-i}(K_n) \geq n + j$.

To prove the reverse inequality, consider a t-coloring of K_{n+j}. The inequality $t \leq \binom{n+j}{2} - i$ implies that there are at least i edges with colors that are used more than once. For $n \geq 4i$, it is possible to select n vertices so that the resulting induced subgraph includes at least i edges that repeat colors. In such a K_n-subgraph, at most $\binom{n}{2} - i$ colors are used. It follows that $r^t_{\binom{n}{2}-i}(K_n) \leq n + j$. □

Theorem 3.3 ([123]) *If $n \geq 3$ and $t > \frac{p(p-1)}{\lfloor p/(n-1)\rfloor(2p-(\lfloor p/(n-1)\rfloor+1)(n-1))}$, then*

$$r^t_{t-1}(K_n) \leq p.$$

Proof Consider a t-coloring of K_p that avoids a K_n-subgraph spanned by edges using at most $t-1$ colors. Some color (say, red), appears on at most $\lfloor \frac{1}{t}\binom{p}{2} \rfloor$ edges, otherwise there would be more than $\binom{p}{2}$ edges in total. So, the other $t-1$ colors appear on at least $\binom{p}{2} - \lfloor \frac{1}{t}\binom{p}{2} \rfloor$ edges. By Theorem 1.7, the subgraph G spanned by the other $t-1$ colors has size at most $\lfloor \frac{(n-2)p^2}{2(n-1)} \rfloor$. If we let $p = (n-1)q + r$, where $q, r \in \mathbb{Z}$ such that $0 \le r < n-1$, then the size of this subgraph satisfies

$$\begin{aligned}
|E(G)| &\le \left\lfloor \frac{(n-2)(p^2-r^2)}{2(n-1)} + \frac{r^2(n-2)}{2(n-1)} \right\rfloor \\
&= \frac{(n-2)(p^2-r^2)}{2(n-1)} + \left\lfloor \frac{r}{2}\left(r - \frac{r}{n-1}\right) \right\rfloor \\
&= \frac{(n-2)(p^2-r^2)}{2(n-1)} + \frac{r}{2}(r-1) \\
&= \frac{(n-2)(p^2-r^2)}{2(n-1)} + \binom{r}{2}.
\end{aligned}$$

It follows that whenever

$$\binom{p}{2} - \left\lfloor \frac{1}{t}\binom{p}{2} \right\rfloor > \frac{(n-2)(p^2-r^2)}{2(n-1)} + \binom{r}{2}, \tag{3.1}$$

a t-coloring of K_p must have a K_n-subgraph spanned by edges using at most $t-1$ colors. Inequality (3.1) is satisfied whenever

$$\begin{aligned}
t &> \frac{p(p-1)(n-1)}{(p-r)(p+r-n+1)} \\
&= \frac{p(p-1)}{((p-r)/(n-1))(p+q(n-1)+r-(q+1)(n-1))} \\
&= \frac{p(p-1)}{q(2p-(q+1)(n-1))}.
\end{aligned}$$

The theorem follows from observing that $q = \lfloor \frac{p}{n-1} \rfloor$. □

Theorem 3.4 ([123]) *If $n \ge 4$ and $t \le 2n-3$, then $r_{t-1}^t(K_n) \ge 2n-1$.*

Proof By Theorem 1.5, K_{2n-2} can be factored into $2n-3$ perfect matchings, each of which consists of $n-1$ disjoint edges. Select t of the perfect matchings and give each one its own distinct color. Regardless of how the other perfect matchings are colored, at most $n-1$ vertices can be chosen to induce a complete subgraph that avoids a given color. Equivalently, every K_n-subgraph includes edges in all t colors, from which it follows that $r_{t-1}^t(K_n) > 2n-2$. □

When $p = 2n-1$, the inequality in the statement of Theorem 3.3 is equivalent to

$$\begin{aligned}
t &> \frac{(2n-1)(2n-2)}{\lfloor (2n-1)/(n-1) \rfloor (2(2n-1) - (\lfloor (2n-1)/(n-1) \rfloor + 1)(n-1))} \\
&= \frac{(2n-1)(n-1)}{n+1} \\
&= 2n-5+\frac{6}{n+1}.
\end{aligned} \tag{3.2}$$

From this observation, Theorems 3.3 and 3.4 imply the following corollary.

Corollary 3.1 ([123]) *If $t = 2n-3$ or $t = 2n-4$ for $n \geq 4$, then*

$$r^t_{t-1}(K_n) = 2n - 1.$$

Theorem 3.5 ([123]) *If $n \geq 4$ is even, then*

$$r^{3n/2-2}_{3n/2-3}(K_n) = 3n - 2.$$

Proof By Theorem 3.3, $r^{3n/2-2}_{3n/2-3}(K_n) \leq 3n-2$. To prove the reverse inequality, let $n = 2a+2$. In 1971, Ray-Chaudhuri and Wilson [191] provided a solution to "Kirkman's schoolgirl problem," showing that it is possible to partition the edges of K_{6a+3} into $3a+1$ sets of $2a+1$ disjoint triangles. If we use one of the $3a+1$ colors for each of these $2a+1$ sets, then every K_{2a+2}-subgraph contains at least one edge of one triangle of every color class. It follows that $r^{3a+1}_{3a}(K_n) \geq 6a+3$, where $3a+1 = 3n/2-2$ and $6a+3 = 3n-3$. □

Next, we consider lower bounds for weakened Ramsey numbers of complete graphs that follow from computational methods. In [49], the authors considered using the clique numbers of complements of cubic residue graphs to obtain lower bounds for certain 3-color weakened Ramsey numbers. In order to describe their results, we first prove a theorem linking such clique numbers to the desired lower bounds.

Theorem 3.6 ([49]) *Let $p \equiv 1 \pmod 6$ be prime and $\psi : \mathbb{F}_p^\times \longrightarrow \mathbb{C}^\times$ be a cubic character (i.e., a character of order 3). If $n = \omega(\overline{\mathrm{Cay}(\mathbb{F}_p; \mathbb{F}_p^{\times 3})})$, then $r^3_2(K_{n+1}) \geq p+1$.*

Proof For a prime $p \equiv 1 \pmod 6$, identify the vertices of K_p with the elements in $\mathbb{F}_p$. Let g be a generator for $\mathbb{F}_p^\times$ and $\psi : \mathbb{F}_p^\times \longrightarrow \mathbb{C}^\times$ be a cubic character. Note that $Ker(\psi) = \mathbb{F}_p^{\times 3}$. For distinct vertices a and b, color edge ab in the following way:

$$f(ab) = \begin{cases} \text{red} & \text{if } b - a \in \mathbb{F}_p^{\times 3} \\ \text{blue} & \text{if } b - a \in g\mathbb{F}_p^{\times 3} \\ \text{green} & \text{if } b - a \in g^2\mathbb{F}_p^{\times 3}. \end{cases}$$

The subgraph G_r spanned by the red edges is isomorphic to the character difference graph $\mathrm{Cay}(\mathbb{F}_p; \mathbb{F}_p^{\times 3})$. Denote by G_b and G_g the subgraphs spanned by the blue edges and the green edges, respectively. Using the isomorphisms $\sigma_1 : V(G_r) \longrightarrow V(G_b)$ and $\sigma_2 : V(G_a) \longrightarrow V(G_g)$ given by $\sigma_1(x) = gx$ and $\sigma_2(x) = g^2x$, it follows that $G_r \cong G_b \cong G_g$. So the subgraph spanned by any two of the colors is necessarily isomorphic to $\overline{G_r}$. Therefore, we obtain the inequality $r^3_2(K_{\omega(\overline{G_r})+1}) > p$. □

In 2002, Su, Li, Luo, and Li [200] developed an algorithm for determining the clique numbers for cubic character graphs and their complements. They applied their algorithm to all primes $p \equiv 1 \pmod 6$ within the range $13 \leq p \leq 1213$, as well as the special values 1303 and 1327. Some of their results are given in the second column of Table 3.1 concerning the clique numbers of the complements of the cubic character graphs. From these clique numbers, Theorem 3.6 implies the lower bounds given in the third column of Table 3.1.

In [40], the authors considered other ways in which character difference graphs could be used to obtain lower bounds for weakened Ramsey numbers for complete graphs in which $t > 3$. For example, when $t = 4$, one can let $p \equiv 1 \pmod 8$ be prime and $\psi : \mathbb{F}_p^\times \longrightarrow \mathbb{C}^\times$ be a *quartic character* (i.e., a character of order 4). The assumed congruence guarantees that $-1 \in \mathbb{F}_p^{\times 4}$. Let g be a generator for $\mathbb{F}_p^\times$. Form a 4-coloring of K_p by identifying the vertices

TABLE 3.1: Lower bounds for some weakened Ramsey numbers $r_2^3(K_n)$, based on the data collected in [200].

p	$\omega(\overline{G_p})$	Lower bound
19	4	$r_2^3(K_5) \geq 20$
31	6	$r_2^3(K_7) \geq 32$
43	7	$r_2^3(K_8) \geq 44$
97	8	$r_2^3(K_9) \geq 98$
109	9	$r_2^3(K_{10}) \geq 110$
163	10	$r_2^3(K_{11}) \geq 164$
229	11	$r_2^3(K_{12}) \geq 230$
349	13	$r_2^3(K_{14}) \geq 350$
439	14	$r_2^3(K_{15}) \geq 440$
487	15	$r_2^3(K_{16}) \geq 488$
673	16	$r_2^3(K_{17}) \geq 674$
769	17	$r_2^3(K_{18}) \geq 770$
1051	18	$r_2^3(K_{19}) \geq 1052$
1303	19	$r_2^3(K_{20}) \geq 1304$
1327	20	$r_2^3(K_{21}) \geq 1328$

with $\mathbb{F}_p$ and coloring the edge ab according to

$$f(ab) = \begin{cases} \text{red} & \text{if } b - a \in \mathbb{F}_p^{\times 4} \\ \text{blue} & \text{if } b - a \in g\mathbb{F}_p^{\times 4} \\ \text{green} & \text{if } b - a \in g^2\mathbb{F}_p^{\times 4} \\ \text{purple} & \text{if } b - a \in g^3\mathbb{F}_p^{\times 4}, \end{cases}$$

where g is a generator for $\mathbb{F}_p^\times$. The authors in [40] used Python to compute the clique numbers for the various subgraphs spanned by edges in two of the four colors, resulting in new lower bounds for weakened Ramsey numbers of the form $r_2^4(K_n)$. This process was carried out for numerous characters $\psi : \mathbb{F}_p \longrightarrow \mathbb{C}^\times$ of order r_ψ, where p is a prime of the form $p \equiv 1 \pmod{2r_\psi}$, leading to the lower bounds given in Theorem 3.7.

Theorem 3.7 ([40]) *The weakened Ramsey number satisfies the following lower bounds:*

$$\begin{array}{lll}
r_2^4(K_5) \geq 18, & r_2^4(K_6) \geq 90, & r_2^4(K_7) \geq 98, \\
r_2^4(K_8) \geq 138, & r_2^4(K_9) \geq 242, & r_2^4(K_{10}) \geq 570, \\
r_2^4(K_{12}) \geq 882, & r_2^4(K_{14}) \geq 1010, & r_2^5(K_5) \geq 72, \\
r_2^5(K_6) \geq 102, & r_2^5(K_8) \geq 402, & r_2^5(K_9) \geq 762, \\
r_2^5(K_{10}) \geq 1052, & r_2^5(K_{11}) \geq 1062, & r_2^6(K_4) \geq 14, \\
r_2^6(K_5) \geq 74, & r_2^6(K_6) \geq 278, & r_2^6(K_7) \geq 734, \\
r_2^6(K_8) \geq 1070, & r_2^6(K_9) \geq 1118, & r_2^6(K_{10}) \geq 1202, \\
r_2^7(K_5) \geq 128, & r_2^7(K_7) \geq 1052, & r_2^7(K_8) \geq 1878, \\
r_2^4(K_{13}) \geq 954, & r_2^4(K_{18}) \geq 1034. &
\end{array}$$

We conclude this section by considering a star-critical analogue of the weakened Ramsey number (cf. Subsection 1.3.2). For $t \geq 3$ and $1 \leq s \leq t-1$, define the *star-critical weakened Ramsey number* $r_*^{s,t}(G)$ to be the least $k \in \{0, 1, \ldots r^{s,t}(G) - 1\}$ such that every t-coloring of $K_{r_s^t(G)-1} \sqcup K_{1,k}$ contains a subgraph isomorphic to G spanned by edges that use at most s colors. When G does not contain isolated vertices, $r_*^{s,t}(G) \geq 1$. These numbers were first suggested in Section 4.1 of [35] and were further developed in [45]. The following theorem determines the case where G is a complete graph, generalizing Theorem 1.21.

Theorem 3.8 **([45])** *For all $n \geq 2$, $t \geq 3$, and $1 \leq s \leq t-1$,*

$$r_*^{s,t}(K_n) = r_s^t(K_n) - 1.$$

Proof This theorem follows from providing a t-coloring of

$$K_{r_s^t(K_n)-1} \sqcup K_{1,r_s^t(K_n)-2} = K_{r_s^t(K_n)} - e$$

that avoids a K_n spanned by edges using at most s colors. Begin with a critical coloring for $r_s^t(K_n)$ and select a vertex x. Introduce a vertex y and join y to all of the vertices in the critical coloring, except for vertex x. Color edge yz, the same color as edge xz, for all $z \notin \{x, y\}$. Viewing xy as the missing edge, consider the resulting $K_{r_s^t(K_n)} - e$. Every K_n-subgraph includes at most one of x and y, and hence, is spanned by edges that use at least $s+1$ colors. □

As we begin to consider weakened Ramsey numbers for graphs that are not complete, we can also consider the corresponding star-critical weakened Ramsey numbers.

3.2 Weakened Ramsey Numbers for Other Graphs

In 2022, Khamseh and Omidi [150] determined the value of $r_{t-1}^t(K_{1,n})$, which can be viewed as the diagonal case of the chromatic Ramsey number given in Theorem 2.30. Recently, the corresponding star-critical weakened Ramsey number was determined in [45]. We combine these two results together in the following theorem.

Theorem 3.9 **([45],[150])** *If $n \geq t \geq 2$, $m = \lfloor \frac{nt-1}{t-1} \rfloor$, and $q = \lfloor \frac{m}{t} \rfloor$, then*

$$r_{t-1}^t(K_{1,n}) = \begin{cases} m & \text{if } m = tq+1 \text{ with } m \text{ odd and } q \text{ odd} \\ m+1 & \text{otherwise,} \end{cases}$$

and

$$r_*^{t-1,t}(K_{1,n}) = \begin{cases} m-1 & \text{if } m = tq+1 \text{ with } m \text{ odd and } q \text{ odd} \\ 1 & \text{otherwise.} \end{cases}$$

Proof Note that the equation $m = \lfloor \frac{nt-1}{t-1} \rfloor$ implies that

$$\frac{nt-1}{t-1} - 1 < m \leq \frac{nt-1}{t-1}.$$

The first inequality is equivalent to $n < m - \frac{m}{t} + 1$ and the second inequality is equivalent to $m - \frac{m}{t} + \frac{1}{t} \leq n$. Let $m = tq + r$, where $q, r \in \mathbb{Z}$ such that $0 \leq r < t$. If $r = 0$, then these inequalities are equivalent to

$$n < m - q + 1 \quad \text{and} \quad m - q + 1 \leq n,$$

which is not solvable. So, $r \neq 0$ and the two inequalities imply that

$$n = m - \left\lfloor \frac{m}{t} \right\rfloor = m - q.$$

Consider the following cases.

Case 1: Suppose that $m = tq + 1$, where m and q are both odd. In this case, $n = (t-1)q + 1$. We must first show that there exists a t-coloring of $K_{m-1} \sqcup K_{1,m-2} = K_m - e$ that avoids a $K_{1,n}$ spanned by edges using at most $t-1$ colors. The assumption that m and q are both odd forces t to be even. By Theorem 1.6, K_m factors into $\frac{tq}{2}$ Hamiltonian cycles. Select a vertex x, which is necessarily contained in every Hamiltonian cycle. For each $i \in \{1, 2, \ldots, t\}$, color $\frac{q-1}{2}$ of the Hamiltonian cycles with color i. There are $\frac{tq}{2} - \frac{t(q-1)}{2} = \frac{t}{2}$ Hamiltonian cycles that remain to be colored. Color each of the remaining Hamiltonian cycles with two colors so that no color appears on more than one of them, and adjacent edges in each Hamiltonian cycle have different colors, except for those incident with x. Note that the two edges incident with x in each Hamiltonian cycle must have the same color since m is odd.

Every vertex other than x is incident with exactly q edges in each color. The largest star spanned by edges using at most $t-1$ colors that has a vertex other than x as its center is isomorphic to $K_{1,(t-1)q}$. So, there does not exist a $K_{1,n}$ spanned by edges using at most $t-1$ colors that has a vertex other than x as its center. Vertex x is incident with exactly $q-1$ edges in each of $\frac{t}{2}$ colors and $q+1$ edges in the other $\frac{t}{2}$ colors. Delete a single edge incident with x that has one of the colors that appears on $q+1$ of the edges that are incident with x. Now, x is incident with $q-1$ edges in $\frac{t}{2}$ of the colors, q edges in one color, and $q+1$ edges in $\frac{t}{2}-1$ of the colors. For example, the case where $m = 13$ and $q = 3$ is shown in Figure 3.2. The maximum number of edges incident with x that can be spanned by $t-1$

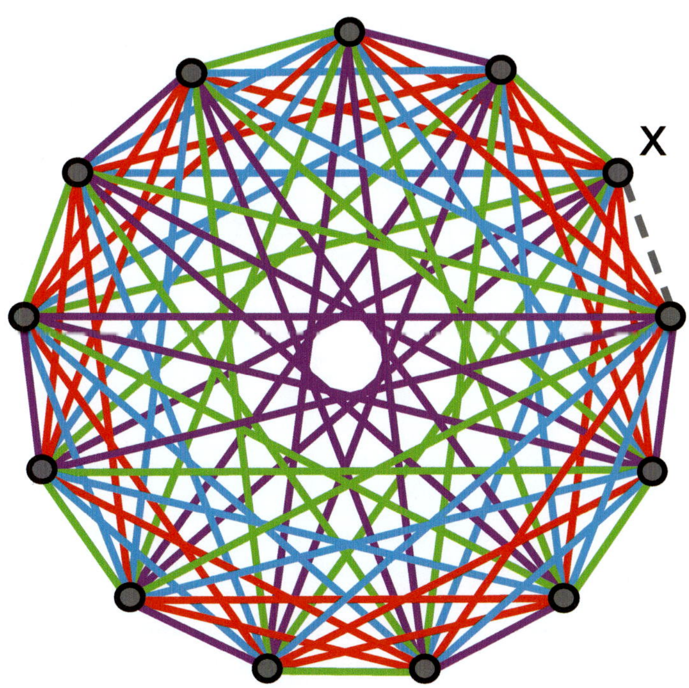

FIGURE 3.2: A 4-coloring of $K_{13} - e$ that lacks a $K_{1,10}$ spanned by edges that use at most three colors.

colors is

$$\left(\frac{t}{2}-1\right)(q-1)+q+\left(\frac{t}{2}-1\right)(q+1)=(t-1)q.$$

Since we have produced a t-coloring of $K_m - e$ that avoids a $K_{1,n}$ spanned by edges using at most $t-1$ colors, it follows that $r_{t-1}^t(K_{1,n}) \geq m$, and once we show that $r_{t-1}^t(K_{1,n}) = m$, it will follow that $r_*^{t-1,t}(K_{1,n}) = m-1$.

Now consider a t-coloring of K_m. If every vertex is incident with exactly q edges in each color, then the subgraph spanned by edges in any given color is q-regular. Of course, this is impossible since q is odd and there are an odd number of vertices. It follows that there exists a vertex x that is incident with at most $q-1$ edges in some color. Then x must be incident with at least $m-1-(q-1) = m-q = n$ edges in the other $t-1$ colors. It follows that $r_{t-1}^t(K_{1,n}) = m$ and $r_*^{t-1,t}(K_{1,n}) = m-1$.

Case 2: Suppose that m is even, q is even, or $m \neq tq+1$. Let $m = tq+r$, where $1 \leq r < t$. Then $n = m-q = (t-1)q+r$. First, we show that in any t-coloring of $K_m \sqcup K_{1,1}$, there exists a $K_{1,n}$ that is spanned by edges using at most $t-1$ colors. Let x be the vertex in a t-coloring of $K_m \sqcup K_{1,1}$ that is incident with $m = tq+r$ edges. Among the edges incident with x, some color must appear on at most q of the edges (otherwise, x would be incident with at least $t(q+1) = tq+t > tq+r = m$ edges). Together, the other $t-1$ colors appear on at least $tq+r-q = (t-1)q+r = n$ of the edges, forming a $K_{1,n}$. Since $K_m \sqcup K_{1,1}$ is a subgraph of K_{m+1}, it follows that $r_{t-1}^t(K_{1,n}) \leq m+1$. Once we show that $r_{t-1}^t(K_{1,n}) \geq m+1$, it will follow that $r_*^{t-1,t}(K_{1,n}) = 1$. To complete the proof of this case, we must show that there exists a t-coloring of K_m that avoids a subgraph isomorphic to $K_{1,n}$ that uses at most $t-1$ colors. Several subcases must be considered.

Subcase 2.1: Suppose that $m = tq+r$ is even, where $1 \leq r < t$. By Theorem 1.5, K_m factors into $m-1 = tq+r-1$ perfect matchings. Color q edges with each of the t colors. For the remaining $r-1$ perfect matchings, color one matching with each of the first $r-1$ colors (for example, the first image in Figure 3.3 shows the case where $m=8$, $q=2$, and $r=2$). In the resulting t-coloring of K_m, every vertex is incident with exactly $q+1$ edges

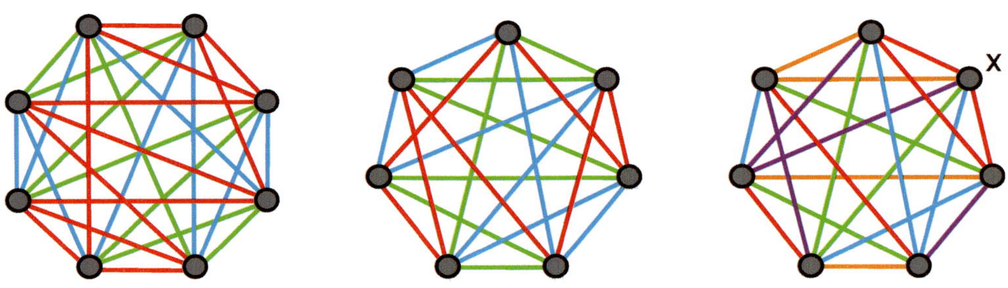

FIGURE 3.3: Critical colorings for $r_2^3(K_{1,6})$, $r_2^3(K_{1,5})$, and $r_4^5(K_{1,6})$.

in colors $1,2,\ldots,r-1$ and q edges in colors $r, r+1, \ldots, t$. The largest star that is spanned by edges using at most $t-1$ of the colors has size $m-1-q < n$. Since a $K_{1,n}$ that uses at most $t-1$ colors does not exist, it follows that $r_{t-1}^t(K_{1,n}) = m+1$ and $r_*^{t-1,t}(K_{1,n}) = 1$.

Subcase 2.2: Suppose that $m = tq+r$ is odd, q is even, and $1 \leq r < t$. By Theorem 1.6, K_m factors into $\frac{m-1}{2} = \frac{tq}{2} - \frac{r-1}{2}$ Hamiltonian cycles. Color $\frac{q}{2}$ of the Hamiltonian cycles with each of the t colors. For the remaining $\frac{r-1}{2}$ Hamiltonian cycles, color one cycle with each of the first $\frac{r-1}{2}$ colors (for example, the second image in Figure 3.3 shows the case where $m=7$, $q=2$, and $r=1$). In the resulting t-coloring of K_m, each vertex is incident with $q+r-1$ edges in each of the colors $1,2,\ldots,\frac{r-1}{2}$ and q edges in each of the colors

$\frac{r+1}{2}, \frac{r+3}{2}, \ldots, t$. The largest star that is spanned by edges using at most $t-1$ colors has size $m-1-q < n$. A $K_{1,n}$ that uses at most $t-1$ colors does not exist, so $r^t_{t-1}(K_{1,n}) = m+1$ and $r_*^{t-1,t}(K_{1,n}) = 1$.

Subcase 2.3: Suppose that $m = tq + r$ is odd, q is odd, and $2 \leq r < t$. Select a vertex x in K_m and consider the complete graph $K_m - \{x\}$, which has even order. By Theorem 1.5, it can be factored into $m-2 = tq+r-2$ perfect matchings. Colors q perfect matchings with each of the t colors, and if $r > 2$, colors the remaining $r-2$ perfect matchings by colorings a single matching with each of the colors $1, 2, \ldots, r-2$. Now, vertex x has degree $tq+r-1$ in the complete graph K_m. Color q of the edges incident with x with each of the t colors and color the remaining $r-1$ edges incident with x arbitrarily (for example, the third image in Figure 3.3 shows the case where $m = 7$, $q = 1$, and $r = 2$). In the resulting t-coloring of K_m, each vertex is incident with at least q edges in each of the t colors. It follows that the largest possible star spanned by edges using at most $t-1$ colors is $m-1-q < n$. So, a $K_{1,n}$ that uses at most $t-1$ colors does not exist. It follows that $r^t_{t-1}(K_{1,n}) = m+1$ and $r_*^{t-1,t}(K_{1,n}) = 1$. □

Besides the value of $r^t_{t-1}(K_{1,n})$, Khamseh and Omidi [150] were also able to determine the value of $r^t_{t-2}(K_{1,n})$. The corresponding star-critical weakened Ramsey numbers were considered in [45]. In order to state these results, let $n \geq t \geq 4$, $\lfloor \frac{t}{2} \rfloor$, $m = \lfloor \frac{nt'-1}{t'-1} \rfloor$, and $m-2 = tq + r$, where $0 \leq r < t$. Consider the following conditions:

(*a*) $r = t-1 > 2q+4$ and m is even,
(*b*) $r = t-1 > 2q+4$ and m and t are both odd,
(*c*) $r = t-1$, m is odd, t is even, and q is odd,
(*d*) $t > 2r+4$ and t is even,
(*e*) $r < t-2$, $t > 2q+2r+5$, and t is odd,
(*f*) $r = t-1$, m is odd, and q is even,
(*g*) $r < t-2$, $t \leq 2r+4$, and t is even,
(*h*) $r < t-2$, $q+r+3 < t \leq 2q+2r+5$, and t is odd,
(*i*) $r = 1$, $\frac{2q+9}{3} < t \leq q+4$, t is odd, and m is even,
(*j*) $r = 1$, $\frac{2q+9}{3} < t \leq q+4$, and m and t are both odd,
(*k*) $1 < r < t-2$, $\frac{2q+2r+7}{3} < t \leq q+r+3$, and t is odd,
(*l*) $r = t-2$ and either t is even or t is odd and $t > \frac{2q+2r+7}{3}$.

Theorem 3.10 ([45], [150]) *Let $t, n \in \mathbb{N}$ such that $t \geq 4$ and $n \geq t-1$. Define $t' = \lfloor \frac{t}{2} \rfloor$, and $m = \lfloor \frac{nt'-1}{t'-1} \rfloor$. If $m-2 = tq+r$, where $q, r \in \mathbb{Z}$ and $0 \leq r < t$, then*

$$r^t_{t-2}(K_{1,n}) = \begin{cases} m+1 & \text{if one of Conditions } (a)\text{-}(e) \text{ hold} \\ m & \text{if one of Conditions } (f)\text{-}(h) \text{ hold} \\ m-1 & \text{if one of Conditions } (i)\text{-}(l) \text{ hold} \\ m-2 & \text{otherwise.} \end{cases}$$

and

$$r_*^{t-2,t}(K_{1,n}) = \begin{cases} 1 & \text{if one of Conditions } (a)\text{-}(e) \text{ or } (g)\text{-}(l) \text{ hold} \\ m-2 & \text{if Condition } (f) \text{ holds.} \end{cases}$$

Proof The assumption that $m = \lfloor \frac{nt'-1}{t'-1} \rfloor$ is equivalent to

$$\frac{nt'-1}{t'-1} - 1 < m \leq \frac{nt'-1}{t'-1}.$$

The first inequality is equivalent to $n < m - \frac{m}{t'} + 1$ and the second inequality is equivalent ot $m - \frac{m}{t'} + \frac{1}{t'} \leq n$. Let $m = t'q' + r'$, where $q', r' \in \mathbb{Z}$ and $0 \leq r' < t'$. If $k = 0$, then these equations are equivalent to

$$n < m - q' + 1 \qquad \text{and} \qquad m - q' + 1 \leq n,$$

which is not solvable. So, $k' \neq 0$ and the two inequalities imply that

$$n = m - \left\lfloor \frac{m}{t'} \right\rfloor.$$

Due to the length and repetitive nature of the proof of Theorem 3.10, we will only prove the theorem for Conditions (a) and (f).

Condition (a) Assume that $r = t - 1 > 2q + 4$ and m is even. Then we can write

$$m = tq + r + 2 = tq + (t-1) + 2 = t(q+1) + 1.$$

It follows that t must be odd, so $t' = \frac{t-1}{2}$. The assumption $r = t - 1$ implies that

$$\begin{aligned} n &= m - \left\lfloor \frac{2m}{t-1} \right\rfloor \\ &= t(q+1) + 1 - \left\lfloor \frac{2(t(q+1)+1)}{t-1} \right\rfloor \\ &= t(q+1) + 1 - \left\lfloor \frac{2((t-1)(q+1)+q+2)}{t-1} \right\rfloor \\ &= t(q+1) + 1 - 2(q+1) - \left\lfloor \frac{2q+4}{t-1} \right\rfloor \\ &= (t-2)(q+1) + 1. \end{aligned}$$

First, we construct a t-coloring of K_m that lacks a $K_{1,n}$ that uses at most $t-2$ colors. By Theorem 1.5, K_m factors into $m-1 = t(q+1)$ perfect matchings. Color $q+1$ of the perfect matchings with each of the t colors. Then every vertex is incident with exactly $q+1$ edges in each color. So, the largest star spanned by edges that use at most $t-2$ colors has size $(t-2)(q+1) < n$. It follows that $r^t_{t-2}(K_{1,n}) \geq m + 1$.

To complete the proof of this case, consider a t-coloring of $K_m \sqcup K_{1,1}$ and let x be the unique vertex of degree $m = t(q+1) + 1$. Some color (assume it is color 1) appears on at most $q+1$ of the edges incident with x. If this were not the case, then x would have to be incident with at least $t(q+2) = t(q+1) + t > m$ edges. Removing the edges incident with x that have color 1, x is now incident with at least $(t-1)(q+1)+1$ edges in colors $2, 3, \ldots, t$. Let E be a set of exactly $t(q+1)+1$ such edges. Among the edges in E, some color (assume it is color 2) must appear on at most $q+1$ of the edges. If this were not the case, then E would contain at least $(t-1)(q+2) = (t-1)(q+1)+t-1$ edges, exceeding the cardinality of E. Remove the edges in color 2 from E and note that at least $(t-2)(q+1)+1 \geq n$ edges remain, which along with vertex x, form a $K_{1,n}$ spanned by edges in colors $3, 4, \ldots, t$. It follows that $r^t_{t-2}(K_{1,n}) = m + 1$ and $r^{t-2,t}_*(K_{1,n}) = 1$.

Condition (f) Assume that $r = t - 1$, m is odd, and q is even. Then we can write

$$m = tq + r + 2 = tq + (t-1) + 2 = t(q+1) + 1.$$

The assumptions on m and q imply that t is even and $t' = \frac{t}{2}$. The assumption $r = t - 1$ implies that

$$n = (t-2)(q+1) + 1.$$

We start by providing a t-coloring of $K_{m-1} \sqcup K_{1,m-3}$ that avoids a $K_{1,n}$ spanned by edges that use at most $t-2$ colors. By Theorem 1.6, K_m can be factored into $\frac{m-1}{2} = \frac{t(q+1)}{2}$ Hamiltonian cycles. Since q is even, we can color $\frac{q}{2}$ of the Hamiltonian cycles with each of the t colors. Color the remaining $\frac{t}{2}$ Hamiltonian cycles, with two colors appearing on each cycle such that no color appear on more than one cycle and adjacent edges of each cycle have different colors, except for the edges incident with a fixed vertex x. Each vertex other than x is incident with exactly $q+1$ edges in each color. The largest star whose edges use at most $t-2$ colors that has a vertex other than x at its center has size $(t-2)(q+1) < n$. Vertex x is incident with q edges in each of $\frac{t}{2}$ of the colors and it is incident with $q+2$ edges in each of the other $\frac{t}{2}$ colors. Delete two edges incident with x that are given distinct colors that each appear on $q+2$ of the edges (see Figure 3.4 for the case where $t=4$ and $m=13$). With these edges removed, x is incident with q edges in $\frac{t}{2}$ of the colors,

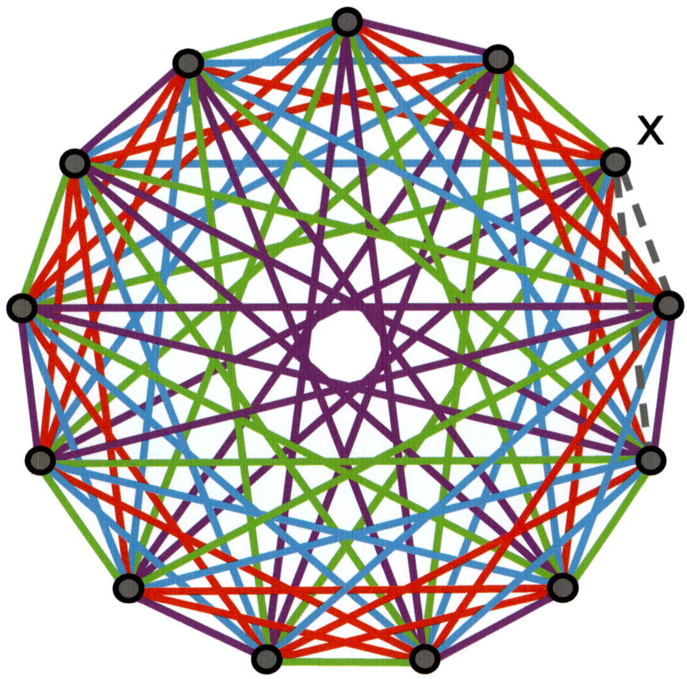

FIGURE 3.4: A 4-coloring of $K_{12} \sqcup K_{1,10}$ that lacks a $K_{1,7}$ spanned by edges that use at most two colors.

$q+1$ edges in two of the colors, and $q+2$ edges in $\frac{t}{2}-2$ of the colors. The largest star that can be formed using at most $t-2$ colors and having vertex x as it s center has size

$$\left(\frac{t}{2}-2\right)q + 2(q+1) + \left(\frac{t}{2}-2\right)(q+2) = (t-2)(q+1) < n.$$

Since we have produced a t-coloring of $K_{m-1} \sqcup K_{1,m-3}$ that avoids a $K_{1,n}$ spanned by edges using at most $t-2$ of the colors, it follows that $r^t_{t-2}(K_{1,n}) \geq m$, and once we show that $r^t_{t-2}(K_{1,n}) \leq m$, it will follow that $r^{t-2,t}_*(K_{1,n}) \geq m-2$.

To complete the proof of this case, consider a t-coloring of $K_{m-1} \sqcup K_{1,m-2} = K_m - e$. Here, $m-2$ of the vertices have degree $m-1$ and two vertices have degree $m-2$. In order to avoids a $K_{1,n}$ that uses at most $t-2$ colors, each of the vertices of degree $m-1$ must be incident with exactly $q+1$ edges in each of the t colors. This is due to the fact that if any two colors appear on at most $2q+1$ of the edges incident with such a vertex, then the other $t-2$ colors muat appear on the remaining

$$m-1-(2q+1)=(t-2)(q+1)+1=n$$

of the edges. Similarly, the two vertices of degree $m-2$ must be incident with $q+1$ edges in each of $t-$ of the colors and q edges in the other color. The subgraph spanned by edges in some color (say, red) must be $(q+1)$-regular and have order m. Since both $q+1$ and m are assumed to be odd, we obtain a contradiction. So, every t-coloring of $K_{m-1} \sqcup K_{1,m-2}$ contains a $K_{1,n}$ spanned by edges using at most $t-2$ of the colors. It follows that $r^t_{t-2}(K_{1,n})=m$ and $r^{t-2,t}_*(K_{1,n})=m-2$. □

Theorem 3.11 ([45]) *If $t \geq 3$, then*

$$r^t_2(K_{1,3}) = \begin{cases} t+2 & \text{if } t \text{ is odd} \\ t+1 & \text{if } t \text{ is even} \end{cases}$$

and

$$r^{2,t}_*(K_{1,3}) = \begin{cases} 1 & \text{if } t \text{ is odd} \\ \frac{t}{2}+1 & \text{if } t \text{ is even.} \end{cases}$$

Proof We separate the proof into cases, based on whether t is even or odd.

Case 1: Assume that t is odd. By Theorem 1.5, K_{t+1} can be factored into t perfect matchings. If we give each perfect matching its own color, then every $K_{1,3}$ subgraph will have distinct colors on its edges. It follows that $r^t_2(K_{1,3}) \geq t+2$.

To finish the proof of this case, consider a t-coloring of $K_{t+1} \sqcup K_{1,1}$. Let x be the unique vertex of degree $t+1$. By the pigeonhole principle, at least two of the edges incident with x must receive the same color. These two edges, along with any other edge incident with x, form a $K_{1,3}$ that uses at most two colors. It follows that $r^t_2(K_{1,3}) \leq t+2$ and $r^{2,t}_*(K_{1,3})=1$.

Case 2: Assume that t is even. By Corollary 1.2, K_t can be factored into $\frac{t}{2}$ Hamiltonian paths, each of which have odd length. Color the edges of each of the Hamiltonian paths with two alternating colors, such that each color appears on at most one Hamiltonian path. The resulting t-coloring of K_t avoids a $K_{1,3}$ that uses at most two colors, so it follows that $r^t_2(K_{1,3}) \geq t+1$. Introduce a vertex x and join it to exactly one of the endpoints of each Hamiltonian path, coloring each such edge so that the colors still alternate with the path the edge extended. For example, see Figure 3.5 for the $t=6$ case. In this example, the Hamiltonian paths are colored red/blue, green/purple, and orange/grey. Every vertex in the resulting $K_t \sqcup K_{1,\frac{t}{2}}$ is incident with edges that have distinct colors. It follows that every $K_{1,3}$-subgraph is rainbow. Once we have shown that $r^t_2(K_{1,3}) \leq t+1$, it will follow that $r^{2,t}_*(K_{1,3}) \geq \frac{t}{2}+1$.

Now consider a t-coloring of $K_t \sqcup K_{1,\frac{t}{2}+1}$, which has size $\binom{t}{2}+\frac{t}{2}+1=\frac{t^2}{2}+1$. By the pigeonhole principle, some color (say, red) must appear on at least $\frac{t}{2}+1=\frac{t+2}{2}$ of the edges. Since there are $t+1$ vertices in total, some vertex y must be incident with at least two red edges. Two such edges, along with any other edge that is incident with y, form a $K_{1,3}$ that uses at most two colors. It follows that $r^t_2(K_{1,3}) \leq t+1$ and $r^{2,t}_*(K_{1,3}) \leq \frac{t}{2}+1$. □

In a very recent preprint, Yi and Luo [212] completed the evaluation of $r^t_s(K_{1,n})$ and $r^{s,t}_*(K_{1,n})$, for all $s \in \{1,2,\ldots,t-1\}$. We state their main results without proof in the following theorem.

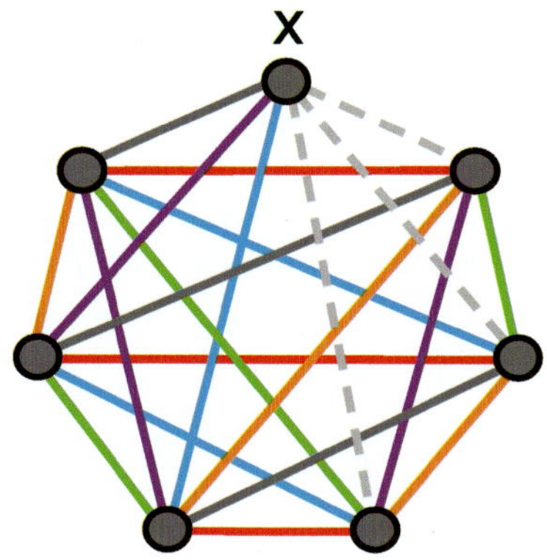

FIGURE 3.5: A Gallai 6-coloring of $K_6 \sqcup K_{1,3}$ that avoids a $K_{1,3}$-subgraph spanned by edges using at most two colors.

Theorem 3.12 (**[212]**) *Let $n, k, t \in \mathbb{N}$, $s \in \{1, 2, \ldots, t-1\}$, and $a \in \{0, 1, \ldots, s-1\}$. Then*

$$r_s^t(K_{1,n}) = \begin{cases} (2k-1)t+s+1 & \text{if } n = 2ks \text{ and } s \neq 1 \\ 2kt+a+1 & \text{if } n = 2ks+a \text{ and } 1 \leq a \leq s-1 \\ 2kt+s+1 & \text{if } n = (2k+1)s \\ (2k+1)t+1 & \text{if } n = (2k+1)s+1 \text{ and } t \text{ is even} \\ (2k+1)t+2 & \text{if } n = (2k+1)s+1 \text{ and } t \text{ is odd} \\ (2k+1)t+a+1 & \text{if } n = (2k+1)s+a \text{ and } 2 \leq a \leq s-1 \end{cases}$$

and

$$r_*^{s,t}(K_{1,n}) = \begin{cases} 2kt+t/2+1 & \text{if } n = (2k+1)s+1,\ t \text{ is even, and } 1 \leq s \leq t/2 \\ 2kt+s+1 & \text{if } n = (2k+1)s+1,\ t \text{ is even, and } t/2+1 \leq s \leq t-1 \\ 1 & \text{otherwise.} \end{cases}$$

3.3 Weakened Gallai-Ramsey Numbers for Complete Graphs

For a graph G and $1 \leq s \leq t$, define the *weakened Gallai-Ramsey number* $gr_s^t(G)$ to be the least $n \in \mathbb{N}$ such that every Gallai t-coloring of K_n contains a subgraph that is isomorphic to G spanned by edges that use at most s colors. The following lemma is similar to Lemma 5 of [18] and will be used to find a relationship between various weakened Gallai-Ramsey numbers for complete graphs.

Lemma 3.1 *Let $n, t \geq 3$, $s, k \geq 1$, and $1 \leq s+k \leq t-1$. If a Gallai t-coloring of K_n contains a K_m-subgraph spanned by edges using at most s colors, then it contains a K_{m+k}-subgraph spanned by edges using at most $s+k$ colors.*

Proof We proceed by induction on $k \geq 0$. The lemma is trivial when $k = 0$. Suppose that the lemma is true for $k-1$, where $k \geq 1$, and consider a Gallai t-coloring of K_n that contains a K_m-subgraph spanned by edges using at most s colors. By the inductive hypothesis, there exists a K_{m+k-1}-subgraph spanned by edges using at most $s+k-1$

colors. Let x be any vertex that is not in the K_{m+k-1}. Consider the possibility that the edges that join x to the K_{m+k-1} contain at least two colors that do not appear in the K_{m+k-1}. Without loss of generality, suppose that xy has color 1 and xz has color 2, where y and z are both in the K_{m+k-1} and the K_{m+k-1} does not contain any edges in colors 1 or 2. Then the subgraph induced by $\{x, y, z\}$ is a rainbow triangle, giving us a contradiction. Hence, the K_{m+k}-subgraph formed by adding in vertex x to the K_{m+k} is spanned by edges using at most $s + k$ colors. □

From Lemma 3.1, it follows that

$$gr^t_{s+k}(K_{m+k}) \leq gr^t_s(K_m),$$

for all $1 \leq s + k \leq t - 1$. The following theorem was proved in 2026 by Jakhar and Moun [136]. It generalizes Corollary 4 from [18].

Theorem 3.13 ([136]) *For all $t \geq 2$ and $1 \leq s < t$, $gr^t_s(K_{s+1}) = s + 1$.*

Proof Since at least $s+1$ vertices are needed to have a K_{s+1}, we have the lower bound $gr^t_s(K_{s+1}) \geq s + 1$. By Lemma 1.1, every Gallai t-coloring of K_{s+1} uses at most s colors, resulting in the upper bound $gr^t_s(K_{s+1}) \leq s + 1$. □

In 2018, it was shown in [18] that $gr^t_2(K_4) \geq t + 2$. The corresponding upper bound $gr^t_2(K_4) \leq t + 2$ was proved in [51] in 2022. Recently, Jakhar and Moun [136] were able to generalize this Gallai-Ramsey number further, resulting in the following theorem (an equivalent result appears in [160]).

Theorem 3.14 ([136]) *For all $t \geq 3$ and $2 \leq s < t$, $gr^t_s(K_{s+2}) = t + 2$.*

Proof For the lower bound, begin with a K_2 in color 1, with vertices x_1 and x_2. Introduce vertex x_3, coloring edges x_1x_3 and x_2x_3 in color 2. Next, add in vertex x_4, coloring edges x_1x_4, x_2x_4, and x_3x_4 with color 3. Continue in this manner, sequentially introducing vertices so that when we introduce vertex x_n, we color edges x_1x_n, x_2x_n, $\ldots$, $x_{n-1}x_n$ with color $n - 1$ (see Figure 3.6 for the $n = 6$ case). Carry out this process until we have a Gallai

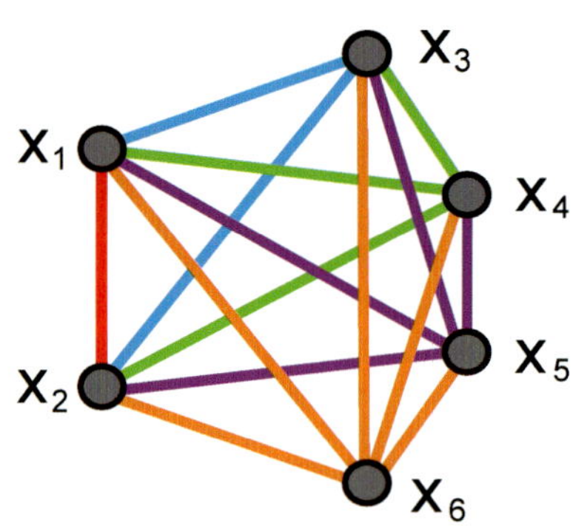

FIGURE 3.6: A Gallai 5-coloring of K_6 that avoids a K_4-subgraph spanned by edges using at most two colors.

t-coloring of K_{t+1}. Note that deleting any single vertex from this coloring removes at most one color. So, if we delete any $t - s - 1$ vertices, the resulting K_{s+2}-subgraph still includes at least $t-(t-s-1)=s+1$ colors, leading to the lower bound $gr_s^t(K_{s+2}) \geq t+2$, for all $t \geq 3$.

To prove the reverse inequality, we use (strong) induduction on $t \geq 3$ to first prove the $s=2$ case. For the base case, consider a Gallai 3-coloring of K_5 and select a vertex x. If deleting x removes all edges in some color, then the result is a K_4 spanned by edges using at most two colors. So, assume that deleting x results in a K_4 whose edges still use all three colors. Since x was selected arbitrarily, assume that this property holds for any vertex selected in the K_5. If we select another vertex y, then deleting it must remove all edges in some color, otherwise, we obtain a K_3 spanned by edges using all three colors, contradicting Lemma 1.1. This is true regardless of which vertex we select. As there are four vertices and three colors, there must exist two distinct vertices, say y and z whose deletion results in the same color (say, red) being removed. This can only occur if the only red edge in the K_4 is yz. Of course, it is possible that vertex x is also incident to a red edge, giving us several cases to consider. If x is not incident with any red edge, then deleting vertex y from the K_5 results in a K_4-subgraph spanned by edges using at most two colors. If x is incident with a red edge which is not adjacent to yz (say, xw is red), then the subgraph induced by $\{w,x,y,z\}$ must be a K_4 spanned by edges using at most two colors. If x is adjacent to exactly one of y or z via a red edge (say, xy is red), then deleting y removes all of the red edges, resulting in a K_4 spanned by edges using at most two colors. If x is adjacent to both y and z via red edges, then there exists a red K_3, and by Lemma 3.1, there exists a K_4 spanned by edges using at most two colors. It follows that $gr_2^3(K_4) \leq 5$, completing the proof of the base case.

Now assume that every Gallai $(t-1)$-coloring of K_{t+1} contains a K_4 spanned by edges using at most two colors and consider a Gallai t-coloring of K_{t+2}. Let $\mathcal{B}$ be the base graph for this Gallai coloring, chosen to have minimal order. If $|V(\mathcal{B})| \geq 4$, then selecting vertices from four distinct blocks results in a K_4 spanned by edges using at most two colors. By Lemma 1.3, $|V(\mathcal{B})| \neq 3$. It follows that $|V(\mathcal{B})| = 2$. Denote the blocks of $\mathcal{B}$ by X and Y and assume that the edges joining them are given the color red. If either X or Y contain a red edge, then there exists a red K_3-subgraph, and by Lemma 3.1, there exists a K_4-subgraph spanned by edges using at most two colors. It follows that X and Y do not contain any red edges. Also, if X and Y have edges that are the same color, then the subgraph induced by the endpoints of those edges form a K_4 spanned by edges using at most two colors. So, assume that X and Y do not have any colors in common. Suppose that X has order k_1 and uses ℓ_1 colors, and Y has order k_2 and uses ℓ_2 colors. Then $k_1+k_2=t+2$ and $\ell_1+\ell_2=t-1$, and by the inductive hypothesis,

$$gr_2^{\ell_1}(K_4) \leq \ell_1+2 \quad \text{and} \quad gr_2^{\ell_2}(K_4) \leq \ell_2+2.$$

If $k_1 \geq \ell_1+2$, then X contains a K_4 spanned by edges using at most two colors. Otherwise,

$$k_2 = t+2-k_1 \geq t+1-\ell_1 = \ell_2+2,$$

and it follows that Y contains a K_4 spanned by edges using at most two colors. It follows that $gr_2^t(K_4) \leq t+2$.

To complete the proof of the upper bound, consider a Gallai t-coloring of K_{t+2}, which we know must contain a K_4-subgraph spanned by edges using at most two colors. By Lemma 3.1, This coloring must also contain a K_{s+2}-subgraph spanned by edges using at most s colors. It follows that $gr_s^t(K_{s+2}) \leq t+2$. □

In 2022, it was shown in [51] that $gr_2^3(K_5)=9$ and $gr_2^t(K_5) \geq 2^t+1$. Working independently at the same time, Li, Broersma, and Wang [160] proved the following more general result. A proof of Theorem 3.15 different from the one give below also appeared in [208], where a recursive partitioning method was employed.

Theorem 3.15 ([160]) *For all $t \geq 2$, $gr_2^t(K_5) = 2^t + 1$.*

Proof We start by constructing a Gallai t-coloring of K_{2^t} that avoids a K_5 spanned by edges using at most two colors by using induction on $t \geq 3$. Denote by $\mathcal{G}_2$ the 2-coloring of K_4 in which color 1 spans a cycle of length 4 and color 2 spans a perfect matching. It is trivially a Gallai coloring and avoids a K_5 spanned by edges using at most two colors. For $t \geq 3$, form $\mathcal{G}_t$ by taking a K_2 in color t and replacing each of its vertices with copies of $\mathcal{G}_{t-1}$ (e.g., the $t = 3$ case is shown in Figure 3.7). The resulting t-coloring of K_{2^t} avoids rainbow

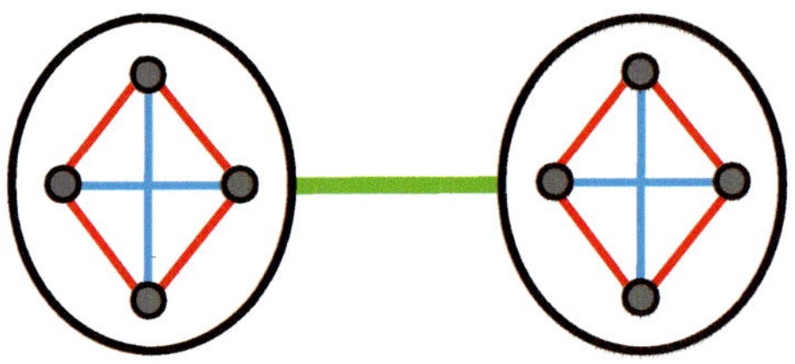

FIGURE 3.7: A Gallai 3-coloring of K_8 that avoids a K_5-subgraph spanned by edges using at most two colors.

triangles by construction. If we assume that $\mathcal{G}_{t-1}$ lacks a K_5 spanned by edges using at most two colors, then $\mathcal{G}_t$ will not have any K_5-subgraphs spanned by edges using at most two colors contained entirely within one of its blocks. For K_5-subgraphs that contain vertices from both blocks, note that by the pigeonhole principle, one of the blocks must contain at least three vertices. Three such vertices span a K_3-subgraph that uses exactly two colors. The edges joining the two blocks then introduce a third color to the K_5-subgraph. It follows that $gr_2^t(K_5) \geq 2^t + 1$.

Next, we use (strong) induction on $t \geq 2$ to prove that $gr_2^t(K_5) \leq 2^t + 1$. For the base case, $gr_2^2(K_5) = 5$ is trivial. Assume that for every $t' \in \{2, 3, \ldots, t-1\}$, $gr_2^{t'}(K_5) \leq 2^{t'} + 1$ and consider a Gallai t-coloring of K_{2^t+1}. Let $\mathcal{B}$ be the base graph for this coloring, chosen to have minimal order. If a K_5-subgraph spanned by edges using at most two colors is avoided, then by Theorem 1.22, $2 \leq |V(\mathcal{B})| \leq 4$. By Lemma 1.3, $|V(\mathcal{B})| \neq 3$. We separate the remainder of the proof into cases based on the value of $|V(\mathcal{B})|$.

Case 1: Assume that $|V(\mathcal{B})| = 2$ and that the edges joining the two blocks are all in color 1. If each block contains at least one edge in color 1, then the endpoints of those edges form a monochromatic K_4. By Lemma 3.1, there exists a K_5 spaned by edges using at most two colors. If neither block contains an edge in color 1, then by the inductive hypothesis, $gr_2^{t-1}(K_5) \leq 2^{t-1} + 1$, and if a K_5 spanned by edges using at most two colors is avoided, there are at most $2 \cdot 2^{t-1} = 2^t$ vertices, leading to a contradiction. Hence, we only need to consider the case where one block contains at least one edge in color 1 and the other block does not contain any edges in color 1. Denote by X the vertex set of the block that contains at least one edge in color 1 and denote by Y the vertex set for the block that does not contain an edge in color 1.

Note that the subgraph induced by X does not contain a K_3-subgraph in color 1, as such a subgraph, along with any vertex from Y, would form a monochromatic K_4. Lemma 3.1 would then imply the existence of a K_5 spanned by edges using at most two colors. Also, the subgraph induced by X does not contain a C_4-subgraph that contains exactly three edges in color 1. This is because if wx, xy, and yz have color 1 and wz has a different color (say, color 2), then edges wy and xz would also have one of the colors 1 or 2. The subgraph induced by $\{w, x, y, z\}$ and any vertex from Y would then be a K_5 spanned by edges using at most two colors. Now, if the subgraph induced by X contains a cycle with odd length given by $x_1x_2 \cdots x_{2n+1}x_1$ ($n \geq 2$) in color 1, then edges $x_1x_4, x_1x_6, \ldots, x_1x_{2n}$

must all have color 1 (since a C_4 with exactly three edges in color 1 is avoided). It follows that the subgraph induced by $\{x_1, x_{2n}, x_{2n+1}\}$ is a K_3 in color 1, which cannot occur. It follows that every cycle in color 1 in the subgraph induced by X has even length. Theorem 1.4 implies that the subgraph induced by X in color 1 is bipartite.

Let E_1 be the set of edges in the subgraph induced by X with color 1 and let X_1 be the vertices in X that are incident with some edge in E_1. Since the subgraph induced by X in color 1 is bipartite, X_1 can be partitioned into two sets A and B such that the subgraphs induced by A and B lack edges in color 1. As the subgraph induced by X contains at least one edge in color 1, $A \neq \emptyset$ and $B \neq \emptyset$. Define $X_1' := X \setminus X_1$ and note that it is possible that $X_1' = \emptyset$.

We already know that color 1 does not appear in the subgraph induced by Y. If $i \in \{2, 3, \ldots, t\}$ is any color that appears in the subgraph induced by X_1 (say, uv has color i with $u, v \in X_1$), then there exists a vertex $w \in X_1 \setminus \{u, v\}$ such that edge uw has color 1. The subgraph induced by $\{u, v, w\}$ is a K_3 in colors 1 and i, and if there exists an edge in color i in Y, then the endpoints of that edge, along with $\{u, v, w\}$, induce a K_5 spanned by edges using at most two colors. So, we have proved the following observation.

Observation 1: If any color $i \neq 1$ appears in the subgraph induced by X_1, then it does not appear in the subgraph induced by Y.

Let $c(S)$ be the number of colors that appear in the subgraph induced by any vertex set S. If $c(Y) = 0$, then $|Y| = 1$. If $c(Y) = 1$, then the subgraph induced by Y is a monochromatic complete subgraph. In this case, $|Y| = 2$ since if $|Y| \geq 3$, then such a K_3, along with the endpoints of an edge in color 1 in the subgraph induced by X, form a K_5 spanned by edges using at most two colors. If $c(Y) \geq 3$, then by the inductive hypothesis, $|Y| \leq 2^{c(Y)}$ Putting together these cases, we obtain the following observation.

Observation 2: $|Y| \leq 2^{c(Y)}$.

If $c(A) \neq 1$, then $|A| \leq 2^{c(A)}$ by an argument similar to the one given above for Y. If $c(A) = 1$, then the subgraph induced by A is a monochromatic complete subgraph. In this case, if $|A| \geq 3$ and $\{u, v, w\} \subseteq A$, then there exists a vertex $x \in B$ such that ux has color 1. The subgraph induced by $\{u, v, w, x\}$, along with any vertex in Y, forms a K_5 spanned by edges using at most two colors. Hence, $|A| \leq 2$, and it follows that $|A| \leq 2^{c(A)}$, regardless of the value of $c(A)$. A similar argument implies that $|B| \leq 2^{c(B)}$, leading to the following observation.

Observation 3: $|A| \leq 2^{c(A)}$ and $|B| \leq 2^{c(B)}$.

Now, consider the possibility that $|A| = 1$ with $A = \{u\}$. Since every vertex in B joins to some vertex in A with an edge in color 1, it follows that all edges joining A and B have color 1. By Observation 1, $c(B) + c(Y) \leq t - 1$. If $X_1' = \emptyset$, then Observations 2 and 3 imply that the total number of vertices satisfies

$$\begin{aligned}
|A| + |B| + |Y| &\leq 1 + 2^c(B) + 2^{c(Y)} \\
&\leq 1 + (2^{c(B)+c(Y)} + 1) \\
&\leq 1 + (2^{t-1} + 1) \\
&< 2^t + 1,
\end{aligned}$$

which is a contradiction. Thus, there exists some vertex $v \in X_1'$. As v is not incident with any edges in color 1, we assume, without loss of generality, that uv has color 2. Avoiding a rainbow triangle forces all of the edges joining v to B to have color 2. If any edge in B has color 2, then its endpoints, along with u, v, and any vertex in Y form a K_5 spanned by edges using at most two colors. If any edge in Y has color 2, then its endpoints, along with u, v, and any vertex in B form a K_5 spanned by edges using at most two colors. Hence, color 2 does not appear in B or Y, and it follows that $c(B) + c(Y) \leq t - 2$. The total number of vertices satisfies

$$\begin{aligned}|A|+|B|+|Y|+|X_1'| &\leq 1+2^{c(B)}+2^{c(Y)}+2^{t-1}\\ &\leq 1+(2^{c(B)+c(Y)}+1)+2^{t-1}\\ &\leq 1+2^{t-2}+1+2^{t-1}\\ &< 2^t+1,\end{aligned}$$

which is a contradiction. Thus, we have proved that $|A| \geq 2$. Using a similar proof for B, we obtain the following observation.

Observation 4: $|A| \geq 2$ and $|B| \geq 2$.

Now we argue that there exists a color that appears in the subgraph induced by B, but not the subgraph induced by A. Recall that the color 1 does not appear in the subgraphs induced by A or B. By Observation 4, $c(B) \geq 1$, and by Corollary 1.4, the subgraph induced by B contains a color (say, color 2) whose subgraph spans B. Suppose that there exist two distinct vertices $u, v \in A$ such that uv has color 2. Then there exists $w, x \in B$ such that wx has color 2 and uw has color 1. Then vw and ux must also have one of the colors 1 or 2. If edge vx also has color 1 or color 2, then the subgraph induced by $\{u, v, w, x\}$, along with any vertex from Y, results in a K_5 spanned by edges using at most 2 colors. So, assume that vx has a color other than 1 or 2. It follows that vw and ux must have color 2. Since vertex v must join to a vertex in B with an edge in color 1, there exists $y \in B \setminus \{w, x\}$ such that vy has color 1. Avoiding a rainbow triangle forces edge uy to have color 1 or color 2 and we find that the subgraph induced by $\{u, v, w, y\}$, along with any vertex from Y, results in a K_5 spanned by edges using at most two colors. Thus, it follows that some color appears in the subgraph induced by B that does not appear in the subgraph induced by A. By symmetry, there exists some color that appears in the subgraph induced by A that does not appear in the subgraph induced by B.

By Observation 1, the subgraphs induced by A and Y do not have any colors in common and the subgraphs induced by B and Y do not have any colors in common. Recall that the subgraph induced by A does not include color 1, nor does the subgraph induced by $B \cup X_1'$. We find that

$$c(A) \leq t-1-c(B-A)-c(Y) \leq t-2-c(Y).$$

Thus, the total number of vertices satisfies

$$\begin{aligned}|A|+|Y|+|B \cup X_1'| &\leq 2^{t-2-c(Y)}+2^{c(Y)}+2^{t-1}\\ &\leq 2^{t-2}+1+2^{t-1}\\ &< 2^t+1,\end{aligned}$$

which is a contradiction. Hence, every Gallai t-coloring of K_{2^t+1} contains a K_5 spanned by edges using at most two colors in this case.

Case 2: Assume that $|V(\mathcal{B})| = 4$. If any vertex in $\mathcal{B}$ is incident with edges in only one color, then the Gallai coloring can be viewed as having only two blocks, contradicting the minimal order of $\mathcal{B}$. So, every vertex in $\mathcal{B}$ is incident with edges in both of the colors (say, colors 1 and 2) that appear in $\mathcal{B}$. If a K_5 spanned by edges using at most two colors is avoided, then no block can contain edges in colors 1 or 2 as the endpoints of such an edge, along with a single vertex chosen from each of the other blocks, would form a K_5 spanned by edges using at most two colors. By the inductive hypothesis, each block has order at most 2^{t-2}, from which it follows that there are at most $4 \cdot 2^{t-2} = 2^t$ vertices, contradicting the assumption that there are $2^t + 1$ vertices. Thus, there exists a K_5 spanned by edges using at most two colors in this case.

In both cases, we find that $gr_2^t(K_5) \leq 2^t + 1$, completing the proof of the theorem. □

In [51], it was proved that when $n \geq 6$, $gr_2^3(K_n) \geq 2n - 1$. This bound was shown to not be optimal when $n = 6$ by Jakhar and Moun [136]. They considered a critical coloring for $r^2(K_3)$ (see Theorem 1.14) in colors red and blue, then replaced four if its vertices with green K_2-subgraphs, and the fifth vertex with a green K_3-subgraph (see Figure 3.8). Here,

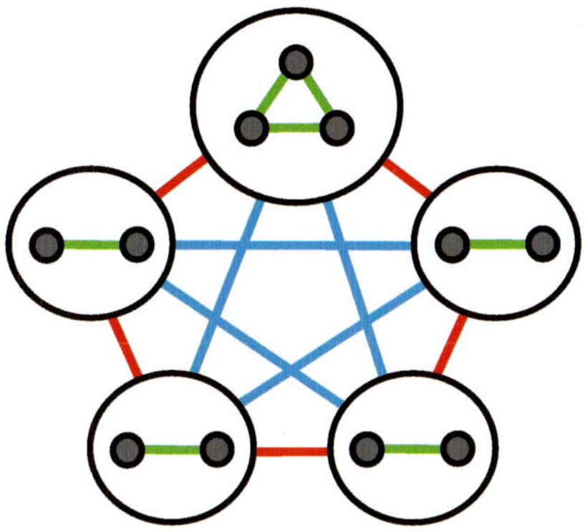

FIGURE 3.8: A Gallai 3-coloring of K_{11} that avoids a K_6-subgraph spanned by edges using at most two colors.

every K_6-subgraph must include at least two vertices from some block and vertices from at least three distinct blocks. Thus, it will contain edges in all three colors, from which it follows that $gr_2^3(K_6) \geq 12$. This lower bound led Jakhar and Moun to conjecture that $gr_2^3(K_6) = 12$ (see Problem 15 in Section 5.1). This construction can easily be generalized to obtain the following theorem.

Theorem 3.16 *For all $n \geq 6$,*

$$gr_2^3(K_n) \geq \begin{cases} \frac{5n-6}{2} & \text{if } n \text{ is even} \\ \frac{5n-3}{2} & \text{if } n \text{ is odd.} \end{cases}$$

Proof We consider the cases where n is even or odd separately.

Case 1: Suppose that $n = 2k$. Consider a critical coloring for $r^2(K_3)$ in colors red and blue (see Theorem 1.14). Replace four of its vertices with green K_{k-1}-subgraphs and the fifth vertex with a green K_k-subgraph (see the first image in Figure 3.9). Every K_n-subgraph includes vertices from at least three distinct blocks and at least two vertices from some block (since $n \geq 6$). So, its edges use all three colors. It follows that

$$gr_2^3(K_n) \geq 4(k-1) + k + 1 = 5k - 3 = \frac{5n-6}{2}.$$

Case 2: Suppose that $n = 2k + 1$. Consider a critical coloring for $r^2(K_3)$ in colors red and blue and replace all of its vertices with green K_k-subgraphs (see the second image in Figure 3.9). Every K_n-subgraph includes vertices from at least three distinct blocks and at least two vertices from some block (since $n \geq 6$). So, its edges use all three colors. It follows that

$$gr_2^3(K_n) \geq 5k + 1 = \frac{5n-3}{2}.$$

□

Besides Theorem 3.15, Li, Broersma, and Wang [160] also proved the following theorem, which we state without proof due to its length.

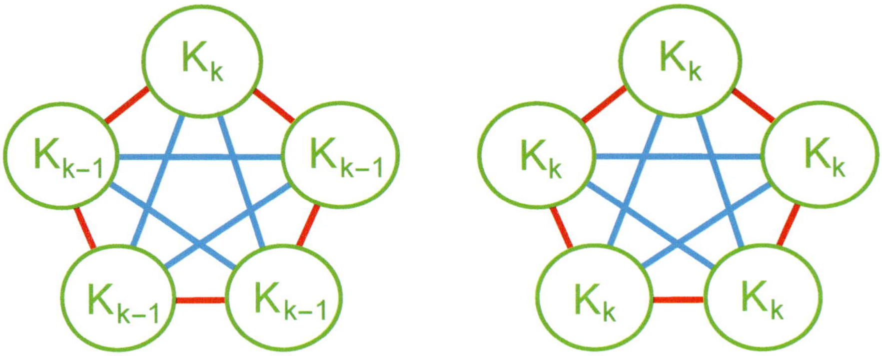

FIGURE 3.9: Gallai 3-colorings of complete graphs that avoid 2-colored complete subgraphs of order at least six.

Theorem 3.17 ([160]) *If $n \geq 8$ and $t \geq n-3$, then $gr^t_{n-3}(K_n) = t+3$.*

Additionally, they offered the following general upper bound:

$$gr^t_s(K_n) \leq 2^{2t(n-2)/s+1},$$

where $n, t \geq 3$ and $1 \leq s \leq \min\{n-2, t\}$.

Finally, we note that for $t \geq 3$ and $1 \leq s \leq t-1$, one can define the *star-critical weakened Gallai-Ramsey number* $gr^{s,t}_*(G)$ to be the least $k \in \{0, 1, \ldots, gr^{s,t}(G)-1\}$ such that every Gallai t-coloring of $K_{gr^{s,t}(G)-1} \sqcup K_{1,k}$ contains a subgraph isomorphic to G that is spanned by edges using at most s colors. As we noted previously, $k \neq 0$ whenever G lacks isolated vertices. A proof similar to that of Theorem 3.8 implies that

$$gr^{s,t}_*(K_n) = gr^t_s(K_n) - 1.$$

3.4 Weakened Gallai-Ramsey Numbers for Other Graphs

Now we turn our attention to weakened Gallai-Ramsey numbers for graphs that are not complete. Over the next four subsections, we consider weakened Gallai-Ramsey numbers for cycles, fans, wheels, and books (along with complete bipartite graphs).

3.4.1 Cycles

In 2010, Faudree, Could, Jacobson, and Magnant [92] proved that

$$gr^t(C_4) = t+4, \quad \text{for all } t \geq 3.$$

Fujita and Magnant [99] proved that

$$gr^t(C_5) = 2^{t+1}+1 \quad \text{and} \quad gr^t(C_6) = 2t+4,$$

for all $t \geq 2$. The complete evaluation of $gr^t(C_n)$ was completed for even cycles by Zhang, Song, and Chen [213] in 2022, where they proved that

$$gr^t(C_n) = \frac{n-2}{2}t + \frac{n+2}{2}, \quad \text{for all even } n \geq 6.$$

In a companion paper [214] in 2023, they also proved that

$$gr^t(C_n) = \frac{n-1}{2} \cdot 2^t + 1, \quad \text{for all odd } n \geq 5,$$

confirming a conjecture of Bondy and Erdős [26] under Gallai colorings. In this subsection, we will consider the weakened Gallai-Ramsey numbers $gr_2^t(C_n)$.

Theorem 3.18 ([51]) *For $n \geq 4$, every Gallai t-coloring of K_n contains a C_4 spanned by edges using at most two colors. In particular,*

$$gr_2^3(C_4) = 4 \qquad \textit{and} \qquad gr_*^{2,3}(C_4) = 3.$$

Proof Let $n \geq 4$ and consider a Gallai t-coloring of K_n. Select four distinct vertices a, b, c, and d. If the cycle $abcda$ uses more than two colors, then by Lemma 1.1, it uses exactly three colors. We obtain two cases, based on whether or not the color that gets repeated appears on adjacent or non-adjacent edges (see Figure 3.10).

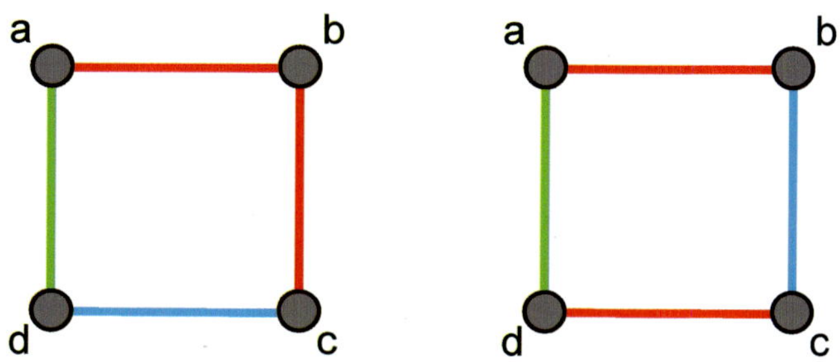

FIGURE 3.10: Two 3-colorings of C_4.

Case 1: Suppose that the color that gets repeated appears on adjacent edges. Without loss of generality, suppose that edges ab and bc are red, cd is blue, and ad is green (see the first image in Figure 3.10). Avoiding a rainbow triangle forces edge bd to be red and edge ac to be blue or green. If ac is blue, then $abdca$ is a red/blue C_4. If ac is green, then $acbda$ is a red/green C_4.

Case 2: Suppose that the color that gets repeated appears on non-adjacent edges. Without loss of generality, suppose that edges ab and cd are red, bc is blue, and ad is green (see the second image in Figure 3.10). Avoiding a rainbow triangle forces edges ac and bd to be red. Then $abdca$ is a red C_4.

So, every Gallai 3-coloring of K_4 contains a C_4 spanned by edges using at most two colors, implying that $gr_2^3(C_4) \leq 4$. The 3-coloring of $K_3 \sqcup K_{1,2} = K_4 - e$ shown in Figure 3.11 lacks a C_4 spanned by edges that use at most two colors since the only such subgraph must include both of the vertices of degree 2, and hence, uses all three colors. It follows that $gr_2^3(C_4) = 4$ and $gr_*^{2,3}(C_4) = 3$. □

In [51], the weakened Gallai-Ramsey number $gr_2^t(C_5)$ was determined. In [36], the corresponding star-critical weakened Gallai-Ramsey number was evaluated. In order to prove these results, we must first prove a couple of lemmas. The following lemma generalizes Lemma 2.1 of [36].

Lemma 3.2 *For $t \geq 3$ and $p \geq n \geq 5$, if a Gallai t-coloring of K_p contains a monochromatic P_{n-1}, then it contains a C_n spanned by edges that use at most two colors.*

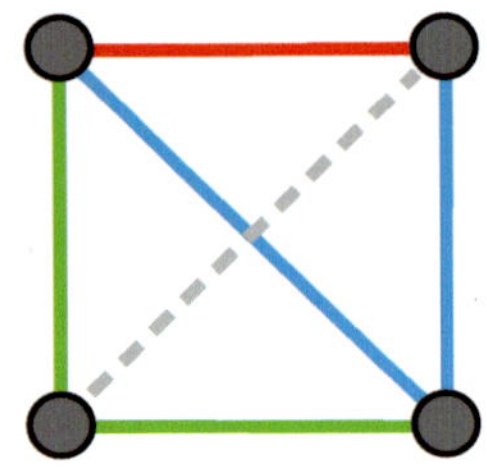

FIGURE 3.11: A Gallai 3-coloring of $K_4 - e$ that lacks a C_4 spanned by edges that use at most two colors.

Proof Consider a Gallai t-coloring of K_p that contains a monochromatic P_{n-1}. Let $x_1x_2\cdots x_{n-1}$ be such a path, and without loss of generality, assume that it is red. Let y be any vertex not contained in the P_{n-1}. If edges yx_1 and yx_{n-1} are the same color, or if one of them is red, then $yx_1x_2\cdots x_{n-1}y$ is a C_n that uses at most two colors. So, assume without loss of generality that yx_1 is blue and yx_{n-1} is green. Since rainbow triangles are avoided, edge x_1x_{n-1} must be either blue or green (assume it is blue). Now edge yx_2 must be red or blue, and either way, $yx_2x_3\cdots x_{n-1}x_1y$ is a red/blue C_n. □

Lemma 3.3 ([**36**]) *For $t \geq 3$, every Gallai t-coloring of K_{t+2} that avoids a C_5 that uses at most two colors contains a vertex that is incident with edges in only one color. Furthermore, the subgraph induced by the leaves of the resulting monochromatic spanning star does not contain any edges in the same color as the edges of the star.*

Proof Consider a Gallai t-coloring of K_{t+2} that avoids a C_5 that uses at most two colors. Let $\mathcal{B}$ be the base graph, which is chosen to have minimal order. By Theorem 1.22 and Lemma 1.3, $2 \leq |V(\mathcal{B})| \leq 4$ and $|V(\mathcal{B})| \neq 3$. Consider the following cases.

Case 1: Suppose that $|V(\mathcal{B})| = 2$. If both blocks have order at least two, then let x_1 and x_2 be vertices in one block and y_1, y_2, and y_3 be vertices in the other block. Regardless of the color of y_1y_3, the cycle $y_1x_2y_2x_2y_3y_1$ is a C_5 that uses at most two colors. So, one block must only contain a single vertex and that vertex is incident with edges that all have the same color.

Case 2: Suppose that $|V(\mathcal{B})| = 4$. Since K_{t+2} has order at least five, by the pigeonhole principle, some block has order at least two. Denote the vertex sets for the blocks by W, X, Y, and Z, and without loss of generality, assume that $w \in W$, $x \in X$, $y \in Y$, and $z_1, z_2 \in Z$. Then wz_1yz_2xw is a C_5 that uses at most two colors.

It follows that every Gallai t-coloring of K_{t+2} that avoids a C_5 that uses at most two colors has a vertex that is incident with edges in only one color. Let x be such a vertex and assume that the other vertices are given by $y_1, y_2, \ldots, y_{t+1}$. Also, assume that the edges incident with x are all red. If some edge in the subgraph induced by $\{y_1, y_2, \ldots, y_{t+1}\}$ is red (say, edge y_1y_2 is red), then $y_1y_2xy_3$ is a red P_4. By Lemma 3.2, this cannot occur. Hence, the subgraph induced by $\{y_1, y_2, \ldots, y_{t+1}\}$ does not contain any red edges. □

With Lemmas 3.2 and 3.3 in place, we are now ready to consider $gr_2^t(C_5)$ and its star-critical analogue. Our method of proof highlights a standard technique employed in star-critical Ramsey theory: we prove the uniqueness of the critical coloring for $gr_2^t(C_5)$, and then build off of that construction by introducing a new vertex and joining it with t edges to the critical coloring.

Theorem 3.19 ([51], [36]) *For all $t \geq 3$,*

$$gr_2^t(C_5) = t + 3 \qquad \textit{and} \qquad gr_*^{2,t}(C_5) = t.$$

Proof First, we will construct a Gallai t-coloring of $K_{t+2} \sqcup K_{1,t-1}$ that avoids a C_5 whose edges use at most two colors. Begin with a K_3 in color 1, introduce a new vertex x_1, and color all edges from x_1 to the K_3 with color 2. Next, introduce vertex x_2 and color all edges from x_2 to the existing graph with color 3. Denote by $\mathcal{G}_3$ the resulting Gallai 3-coloring of K_5. Continue in this manner, sequentially introducing vertices $x_3, x_4, \ldots, x_{t-1}$, coloring all edges from x_i to the existing graph with color $i+1$ (the $t = 5$ case is shown in Figure 3.12). Denote the resulting Gallai t-coloring of K_{t+2} by $\mathcal{G}_t$. Every C_5-subgraph

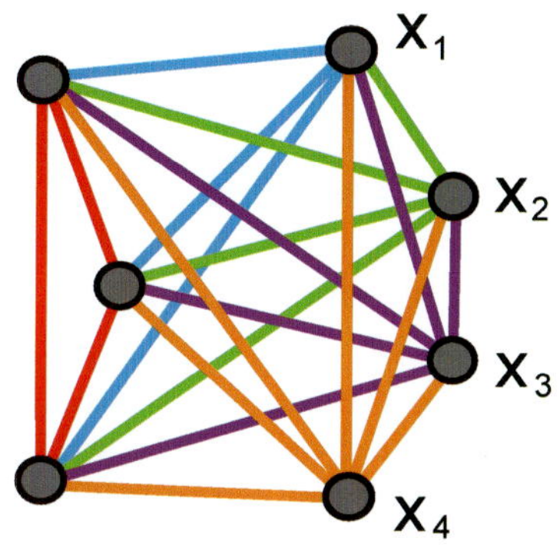

FIGURE 3.12: A Gallai 5-coloring of K_7 that avoids a C_5-subgraph spanned by edges using at most two colors.

includes at least two of the vertices $x_1, x_2, \ldots, x_{t-1}$. For the two x_i that are included for which i is greatest, the C_5-subgraph will use at least one, and no more than two edges in color $i+1$. It follows that every C_5-subgraph uses at least three colors, leading to the lower bound $gr_2^t(C_5) \geq t + 3$.

Introduce a vertex v and join it to x_i with an edge in color $i+1$, for all $i \in \{1, 2, \ldots, t-1\}$. Denote the resulting Gallai t-coloring of $K_{t+2} \sqcup K_{1,t-1}$ by $\mathcal{G}'_t$ (e.g., Figure 3.13 shows $\mathcal{G}'_3$). Every C_5-subgraph that includes vertex v must include two edges vx_i and vx_j (where $i \neq j$)

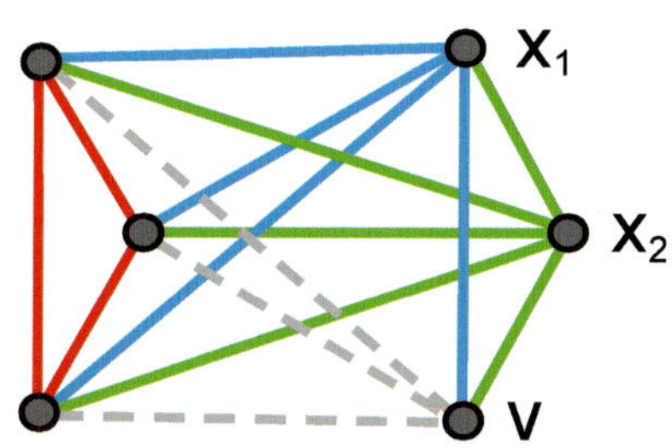

FIGURE 3.13: A Gallai 3-coloring of $K_5 \sqcup K_{1,2}$ that avoids a C_5-subgraph spanned by edges using at most two colors.

in colors $i+1$ and $j+1$. Regardless of which two other vertices are included in the C_5-subgraph, the edge joining them must include a third color. Hence, a C_5-subgraph that uses at most two colors is avoided. It follows that $gr_2^t(C_5) \geq t+3$, and once we show that $gr_2^t(C_5) \leq t+3$, it will follow that $gr_*^{2,t}(C_5) \geq t$.

To prove the upper bounds, consider a Gallai t-coloring of $K_{t+2} \sqcup K_{1,t}$ and denote by v the unique vertex of degree t. We will now prove that if v is deleted, and a C_5 spanned by edges using at most two colors is avoided, then the resulting Gallai t-coloring of K_{t+2} is isomorphic to $\mathcal{G}_t$, up to a reordering of the colors. This is accomplished by induction on $t \geq 3$. For the base case, consider a Gallai 3-coloring of K_5 that avoids a C_5 that uses at most two colors. By Lemma 3.3, some vertex is incident with edges in only one color. Without loss of generality, suppose that all of the edges incident with vertex x receive color 3, and denote the other vertices by y_1, y_2, y_3, and y_4. Lemma 3.3 also implies that the subgraph induced by $\{y_1, y_2, y_3, y_4\}$ only contains edges in colors 1 and 2. Figure 3.14 shows all 2-colorings of K_4, up to isomorphism or a reordering of the colors. By Lemma 3.2,

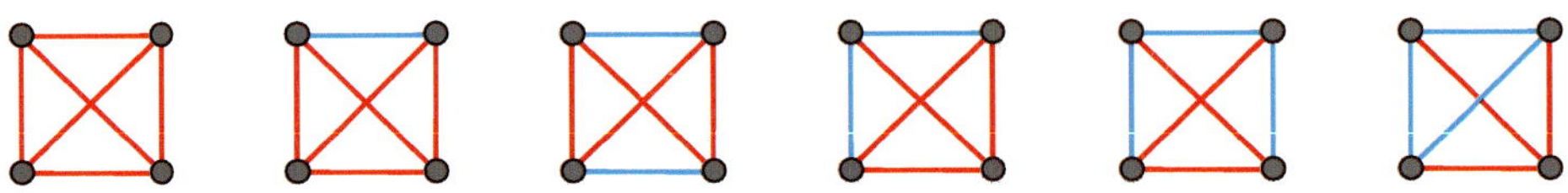

FIGURE 3.14: All 2-colorings of K_4, up to isomorphism or reordering of the colors.

the subgraph induced by $\{y_1, y_2, y_3, y_4\}$ does not contain a monochromatic P_4. The only 2-coloring of K_4 in Figure 3.14 is the last one. It follows that the Gallai 3-coloring of K_5 must be isomorphic to $\mathcal{G}_3$, up to a reordering of the colors.

Assume that up to a reordering of the colors, every critical coloring for $gr_2^t(C_5)$ is isomorphic to $\mathcal{G}_t$ and consider a Gallai $(t+1)$-coloring of K_{t+3} that avoids a C_5 spanned by edges that use at most two colors. By Lemma 3.3, there exists a vertex that is only incident with edges in one color. Without loss of generality, assume that x is only incident with edges in color $t+1$ and the other vertices are given by $y_1, y_2, \ldots, y_{t+2}$. Lemma 3.3 also implies that the subgraph induced by $\{y_1, y_2, \ldots, y_{t+2}\}$ only uses colors $1, 2\ldots, t$. By the inductive hypothesis, this subgraph must be isomorphic to $\mathcal{G}_t$, up to a reordering of the colors. It follows that the Gallai $(t+1)$-coloring of K_{t+3} being considered is isomorphic to $\mathcal{G}_{t+1}$, up to a reordering of the colors.

Returning to the Gallai t-coloring of $K_{t+2} \sqcup K_{1,t}$ with vertex v having degree t, we have shown that deleting v results in a Gallai t-coloring of K_{t+2} that is isomorphic to $\mathcal{G}_t$, up to a reordering of the colors. In this $\mathcal{G}_t$, denote the vertices in the K_3 in color 1 by y_1, y_2, and y_3, and assume that the other vertices $x_1, x_2, \ldots, x_{t-1}$ were sequentially added in, with vertex x_i joining to the existing graph with edges in color $i+1$, for each $i \in \{1, 2, \ldots, t-1\}$. Since v has degree t, it must join to at least one of y_1, y_2 or y_3. Consider the following cases.

<u>Case 1:</u> Assume that v joins to only one of y_1, y_2, and y_3 (say, y_1) and all of $x_1, x_2, \ldots, x_{t-1}$. If vy_1 has color 1, then vx_1 must receive either color 1 or color 2. Either way, $vy_1y_2y_3x_1v$ is a C_5 whose edges only use colors 1 and 2. If vy_1 has a color other than 1 (say, color 2), then edge vx_2 must receive one of the colors 2 or 3. Either way, $vy_1x_1y_2x_2v$ is a C_5 whose edges only use colors 2 and 3.

<u>Case 2:</u> Assume that v joins to two or more of y_1, y_2, and y_3. Without loss of generality, suppose that v joins to both y_1 and y_2. If both of the edges vy_1 and vy_2 have color 1, then $vy_1y_3x_1y_2c$ is a C_5 whose edges use only colors 1 and 2. If one of vy_1 and vy_2 use a color other than 1, note that they cannot both receive distinct colors other than 1 or a rainbow triangle would be formed. So, assume that color i (where $2 \leq i \leq t$) appears on at least one of vy_1 and vy_2. Then $vy_1y_3x_{i-1}y_2v$ is a C_5 whose edges only use the colors 1 and i.

In both cases, we have seen that a Gallai t-coloring of $K_{t+2} \sqcup K_{1,t}$ contains a C_5 that uses at most two colors. It follows that $gr_2^t(C_5) = t+3$ and $gr_*^{2,t}(C_5) = t$. □

One consequence of Lemma 3.2 is that for all $t \geq 3$ and $n \geq 5$,

$$gr_2^t(C_n) \leq gr^t(P_{n-1}). \tag{3.3}$$

The Gallai-Ramsey number $gr^t(P_n)$ was determined for $n \in \{4,5,6\}$ in [92], for $n = 8$ in [199], and for $n \in \{7,9,10,11,12,13\}$ in [157]. The general evaluation was completed in 2022 by Zhang, Song, and Chen [213], where they proved that for all $t \geq 2$ and $n \geq 6$,

$$gr^t(P_n) = \begin{cases} \frac{n-2}{2} \cdot t + \frac{n+2}{2} & \text{if } n \text{ is even} \\ \frac{n-3}{2} \cdot t + \frac{n+3}{2} & \text{if } n \text{ is odd.} \end{cases}$$

Inequality (3.3) implies that

$$gr_2^t(C_n) \leq \begin{cases} \frac{n-4}{2} \cdot t + \frac{n+2}{2} & \text{if } n \text{ is even} \\ \frac{n-3}{2} \cdot t + \frac{n+1}{2} & \text{if } n \text{ is odd,} \end{cases} \tag{3.4}$$

for all $t \geq 3$ and $n \geq 7$. The next theorem offers a general lower bound for $gr_2^t(C_n)$.

Theorem 3.20 *If $t \geq 3$ and $n \geq 5$, then $gr_2^t(C_n) \geq t + n - 2$.*

Proof Begin with a K_{n-2} in color 1, add in vertex x_1, and join x_1 to the K_{n-2} with edges in color 2. Introduce vertex x_2 and join it to all of the other vertices with edges in color 3. Continue in this manner, sequentially introducing vertices $x_3, x_4, \ldots, x_{t-1}$, coloring all of the edges from x_i to the existing graph with color $i+1$. Figure 3.15 show the $t = 4$ case. In the resulting Gallai t-coloring of K_{t+n-3}, every C_n-subgraph includes at least two

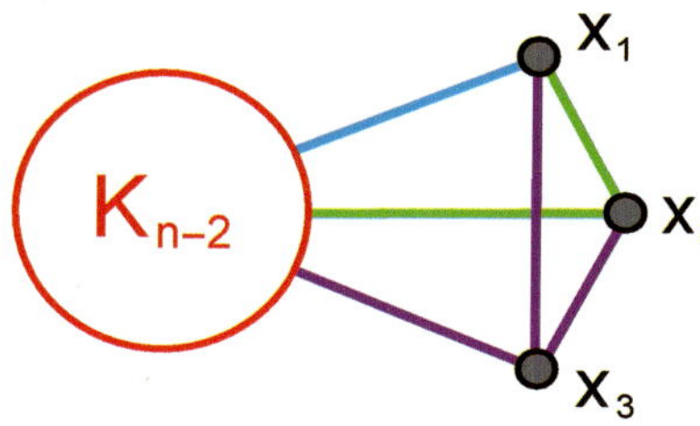

FIGURE 3.15: A Gallai 4-coloring of K_{t+n-3} that avoids a C_n-subgraph spanned by edges using at most two colors.

of the vertices $x_1, x_2, \ldots, x_{t-1}$. For the two x_i that are included for which i is greatest, the C_n-subgraph will use at least one, and no more than two edges in color $i+1$. So, every C_n-subgraph uses at least three colors, and it follows that $gr_2^t(C_n) > t + n - 3$. □

Combining Theorem 3.20 with Inequality (3.3) leads to the following theorem.

Theorem 3.21 *For all $t \geq 3$, $gr_2^t(C_6) = t + 4$.*

Proof From Theorem 3.20, it follows that $gr_2^t(C_6) \geq t + 4$. In [92], Faudree, Gould, Jacobson, and Magnant showed that $gr^t(P_5) = t + 4$, for all $t \geq 3$. Combining this with Inequality (3.3), we obtain the upper bound $gr_2^t(C_6) \leq t + 4$. □

It is not immediately clear whether or not Inequality (3.4) is sharp for $n \geq 7$. The values of the corresponding star-critical Ramsey numbers $gr_*^{2,t}(C_n)$, for $n \geq 6$ are also currently unknown (see Problem 10 in Section 5.1).

3.4.2 Fans

Gallai-Ramsey numbers for fans were first considered by Mao, Wang, Magnant, and Schiermeyer [174] in 2023, where they proved that

$$gr^t(F_2) = \begin{cases} 9 & \text{if } t = 2 \\ \frac{83}{2} \cdot 5^{(t-4)/2} + \frac{1}{2} & \text{if } t \geq 4 \text{ is even} \\ 4 \cdot 5^{(t-1)/2} + 1 & \text{if } t \text{ is odd.} \end{cases}$$

and

$$gr^t(F_3) \geq \begin{cases} 14 \cdot 5^{(t-2)/2} - 1 & \text{if } t \text{ is even} \\ 33 \cdot 5^{(t-3)/2} & \text{if } t \text{ is odd,} \end{cases}$$

when $t \geq 2$. In this subsection, we consider the weakened version of Gallai-Ramsey numbers for fans.

One useful result for working with fans is the Ramsey number for matchings, which was determined by Cockayne and Lorimer [71] in 1975. They proved that if $n_1, n_2, \ldots, n_t \in \mathbb{N}$ such that $n_1 = \max\{n_1, n_s, \ldots, n_t\}$, then

$$r(n_1K_2, n_2K_2, \ldots, n_tK_2) = n_1 + 1 + \sum_{i=1}^{t}(n_i - 1).$$

In particular, this implies that $r^t(nK_2) = n(t+1) - t + 1$.

Theorem 3.22 ([31]) *For all $t \geq 3$, $gr_2^t(F_2) = t + 3$.*

Proof The lower bound follows from the critical colorings for $gr_2^t(C_5)$ described in the proof of Theorem 3.19. (see Figure 3.12 for a Gallai 5-coloring of K_7 that avoids an F_2 that uses at most two colors). In this construction, every F_2-subgraph in which the center of the fan is a vertex in the K_3 in color 1 uses at least three colors on its four edges to the blades of the fan. If an F_2-subgraph has x_i as its center and the edges to its blades are chosen to use at most two colors, then at least one of the blades must use a third color. So, no F_2 exists that uses at most two colors. It follows that $gr_2^t(F_2) \geq t + 3$.

To prove the upper bound, consider a Gallai t-coloring of K_{t+3} and let $\mathcal{B}$ be the base graph, chosen to have minimal order. Theorem 1.22 and Lemma 1.3 imply that $2 \leq |V(\mathcal{B})| \leq 4$ and $|V(\mathcal{B})| \neq 3$. Consider the following cases.

Case 1: Suppose that $|V(\mathcal{B})| = 2$ and assume that the edges joining the blocks are red (color 1). Consider two subcases.

Subcase 1.1: Suppose that one block only contains a single vertex x and the larger block has order $t + 2$. If we view the colors red and blue (colors 1 and 2, respectively) as if they are the same color, then since $gr^{t-1}(2K_2) \leq t + 2$ [71], there exists a red/blue $2K_2$ or a monochromatic copy of $2K_2$ in one of the colors $3, 4, \ldots, t$. In both cases, an F_2 that uses at most two colors is formed with x as its center and the $2K_2$ forming the blades.

Subcase 1.2: Suppose that both blocks have order at least 2. Note that one block must have order at least 3. Let x_1 and x_2 be in one block and y_1, ay_2, and y_3 be in the other block. Since rainbow triangles are avoided, at least two of the edges of the subgraph induced by $\{y_1, y_2, y_3\}$ use the same color. Without loss of generality, suppose that y_1y_2 and y_1y_3 are both blue. Then a red/blue F_2 is formed with center y_1 and blades x_1y_2 and x_2y_3.

Case 2: Suppose that $|V(\mathcal{B})| = 4$ and assume that the edges joining the blocks are red and blue. Denote the vertex sets for the blocks by W, X, Y, and Z. Note that one block must contain at least two vertices. Without loss of generality, assume that $w \in W$, $x \in X$,

$y \in Y$, and $z_1, z_2 \in Z$. Then a red/blue F_2 can be formed with center w and blades xz_1 and yz_2.

In both cases, we find that there exists an F_2 whose edges use at most two colors. It follows that $gr_2^t(F_2) \leq t + 3$. □

Theorem 3.23 ([31]) $gr_2^3(F_3) = 9$.

Proof To prove the lower bound, start by replacing one vertex in a blue K_2 with a green K_5. Denote by $\mathcal{G}$ the resulting blue/green coloring of K_6, which is a critical coloring for $r(2K_2, 3K_2)$ (see [71]). Next, take a red K_2 and replace one of its vertices with a blue K_2 and the other vertex with a copy of $\mathcal{G}$ (see Figure 3.16). The resulting Gallai 3-coloring

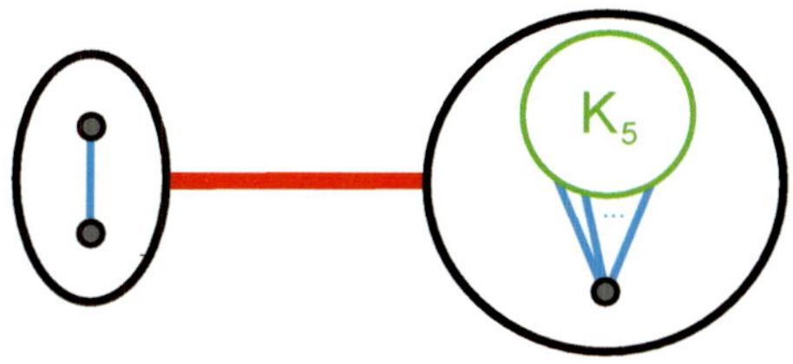

FIGURE 3.16: A Gallai 3-coloring of K_8 that avoids an F_3-subgraph spanned by edges using at most two colors.

of K_8 does not have a large enough blue/green connected component to have a blue/green F_3. It lacks a red/blue F_3 as such a subgraph would require the existence of a blue $2K_2$ in the larger block. It also lacks a red/green F_3 as it would require there to be a green $3K_2$ in the larger block. It follows that $gr_2^3(F_3) \geq 9$.

To prove the reverse inequality, consider a Gallai 3-coloring of K_9 and let $\mathcal{B}$ be the base graph, chosen to have minimal order. Theorem 1.22 and Lemma 1.3 imply that $2 \leq |V(\mathcal{B})| \leq 6$ and $|V(\mathcal{B})| \neq 3$. Consider the following cases, based on the order of $\mathcal{B}$.

Case 1: Suppose that $|V(\mathcal{B})| = 2$ and that the edges joining the two blocks are red. Consider the following three subcases.

Subcase 1.1: Assume that one block has order at most 2. Denote its vertex set by X and denote the vertex set for the larger block by Y, noting that $|Y| \geq 7$. If the subgraph induced by Y does not contain any red edges, then contains a blue/green K_7, which contains F_3 as a subgraph. So, without loss of generality, assume that y_1y_2 is red, where $y_1, y_2 \in Y$. If the subgraph induced by $Y \setminus \{y_1, y_2\}$ lacks red edges, then $r^2(2K_2) = 5$ [71] implies that it contains a monochromatic $2K_2$ that is blue or green. This $2K_2$, along with edge y_1y_2 and a vertex $x \in X$, forms an F_3 spanned by edges using at most two colors in which x is the center. So, assume that the subgraph induced by $Y \setminus \{y_1, y_2\}$ does contain a red edge, and without loss of generality, assume that y_3y_4 is one such edge. Then an F_3 that uses at most two colors can be formed with center x and blades y_1y_2, y_3y_4, and y_5y_6, where $y_5, y_6 \in Y \setminus \{y_1, y_2, y_3, y_4\}$.

Subcase 1.2: Assume that one block has order 3. Denote its vertex set by $X = \{x_1, x_2, x_3\}$ and denote the vertex set for the block of order 6 by $Y = \{y_1, y_2, \ldots, y_6\}$. If the subgraph induced by X contains a red edge (say, x_1x_2 is red), then $r^3(2K_2) = 6$ [71] implies that the subgraph induced by Y contains a monochromatic $2K_2$. If the edges in this $2K_2$ are given by y_1y_2 and y_3y_4, then an F_3 whose edges use at most two colors is formed with center x_1 and blades y_1y_2, y_3y_4, and x_2y_5. So, assume that the subgraph induced by X is blue/green. Since rainbow triangles are avoided, at least two of the edges in this blue/green K_3 must be the same color. Without loss of generality, assume that x_1x_2 and x_1x_3 are both blue. If the subgraph induced by Y contains a red or blue edge (say,

y_1y_2 is red or blue), then a red/blue F_3 is formed with x_1 as the center and blade y_1y_2, x_2y_3, adn x_3y_4. So, assume that the subgraph induced by Y only contains green edges. Then a red/green F_3 is formed with center x_1 and blades y_1y_2, y_3y_4, and y_5y_6.

Subcase 1.3: Assume that one block has order 4 and the other block has order 5. Without loss of generality, assume that $X = \{x_1, x_2, x_3, x_4\}$ and $Y = \{y_1, y_2, y_3, y_4\}$ are the vertex sets for the blocks. If the subgraph induced by X contains a red edge (say, x_1x_2 is red), then if we view red and blue as the same color, $r^2(2K_2) = 5$ [71] implies that the subgraph induced by Y contains a red/blue $2K_2$ or a green $2K_2$. If y_1y_2 and y_3y_4 form a red/blue $2K_2$, then a red/blue F_3 is formed with center x_1 and blades y_1y_2, y_3y_4, and x_2y_5. If y_1y_2 and y_3y_4 form a green $2K_2$, then a red/green F_3 can be formed with center x_1 and blades y_1y_2, y_3y_4, and x_2y_5.

Consider the possibility that X does not contain any red edges. Since $r^2(P_3) = 3$ [105], it must contain a monochromatic P_3. Without loss of generality, assume that $x_1x_2x_3$ is a blue P_3. If the subgraph induced by Y contains a red or blue edge (say, y_1y_2 is red or blue), then a red/blue F_3 is formed with center x_2 and blades y_1y_2, x_1y_3, and x_3y_4. So, assume that all of the edges in the subgraph induced by Y are green. Then a red/green F_3 is formed with center y_1 and blades x_1y_2, x_2y_3, and x_3y_4.

Case 2: Suppose that $|V(\mathcal{B})| = 4$ and that the edges in $\mathcal{B}$ are red and blue. Denote the vertex sets for the blocks by W, X, Y, and Z. By the pigeonhole principle, some block must contain at least three vertices, and we assume it is the block with vertex set Z. Consider the following two subcases.

Subcase 2.1: Assume that some block only contains a single vertex (say, $W = \{w\}$). If at least one of X and Y has order at least 2 (say, $x_1x_2 \in X$, $y \in Y$, and $z_1, z_2, z_3 \in Z$), then a red/blue F_3 is formed with center w and blades x_1z_1, x_2z_2, and yz_3. So, assume that X and Y both have order 1 and that $X = \{x\}$, $Y = \{y\}$, and $Z = \{z_1, z_2, \ldots, z_6\}$. If the subgraph induced by Z contains a red or blue edge (say, z_1z_2 is red or blue), then a red/blue F_3 is formed with center w and blades z_1z_2, xz_3, and yz_4. So, assume that Z forms a green K_6. Then an F_3 using at most two colors (one of which is green) is formed with center w and blades z_1z_2, z_3z_4, and z_5z_6.

Subcase 2.2: Assume that every block has order at least 2. Without loss of generality, suppose that the vertex sets for the blocks are given by $W = \{w_1, w_2\}$, $X = \{x_1, x_2\}$, $Y = \{y_1, y_2\}$, and $Z = \{z_1, z_2, z_3\}$. Then a red/blue F_3 is formed with center w_1 and blades x_1z_1, x_2z_2, and y_1z_3.

Case 3: Suppose that $|V(\mathcal{B})| = 5$ and that the edges in $\mathcal{B}$ are red and blue. Denote the vertex sets for the blocks by V, W, X, Y, and Z. By the pigeonhole principle, some block has order at least 2, and since there are only nine vertices in total, some block has order 1. Without loss of generality, assume that $V = \{v\}$. If some block has order at least 3 (say, $w \in W$, $x \in X$, $y \in Y$, and $z_1, z_2, z_3 \in Z$), then a red/blue F_3 is formed with center v and blades wz_1, xz_2, and yz_3. So, assume that all blocks other than V have order 2. If $w_1, w_2 \in W$, $x_1, x_2 \in X$, $y_1y_2 \in Y$, and $z_1z_2 \in Z$, then a red/blue F_3 is formed with center v and blades w_1x_1, w_2x_2, and y_1z_1.

Case 4: Suppose that $|V(\mathcal{B})| = 6$ and that the edges in $\mathcal{B}$ are red and blue. Denote the vertex sets for the blocks by U, V, W, X, Y, and Z. By the pigeonhole principle, some block must have order at least 2. Without loss of generality, assume that $u \in U$, $v \in V$, $w \in W$, $x \in X$, $y \in Y$, and $z_1, z_2 \in Z$. Then a red/blue F_3 is formed with center u and blades vw, xz_1, and yz_2.

In all cases, we have shown that a Gallai 3-coloring of K_9 contains an F_3 spanned by edges that use at most two colors. It follows that $gr^3_2(F_3) \leq 9$. □

In the next theorem, we give a more general evaluation of certain weakened Gallai-Ramsey numbers for fans.

Theorem 3.24 ([31]) *For all $n \geq 2$ and $t \geq 2n$, $gr^t_{2n-1}(F_n) = 2n+1$.*

Proof At least $2n+1$ vertices are required to have a fan F_n, so, trivially, it follows that $gr^t_{2n-1}(F_n) \geq 2n+1$. We will prove the reverse inequality by using induction on $n \geq 2$. For the base case $n = 2$, consider a Gallai t-coloring of K_5 and let $\mathcal{B}$ be the base graph, chosen to have minimal order. Theorem 1.22 and Lemma 1.3 imply that $|V(\mathcal{B})| \in \{2,4\}$. In the latter case, some block must contain at least two vertices by the pigeonhole principle. If we select two vertices from one block and a single vertex from each of the other three blocks, the subgraph induced by these five vertices is a K_5 (which contains F_2 as a subgraph) that uses at most three colors.

Next, assume that $|V(\mathcal{B})| = 2$ with the edges joining the two blocks being red. By the pigeonhole principle, one of the blocks must contain at least three vertices. An F_2 that uses at most three colors can then be formed by selecting a single vertex from the smaller block to serve as its center. We then pick four other vertices (at least three of which come from the larger block) and pair them up to form the blades. It follows that $gr^t_3(F_2) \leq 5$, completing the proof of the base case.

Assume that $gr^t_{2n-3}(F_{n-1}) \leq 2n-1$ and consider a Gallai t-coloring of K_{2n+1}. By the inductive hypothesis, there exists an F_{n-1} whose edges use at most $2n-3$ colors. If we select two vertices not contained in the F_{n-1} and let them form and n^{th} blade, at most two more colors can be introduced since rainbow triangles are avoided. The resulting F_n is spanned by edges using at most $2n-1$ colors, implying that $gr^t_{2n-1}(F_n) \leq 2n+1$. □

In particular, Theorem 3.24 implies that $gr^t_3(F_2) = 5$ when $t \geq 4$, and $gr^t_5(F_3) = 7$ when $t \geq 6$. At present, the star-critical weakened Ramsey numbers $gr^{s,t}_*(F_n)$ have not been considered (see Problem 11 in Section 5.1).

3.4.3 Wheels

The smallest wheel that is not a complete graph is W_5. In 2020, Song, Wei, Zhang, and Zhao [198] proved that

$$gr^t(W_5) = \begin{cases} 14 \cdot 5^{(t-2)/2} + 1 & \text{if } t \text{ is even} \\ 28 \cdot 5^{(t-3)/2} + 1 & \text{if } t \text{ is odd,} \end{cases}$$

for all $t \geq 2$. A different proof of this result is also given in [173]. We now consider the weakened Gallai-Ramsey number $gr^t_2(W_5)$, which was determined in 2026 by Jakhar and Moun [136]. In the following theorems, the original proofs omitted some important details, which we fill in below.

Theorem 3.25 ([136]) *For all $t \geq 3$, $gr^t_2(W_5) = t+3$.*

Proof To prove the lower bound, begin with a 2-coloring of K_4 (using colors 1 and 2) in which the subgraph spanned by each color is isomorphic to P_4. Introduce vertex x_1 and join it to the other four vertices with edges in color 3. Denote the resulting 3-coloring of K_5 by $\mathcal{G}_3$. Continue in this manner, sequentially introducing vertices $x_2, x_3, \ldots, x_{t-2}$, coloring all edges from x_i to the existing graph with color $i+2$ (the $t = 5$ case is shown in Figure 3.17). After vertex x_{t-2} has been added in, denote the resulting Gallai t-coloring of K_{t+2} by $\mathcal{G}_t$. If a W_5-subgraph uses at least three vertices from $X := \{x_1, x_2, \ldots, x_{t-2}\}$ then it necessarily has edges in at least three colors. If a W_5-subgraph uses at most two vertices from X, then it must use at least three vertices from the original K_4, and hence includes colors 1 and 2. As there are only four vertices in the original K_4, it must include at least

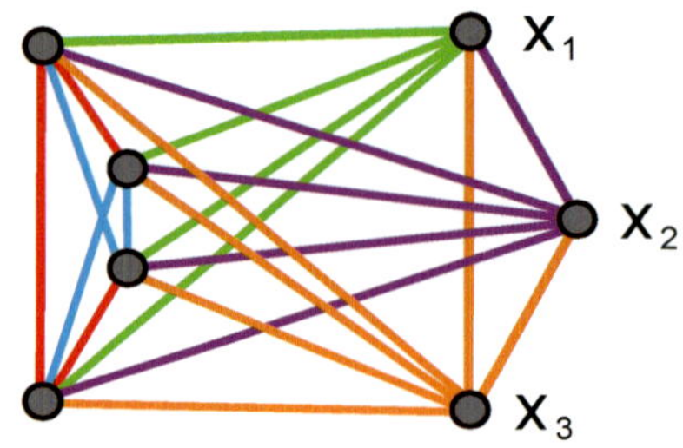

FIGURE 3.17: A Gallai 5-coloring of K_7 that avoids a W_5-subgraph spanned by edges using at most two colors.

one vertex from X, forcing a third color to be included. Hence, there does not exist a W_5 spanned by edges using at most two colors, and so it follows that $gr_2^t(W_5) \geq t + 3$.

To prove the reverse inequality, we will first prove that if a Gallai t-coloring of K_{t+3} avoids a W_5 spanned by edges using at most two colors, then it contains a vertex that is only incident with edges in one color. Consider such a coloring and let $\mathcal{B}$ be its base graph, chosen to have minimal order. By Theorem 1.22, $2 \leq |V(\mathcal{B})| \leq 5$ and by Lemma 1.3, $|V(\mathcal{B})| \neq 3$. We handle the cases $|V(\mathcal{B})| = 2, 4$ separately.

Case 1: Suppose that $|V(\mathcal{B})| = 2$ and that the edges joining the two blocks are red. If either block contains only one vertex, then that vertex is only incident with red edges, proving the claim. So, suppose that each block contains at least two vertices (and one of the blocks contains at least three vertices). Let a, b, and c be in one block and d and e be in the other block. Since rainbow triangles are avoided, the subgaph induced by $\{a, b, c\}$ must contain a monochromatic P_3, which we assume is given by abc. Then a W_5 spanned by edges using at most two colors can be formed with b as the center and $adcea$ as the C_4-subgraph.

Case 2: Suppose that $|V(\mathcal{B})| = 4$ and that the edges in $\mathcal{B}$ are red and blue. Also, note that every vertex in $\mathcal{B}$ is incident with edges in both colors since $\mathcal{B}$ was chosen to have minimal order. Label vertex sets for the blocks W, X, Y, and Z. By the pigeonhole principle, some block (say, W) contains at least two vertices. Without loss of generality, suppose that $v, w \in W$, $x \in X$, $y \in Y$, and $z \in Z$. Then a W_5spanned by edges in colors red and blue can be formed with center z and C_4-subgraph $vxwyv$.

Hence, we have shown that if a Gallai t-coloring of K_{t+3} avoids a W_5 spanned by edges using at most two colors, then it contains a vertex that is only incident with edges in one color. Now, we complete the proof of the upper bound by using induction on $t \geq 3$. For the base case $t = 3$, consider a Gallai 3-coloring of K_6 and suppose that vertex a is only incident with red edges. Let b, c, d, e, and f be the other vertices. If the subgraph induced by $X := \{b, c, d, e, f\}$ does not have any red edges, then it is a K_5 spanned by edges usng at most two colors, which includes a W_5-subgraph. So, assume that it contains at least one red edge, and without loss of generality, suppose that bc is red. If we consider the edges that join $\{b, c\}$ to $\{d, e, f\}$, by the pigeonhole principle, there exists two vertices (say, d and e) such that $bdceb$ is a C_4 spanned by edges using at most two colors, one of which is red. This C_4-subgraph, along with vertex a as the center, form a W_5 spanned by edges using at most two colors.

Now suppose that every Gallai $(t-1)$-coloring of K_{t+2} contains a W_5 spanned by edges using at most two colors and consider a Gallai t-coloring of K_{t+3}. Suppose that vertex a is only incident with red edges and denote the other vertices by $b_1, b_2, \ldots, b_{t+2}$. If the subgraph induced by $\{b_1, b_2, \ldots, b_{t+2}\}$ does not contain any red edges, then by the inductive hypothesis, there exists a W_5 spanned by edges using at most two colors. Otherwise, suppose that b_1b_2 is red. By the pigeonhole principle, there exist two vertices (say, b_3 and b_4) such that $b_1b_3b_2b_4b_1$ is a C_4 spanned by edges using at most two colors, one of which is red. This C_4-subgraph, along with vertex a as the center, form a W_5 spanned by edges using at most two colors. It follows that $gr_2^t(W_5) \leq t+3$. □

In addition to evaluating $gr_2^t(W_5)$, Jakhar and Moun [136] also proved that $gr_2^t(W_6) = 2t+2$, for all $t \geq 3$. Due to its length, we omit the complete proof, and instead, provide a proof of the $t = 3$ case.

Theorem 3.26 ([136]) $gr_2^3(W_6) = 8$.

Proof For the lower bound, consider the Gallai 3-coloring of K_7 shown in Figure 3.18. Note that the hub of W_6 has degree 5, and hence, must be incident with at least one green

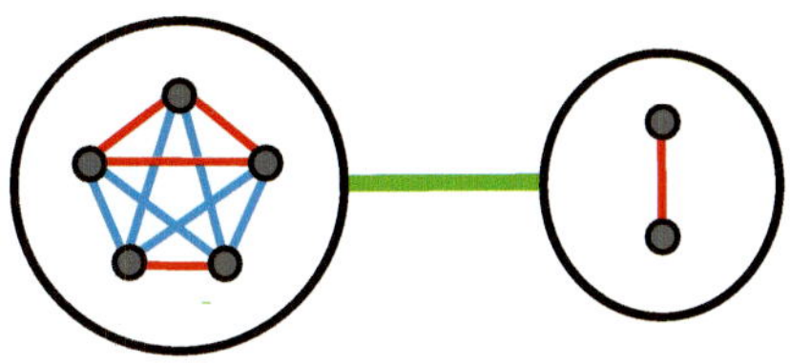

FIGURE 3.18: A Gallai 3-coloring of K_7 that avoids a W_6-subgraph spanned by edges using at most two colors.

edge in any W_6-subgraph. Regardless of which block the hub is in, every W_6-subgraph must use at least one red edge and at least one blue edge from the larger block. It follows that $gr_2^3(W_6) \geq 8$.

To prove the reverse inequality, consider a Gallai t-coloring of K_8 and let $\mathcal{B}$ be its base graph, chosen to have minimal order. By Theorem 1.22 and Lemma 1.3, it follows that $2 \leq |V(\mathcal{B})| \leq 5$ and $|V(\mathcal{B})| \neq 3$. Consider the following cases.

Case 1: Suppose that $|V(\mathcal{B})| = 2$ and the edges joining the two blocks are red. We consider several subcases, based on the orders of the blocks.

Subcase 1.1: Assume that one block contains a single vertex (call it x) and the other block (call its vertex set $Y = \{y_1, y_2, \ldots, y_7\}$) has order 7. Let G_r, G_b, and G_g denote the subgraphs of the larger block spanned by the red, blue, and green edges, respectively. If there exists a red/blue or red/green C_5 in the larger block, then that cycle, along with vertex x, forms a W_6 spanned by edges using at most two colors (one of which is red). It is known that the Turán number for C_5 satisfies $ex(7, C_5) = \lfloor \frac{7^2}{4} \rfloor = 12$ (see Corollary 9 of [80]). So, if the larger block lacks a red/blue C_5 and a red/green C_5, then

$$|E(G_r)| + |E(G_b)| \leq 12 \quad \text{and} \quad |E(G_r)| + |E(G_g)| \leq 12.$$

Adding these inequalities together gives

$$2|E(G_r)| + |E(G_b)| + |E(G_g)| \leq 24,$$

and since there are a total of 21 edges in the larger block, it follows that $|E(G_r)| \leq 3$.

We now work through all of the possibilities for G_r. If $|E(G_r)| \leq 1$, then it is possible to select six vertices from Y that induce a blue/green K_6, which contains W_6 as a subgraph. If $|E(G_r)| = 2$, then the two red edges are either adjacent or disjoint. If the two edges in G_r are adjacent, then without loss of generality, assume that y_1y_2 and y_2y_3 are red. It follows that the subgraph induced by $\{y_1, y_3, y_4, y_5, y_6, y_7\}$ is a blue/green K_6, which contains W_6

as a subgraph. If the two edges in G_r are disjoint, then suppose that they are given by y_1y_2 and y_3y_4. The subgraph induced by $\{y_2, y_3, \ldots, y_7\}$ that is spanned by the blue and green edges is a blue/green $K_6 - e$, which contains W_6 as a subgraph.

Now consider the possiblity that $|E(G_r)| = 3$. If the red edges form a C_3 given by $y_1y_2y_3y_1$, then the subgraph induced by $\{y_2y_3 \ldots, y_7\}$ that is spanned by the blue and green edges is a blue/green $K_6 - e$, which contains W_6 as a subgraph. If the three red edges in G_r form a $K_{1,3}$, then assume that y_1y_2, y_1y_3, and y_1y_4 are red. The subgraph induced by $\{y_2, y_3, y_4, y_5, y_6, y_7\}$ is a blue/green K_6, which contains W_6 as a subgraph. If the three red edges in G_r forma a P_4, then assume that y_1y_2, y_2y_3, and y_3y_4 are red. The subgraph induced by $\{y_1, y_3, y_4, \ldots, y_7\}$ that is spanned by the blue and green edges is a blue/green $K_6 - e$, which contains W_6 as a subgraph.

Next, consider the cases where $|E(G_r)| = 3$ and more than one connected component of G_r has order greater than 1. If G_r contains a red $K_2 \cup P_3$, then assume that y_1y_2, y_3y_4, and y_4y_5 are red. The subgraph induced by $\{y_1, y_2, y_3, y_5, y_6, y_7\}$ that is spanned by the blue and green edges is a blue/green $K_6 - e$, which contains a W_6 as a subgraph. If G_r contains a red $3K_2$, then assume that y_1y_2, y_3y_4, and y_5y_6 are red. The subgraph induced by $\{y_2, y_3, \ldots, y_7\}$ that is spanned by the blue and green edges is a blue/green $K_6 - E(2K_2)$ (i.e., a K_6 with two disjoint edges deleted), which contains W_6 as a subgraph.

Subcase 1.2: Assume that one block has order 2 and the other block has order 6. Let the vertex sets for the blocks be given by $X = \{x_1, x_2\}$ and $Y = \{y_1, y_2, \ldots, y_6\}$. Let G_r, G_b, and G_g denote the subgraphs of the larger block spanned by the red, blue, and green edges, respectively. Without loss of generality, suppose that x_1x_2 is either red or blue. If the subgraph induced by Y contains a red/blue C_4, then a red/blue W_6 exists. For example, if $y_1y_2y_3y_4y_1$ is a red/blue C_4, then a red/blue W_6 is formed with hub x_1 and cycle $x_2y_1y_2y_3y_4x_2$. It is known that $ex(6, C_4) = 7$ [70], from which it follows that

$$|E(G_r)| + |E(G_b)| \leq 7. \tag{3.5}$$

If the subgraph induced by Y contains a re/green C_5, then a red/green W_6 exists. For example, if $y_1y_2y_3y_4y_5y_1$ is a red/green C_5, then this cycle, along with vertex x_1 as the hub, form a red/green W_6. It is known that $ex(6, C_5) \leq \lfloor \frac{6^2}{4} \rfloor = 9$ [80], from which it follows that

$$|E(G_r)| + |E(G_g)| \leq 9. \tag{3.6}$$

Adding Inequalities (3.5) and (3.6) together results in

$$2|E(G_r)| + |E(G_b)| + |E(G_g)| \leq 16.$$

Since the subgraph induced by Y has size 15, it follows that $|E(G_r)| \leq 1$. So, the subgraph induced by Y that is spanned by the blue and green edges is isomorphic to one of K_6 or $K_6 - e$, both of which contain W_6 as a subgraph.

Subcase 1.3: Assume that each block contains at least three vertices and note that one block must contain at least four vertices. Suppose that one block contains vertices x_1, x_2, and x_3, with x_1x_2 red, and the other block contains vertices y_1, y_2, and y_3. Since rainbow triangles are avoided, the subgraph induced by $\{y_1, y_2, y_3\}$ contains a monochromatic P_3. Suppose that the path $y_1y_2y_3$ is monochromatic. Then a W_6 is formed that is spanned by edges using at most two colors with hub y_2 and cycle $x_1y_1x_2y_3x_3x_1$. Therefore, we can assume that the blocks do not contain any red edges.

A similar argument can be given to handle the case where one block contains a monochromatic P_3 (which is necessarily blue or green) and the other block contains a single edge of the same color. The only way this scenario can be avoided is if one block is a blue complete graph and the other block is a green complete graph. Without loss of generality, suppose that $\{x_1, x_2, x_3\}$ is the vertex set for a blue K_3 in one block and $\{y_1, y_2, y_3, y_4\}$ is the vertex set for a green K_4 in the other block. Then a red/green W_6 is formed with hub y_1 and cycle $y_2x_1y_3x_2y_4y_2$.

Case 2: Suppose that $|V(\mathcal{B})| = 4$ and the edges joining the blocks are red and blue. Consider the case where there exist two blocks having order at least 2. Suppose that W, X, Y, and Z are the vertex sets for the blocks and $w_1, w_2 \in W$, $x_1, x_2 \in X$, $y \in Y$, and $z \in Z$. Then a red/blue W_6 is formed with hub z and cycle $yw_1x_1w_2x_2y$. So, assume that only one block has more than one vertex. Without loss of generality, assume that $W = \{w_1, w_2, w_3, w_4, w_5\}$, $X = \{x\}$, $Y = \{y\}$, and $Z = \{z\}$ are the vertex sets for the blocks. If the subgraph induced by W contains a red or blue edge (say, w_1w_2 is red or blue), then a red/blue W_6 is formed (with hub z and cycle $w_1xw_3yw_2w_1$). So, the subgraph induced by W must be a green K_5. Then a W_6 spanned by edges using at most two colors (one of which is green) is formed with hub w_1 and cycle $w_2w_3w_4w_5xw_2$.

Case 3: Suppose that $|V(\mathcal{B})| = 5$ and the edges joining the blocks are red and blue. By the pigeonhole principle, some block contains at least two vertices. Assume that U, W, X, Y, and Z are the vertex sets for the blocks and that $u_1, u_2 \in U$, $w \in W$, $x \in X$, $y \in Y$, and $z \in Z$. Then a red/blue W_6 is formed with hub z and cycle $u_1wu_2xyu_1$.

In all cases, we find that there is a W_6 spanned by edges using at most two colors. It follows that $gr_2^3(W_6) \leq 8$. □

The star-critical analogue of the weakened Gallai-Ramsey number for wheels has not yet been studied (see Problem 12 in Section 5.1).

3.4.4 Books and Complete Bipartite Graphs

Next, we consider weakened Gallai-Ramsey numbers for books. The smallest book that is not a complete graph is $B_2 \cong K_4 - e$. In 2019, Zou, Mao, Magnant, Wang, and Ye [215] proved that for all $t \geq 2$ and $2 \leq m \leq 5$,

$$gr^t(B_m) = \begin{cases} (r^2(B_m) - 1) \cdot 5^{(t-2)/2} + 1 & \text{if } t \text{ is even} \\ 2(r^2(B_m) - 1) \cdot 5^{(t-3)/2} + 1 & \text{if } t \text{ is odd.} \end{cases}$$

They also proved general upper and lower bounds for $gr^t(B_m)$ for larger values of m. In this subsection, we consider $gr_2^3(B_m)$ for a few small values of m. We also determine the values of $gr_2^3(K_{2,3})$, $gr_2^3(K_{2,4})$, and $gr_2^3(K_{3,3})$. First, determining $gr_2^3(B_2)$ requires the following theorem.

Theorem 3.27 *Up to isomorphism or a reordering of the colors, the only two Gallai 3-colorings of K_4 that use all three colors are shown in Figure 3.19.*

Proof Consider an exact Gallai 3-coloring of K_4 with base graph $\mathcal{B}$, chosen to have minimal order. By Theorem 1.22 and Lemma 1.3, $|V(\mathcal{B})| = 2$. Suppose that the edges joining the two blocks are red. If both blocks have order 2, then each block contains a single edge, one of which is blue and one of which is green. This corresponds with the first image in Figure 3.19. If one block has a single vertex and the other block contains the other three vertices, then the K_3-subgraph in the larger block must be blue/green. Without loss of generality, suppose that it contains two blue edges and one green edge. This corresponds with the second image in Figure 3.19. □

Note that the two Gallai 3-colorings shown in Figure 3.19 have base graphs isomorphic to a red K_2. In both graphs, removing the green edge produces a B_2 spanned by edges using at most two colors. Since at least four vertices are required to have a B_2, and every Gallai

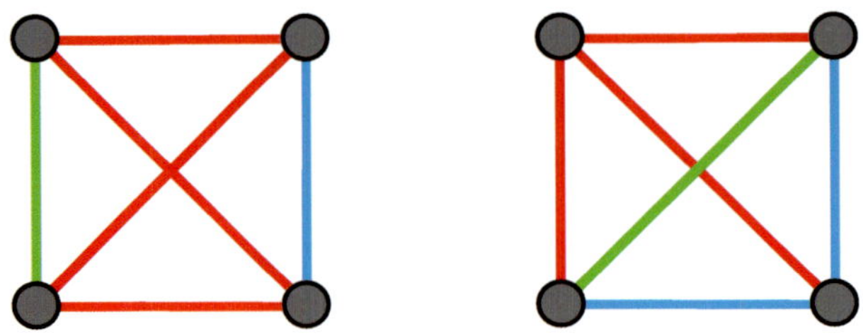

FIGURE 3.19: Two Gallai 3-colorings of K_4.

t-coloring of K_4 uses at most three colors by Lemma 1.1 (t-colorings are not assumed to be surjective), we obtain the following corollary.

Corollary 3.2 *For all* $t \geq 3$, $gr_2^t(B_2) = 4$.

Now we consider larger books.

Theorem 3.28 ([136]) $gr_2^3(B_3) = 6$.

Proof The lower bound follows from the Gallai 3-coloring of K_5 given in Figure 3.20. Every B_3-subgraph can be formed by removing three edges in the shape of a triangle from

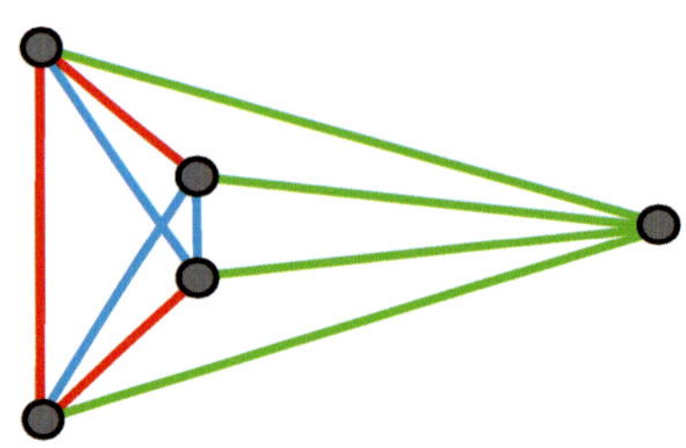

FIGURE 3.20: A Gallai 3-coloring of K_5 that avoids a B_3-subgraph spanned by edges using at most two colors.

K_5. Regardless of which three edges in the shape of a triangle are removed, all three colors still appear in the resulting B_3. Hence, this Gallai 3-coloring of K_5 lacks a B_3 spanned by edges using at most two colors. It follows that $gr_2^3(B_3) \geq 6$.

To prove the reverse inequality, consider a Gallai 3-coloring of K_6 and let $\mathcal{B}$ be its base graph, chosen to have minimal order. By Theorem 1.22, $2 \leq |V(\mathcal{B})| \leq 4$, and by Lemma 1.3, $|V(\mathcal{B})| \neq 3$. We consider the following cases.

Case 1: Suppose that $|V(\mathcal{B})| = 2$ and that the edges joining the two blocks are red. If each block has order at least two, then one block will have order at least three. Let a and b be in one block and c, d, and e be in the other block. Then regardless of the color of edge ab, a B_3 spanned by edges using at most two colors is formed with spine ab and pages c, d, and e.

So, suppose that one block contains a single vertex (call it a) and the other block contains the other five vertices (call them b, c, d, e, and f). Since $r^3(P_3) = 5$, the larger block contains a monochromatic P_3. If the P_3-subgraph is red (say, bcd is a red P_3), then a

B_3 can be formed with ac as the spine and b, d, and e as the pages. Regardless of the color of edge ce, the B_3 is spanned by edges using at most two colors. If the P_3-subgraph is not red (say, bcd is a blue P_3), then if either or ce or cf is blue, a B_3 spanned by edges using at most two colors exists. For example, if ce is blue, the the B_3 with ac as the spine and b, d, and e uses at most two colors. So, assume that ce and cf are green. In order to avoid a rainbow triangle, edges be, bf, de, and df must be blue or green. If bd is blue or green, then a blue/green B_3 is formed with bd as the spine and c, e, and f as the pages. So, assume that bd is red. If be is blue, then de must also be blue, and a red/blue B_3 can be formed with spine bd and pages a, c, and e. A similar argument holds if bf is blue. So, assume that be, bf, de, and df are all green. Then a red/green B_3 can be formed with spine bd and pages a, e, and f.

Case 2: Suppose that $|V(\mathcal{B})| = 4$ and that the edges in $\mathcal{B}$ are red and blue. Each vertex in $\mathcal{B}$ must be incident with edges in both colors due to the minimal assumption on $\mathcal{B}$. If any block contains a red or blue edge, then that edge is the spine of a red/blue B_3 with pages formed by selecting a single vertex from each of the other three blocks. So, all edges within the blocks must be green. We consider two subcases.

Subcase 2.1: Suppose that one block (call its vertex set W) has three vertices and each of the other blocks (call their vertex sets X, Y, and Z) contain a single vertex. By the pigeonhole principle, the edges joining the vertices in W to two of the other blocks must have the same color. Without loss of generality, suppose that the edges joining the vertices in W to the vertices in X and Y are all red. Then a red/green B_3 can be formed with the spine being contained in W, and the pages consisting of the other three vertices in W, X, and Y.

Subcase 2.2: Suppose that two of the blocks (call their vertex sets W and X) have two vertices each and the other two blocks (call their vertex sets Y and Z) each contain a single vertex. Without loss of generality, suppose that the edges joining the vertices in W and X are red. If W or X join to Y or Z with a red edge, then a red/green B_3 can be formed. For example, if W joins to Y with red edges, then a red/green B_3 can be formed with the vertices in W making up the spine and the pages consist of the vertices in X and Y. So, suppose that all of the edges joining W and X to Y and Z are blue. Regardless of the color of the edge joining Y and Z, this edge then serves as the spine of a B_3 spanned by edges using at most two colors with any of the other three vertices as pages.

In all cases, we have shown that a Gallai 3-coloring of K_6 contains a B_3 spanned by edges using at most two colors. It follows that $gr_2^3(B_3) \leq 6$. □

Note that Figure 3.20 also avoids a $K_{2,3}$ spanned by edges using at most two colors. Since $K_{2,3}$ is a subgraph of B_3, it follows from Theorem 3.28 that $gr_2^3(K_{2,3}) \leq gr_2^3(B_3)$, leading to the following corollary.

Corollary 3.3 ([136]) $gr_2^3(K_{2,3}) = 6$.

In the case of the book B_4, the proof of the following theorem in [136] contains a mistake. The mistake is corrected in the proof given below by appealing to Theorem 1.22 a second time.

Theorem 3.29 ([136]) $gr_2^3(B_4) = 7$.

Proof To prove the lower bound, begin with a critical coloring for $r(K_3, K_3)$ in colors red and blue. That is, consider a K_5 in which the subgraph spanned by the edges in each of the two colors are C_5-subgraphs. Introduce a sixth vertex and join it to all of the vertices in the K_5 with green edges (see Figure 3.21). Every B_4-subgraph can be formed by removing four edges in the shape of a K_4-subgraph from K_6. Regardless of which six edges in the

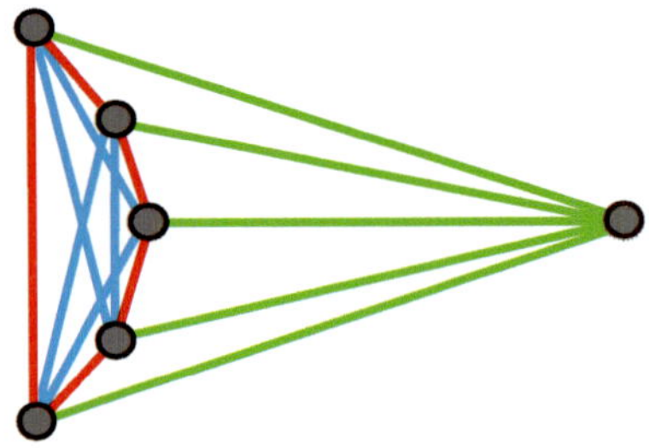

FIGURE 3.21: A Gallai 3-coloring of K_6 that avoids a B_4-subgraph spanned by edges using at most two colors.

shape of a K_4 are removed, all three colors still appear in the resulting B_4. Hence, this Gallai 3-coloring of K_6 lacks a B_4 spanned by edges using at most two colors. It follows that $gr_2^3(B_4) \geq 7$.

To prove the reverse inequality, consider a Gallai 3-coloring of K_7 and let $\mathcal{B}$ be its base graph, chosen to have minimal order. By Theorem 1.22, $2 \leq |V(\mathcal{B})| \leq 5$, and by Lemma 1.3, $|V(\mathcal{B})| \neq 3$. We consider the following cases.

Case 1: Suppose that $|V(\mathcal{B})| = 2$ and the edges joining the two blocks are red. If each block has order at least two, then one block will have order at least four. Let a and b be in one block and c, d, e, and f be in the other block. Then regardless of the color of edge ab, a B_4 spanned by edges using at most two colors is formed with spine ab and pages c, d, e, and f.

So, suppose that one block contains a single vertex and the other block contains the other six vertices. Denote the vertex sets for the two blocks by $X = \{a\}$ and $Y = \{b, c, d, e, f, g\}$. Of course, the subgraph induced by Y is a Gallai 3-coloring of K_6 using all three colors (otherwise, it is a K_6 spanned by edges using at most two colors, and B_4 is a subgraph of K_6). Let $\mathcal{B}'$ be the base graph for this Gallai coloring, chosen to have minimal order. By Theorem 1.22 and Lemma 1.3, $2 \leq |V(\mathcal{B}')| \leq 5$ and $|V(\mathcal{B}')| \neq 3$, leading to the following subcases.

Subcase 1.1: Suppose that $|V(\mathcal{B}')| = 2$, and the edges joining the two blocks in $\mathcal{B}'$ are red. Without loss of generality, suppose that b is in the smaller of the two blocks in $\mathcal{B}'$ and that c, d, and e are in the other block. Then regardless of the color of the edge bf, a B_4 spanned by edges using at most two colors (one of which is red) is formed with spine ab and pages c, d, e, and f.

If the edges joining the two blocks in $\mathcal{B}'$ are blue or green (say, blue), then consider the possibility that one block has order at most 2. Without loss of generality, suppose that b is in the smaller block and c, d, e, and f are in the other block. Then a red/blue B_4 is formed with spine ab and pages c, d, e, and f. So, assume that each of the blocks in $\mathcal{B}'$ have order three. Without loss of generality, assume that b, c, and d are in one block and e, f, and g are in the other block. If any edge in either block is red or blue, then a red/blue B_4 is formed. For example, if bc is red or blue, then a red/blue B_4 is formed with spine ab and pages c, e, f, and g. So, all of the edges in the blocks in $\mathcal{B}'$ must be green. However, a blue/green B_4 can then be formed with spine bc and pages d, e, f, and g.

Subcase 1.2: Suppose that $|V(\mathcal{B}')| = 4$. Denote the vertex sets for the blocks by W, X, Y, and Z. Two blocks (say, W and X) contain only one vertex each (otherwise, there would be at least seven vertices). Let $b \in W$, $c \in X$, and $d, e, f, g \in Y \cup Z$. Then a B_4 spanned by edges using at most two colors is formed with spine bc and pages d, e, f, and g.

Subcase 1.3: Suppose that $|V(\mathcal{B}')| = 5$. Four of the blocks must contain a single vertex, and one block contains two vertices. Suppose that f and g are in the same block, while b, c,

d, and e appear in their own blocks. Then a B_4 spanned by egdes using at most two colors is formed with spine bc and pages c, d, e, and f.

Case 2: Suppose that $|V(\mathcal{B})| = 4$. The proof is similar to the proof of Subcase 1.2, so the details are left to the reader.

Case 3: Suppose that $|V(\mathcal{B})| = 5$. The proof is similar to the proof of Subcase 1.3, so the details are left to the reader.

In all cases, we have shown that a Gallai 3-coloring of K_7 contains a B_4 spanned by edges using at most two colors. It follows that $gr_2^3(B_4) \leq 7$. □

If we remove the edges from any K_4-subgraph of the 3-coloring of K_6 given in Figure 3.21, there exists at least two edges remaining in each color. Removing one more edge does not delete one of the colors, so every $K_{2,4}$-subgraph still uses all three colors. Since $K_{2,4}$ is a subgraph of B_4, the following corollary is an immediate consequence of Theorem 3.29.

Corollary 3.4 ([136]) $gr_2^3(K_{2,4}) = 7$.

Despite the similarity of the proofs of Theorems 3.28 and 3.29, they do not easily generalize to a full evalation of $gr_2^3(B_n)$ (see Problem 13 in Section 5.1). We conclude this section by evaluating $gr_2^3(K_{3,3})$.

Theorem 3.30 ([136]) $gr_2^3(K_{3,3}) = 8$.

Proof For the lower bound, consider a red K_2 and replace one of its vertices with a red K_2 and its other vertex with a critical coloring for $r(K_3, K_3)$ in colors blue and green (see Figure 3.22). The blue/green edges only span connected graph of order five and if a

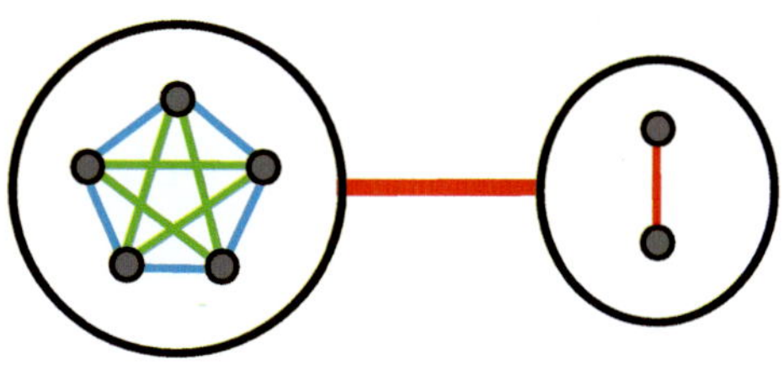

FIGURE 3.22: A Gallai 3-coloring of K_7 that avoids a $K_{3,3}$-subgraph spanned by edges using at most two colors.

$K_{3,3}$-subgraph existed that was either red/blue or red/green, one of its partite sets would include three vertices from the larger block. Any such selection uses both blue and green edges. It follows that $gr_2^3(K_{3,3}) \geq 8$.

To prove the reverse inequality, consider a Gallai 3-coloring of K_8. Let $\mathcal{B}$ be the base graph, chosen to have minimal order. Theorem 1.22 and Lemma 1.3 imply that $2 \leq |V(\mathcal{G})| \leq 5$ and $|V(\mathcal{B})| \neq 3$. Consider the following cases.

Case 1: Suppose that $|V(\mathcal{B})| = 2$ and that the edges joining the two blocks in $\mathcal{B}$ are red. If each block has order at least three, then selecting three vertices from each block produces a red $K_{3,3}$. So, assume that one block has at most two vertices. Without loss of generality, let X and Y be the vertex sets for the two blocks and suppose that $|X| \leq 2$. We consider two subcases.

Subcase 1.1: Assume $|X| = 2$ and $|Y| = 6$. In [116], it was proved that $gr^3(K_{1,3}) = 6$, so Y contains a monochromatic $K_{1,3}$. Suppose that $X = \{a, b\}$, c is the center of the $K_{1,3}$, and d, e, and f are the leaves of the $K_{1,3}$. Then a $K_{3,3}$ spanned by edges using at most two colors is formed with partite sets $\{a, b, c\}$ and $\{d, e, f\}$.

Subcase 1.2: Assume $|X| = 1$ and $|Y| = 7$. Let $X = \{a\}$ and $Y = \{b, c, d, e, f, g, h\}$. Some edge in Y must be red (otherwise, the subgraph induced by Y is a blue/green K_7, which contains $K_{3,3}$ as a subgraph). Without loss of generality, suppose that bc is red. By the pigeonhole principle, at least three of the vertices d, e, f, g, and h join to b and c via either red/blue edges or red/green edges. Without loss of generality, assume that all edges joining b and c to d, e, and f are red/blue. Then a red/blue $K_{3,3}$ is formed with partite sets $\{a, b, c\}$ and $\{d, e, f\}$.

Case 2: Suppose that $|V(\mathcal{B})| \in \{4, 5\}$. If some block has order at least 3, then selecting three vertices from that block and three vertices from other blocks produces a $K_{3,3}$ spanned by edges using at most two colors. So, assume that each block has order at most 2, and note that at least three blocks have order at least 2. From three such blocks, choose a single vertex for one partite set and the other vertex for the other partite set. The result is a $K_{3,3}$ spanned by edges using at most two colors.

In all cases, we have shown that a Gallai 3-coloring of K_8 contains a $K_{3,3}$ spanned by edges using at most two colors. It follows that $gr_2^3(K_{3,3}) \leq 8$. □

The star-critical analogues of the weakened Gallai-Ramsey numbers for books and complete bipartite graphs have not yet been studied (see Problem 13 in Section 5.1).

3.5 Weakened Size Ramsey Numbers

Given a graph G and $1 \leq s < t$, let $\mathcal{C}_s^t(G)$ be the set of all graphs H such that every t-coloring of H contains a subgraph that is isomorphic to G that uses at most s colors. The *weakened size Ramsey number* $\hat{r}_s^t(G)$ is defined by

$$\hat{r}_s^t(G) := \min\{|E(H)| \mid H \in \mathcal{C}_s^t(G)\}.$$

For any $s_1 \leq s_2$, note that

$$\hat{r}_{s_1}^t(G) \geq \hat{r}_{s_2}^t(G),$$

from which it follows that such numbers are finite since $\hat{r}_1^t(G) = \hat{r}^t(G)$. Also, observe that

$$\hat{r}_s^t(G) \leq \binom{r_s^t(G)}{2} = |E(K_{r_s^t(G)})|, \tag{3.7}$$

since every t-coloring of $K_{r_s^t(G)}$ is guaranteed to have a copy of G whose edges use at most s colors. The following theorem generalizes Theorem 1 of [84], where the result is originally credited to Vašek Chvátal.

Theorem 3.31 *For all* $t \geq 2$, $1 \leq s < t$, *and* $n \geq 2$,

$$\hat{r}_s^t(K_n) = \binom{r_s^t(K_n)}{2}.$$

Proof Let $p = r_s^t(K_n)$. The upper bound $\hat{r}_2^t(K_5) \leq \binom{p}{2}$ follows from Inequality (3.7). To prove the lower bound let $H \in \mathcal{C}_s^t(K_n)$. If $|V(H)| < p$, then H is a subgraph of K_{p-1}. Since $r_2^t(K_n) = p$, there exists a t-coloring of K_{p-1} in which every K_n-subgraph uses at least $s+1$ colors. This property is also true of the subgraph H. Hence, we see that $|V(H)| \geq p$.

Now assume that $|V(H)| \geq p$ and $|E(H)| < \binom{p}{2}$. Writing $|V(H)| = p + k$, we will prove by induction on $k \geq 0$ that there exists a t-coloring of H in which every K_n-subgraph uses at least $s+1$ colors. For the $k=0$ case, $H \not\cong K_p$, from which it follows that H is a subgraph of $K_p - e$. In the proof of Theorem 3.8, we constructed a t-coloring of $K_p - e$ in which every K_n-subgraph used at least $s+1$ colors. This property is also true of the subgraph H.

Now assume that for some $k \geq 0$, every graph with order $p+k$ and size less than $\binom{p}{2}$ has a t-coloring in which every K_n-subgraph uses at least $s+1$ colors. Let H be a graph of order $p+k+1$ such that $|E(H)| < \binom{p}{2}$ and note that $H \not\cong K_{p+k+1}$. Select two vertices $x, y \in V(H)$ that are not adjacent. By the inductive hypothesis, the subgraph H_x induced by $V(H)\setminus\{x\}$ has a t-coloring in which every K_n-subgraph uses at least $s+1$ colors. Extend such a t-coloring to H by giving edge xz the same color as yz, for all $z \in V(H)\setminus\{x, y\}$. The subgraph H_y induced by $V(H)\setminus\{y\}$ is isomorphic to H_x, and hence, every K_n-subgraph uses at least $s+1$ colors. Since x and y are not adjacent in H, every K_n-subgraph is contained in H_x or H_y. It follows that $\hat{r}_s^t(K_n) \geq \binom{p}{2}$. □

In 2023, Mao [170] introduced *size Gallai-Ramsey numbers* as an analogue of size Ramsey numbers in which only Gallai t-colorings were considered. This concept was further generalized by Wang and Zhang [208] in 2025 when they defined weakened size Gallai-Ramsey numbers. For $1 \leq s < t$, let $\mathcal{GC}_s^t(G)$ be the set of all graphs H such that every Gallai t-coloring of H contains a subgraph that is isomorphic to G that uses at most s colors. The *weakened size Gallai-Ramsey number* $\widehat{gr}_s^t(G)$ is then defined by

$$\widehat{gr}_s^t(G) = \min\{|E(H)| \mid H \in \mathcal{GC}_s^t(G)\}.$$

For any $s_1 \leq s_2$, note that

$$\widehat{gr}_{s_1}^t(G) \geq \widehat{gr}_{s_2}^t(G),$$

from which it follows that such numbers are finite since $\widehat{gr}_1^t(G) = \widehat{gr}^t(G)$. Of course, their finiteness also follows from the observation

$$\widehat{gr}_s^t(G) \leq \binom{gr_s^t(G)}{2} = |E(K_{gr_s^t(G)})|.$$

The proof given for Theorem 3.31 also holds for Gallai colorings, from which it follows that for all $t \geq 2$, $1 \leq s < t$, and $n \geq 2$,

$$\widehat{gr}_s^t(K_n) = \binom{gr_s^t(K_n)}{2}.$$

At present, weakened size Ramsey numbers and weakened size Gallai-Ramsey numbers have not been studied for graphs that are not complete (see Problem 21 in Section 5.1).

3.6 Weakened Hypergraph Ramsey Numbers

Define the *weakened (k-uniform) hypergraph Ramsey number* $r_s^t(H;k)$ to be the least $n \in \mathbb{N}$ such that every t-coloring of $K_n^{(k)}$ contains a subhypergraph isomorphic to H that is spanned by hyperedges that use at most s colors, where $1 \leq s \leq t-1$. In particular, note that $r_s^t(H;2) = r_s^t(H)$.

In [49], it was shown that

$$7 \leq r_2^3(K_4^{(3)};3) \leq 8,$$

and in [114], Gyárfás completed the evaluation, resulting in the following theorem. In the proof, we will need to use a known value of a variation of the Turán number introduced in Section 1.1. Define the *generalized Turán number* $T(n)$ to be the least positive integer m such that there exist m 3-tuples in an n-element set such that every 4-element subset contains at least one of the m 3-tuples.

Theorem 3.32 ([49], [114]) $r_2^3(K_4^{(3)};3) = 7$.

Proof First we describe a 3-coloring of $K_6^{(3)}$ in which every $K_4^{(3)}$-subhypergraph contains hyperedges in all three colors. Denote the vertices by a, b, c, d, e, and f and color the hyperedges as follows:

Red: $abc,\ abd,\ aef,\ cde,\ adf,\ bcf,\ bef,$

Blue: $bcd,\ ade,\ abe,\ acf,\ bdf,\ cef,$

Green: $acd,\ ace,\ bce,\ bde,\ abf,\ cdf,\ def.$

One can manually check that each $K_4^{(3)}$-subhypergraph contains red, blue, and green hyperedges. It follows that $r_2^3(K_4^{(3)};3) \geq 7$.

To prove the reverse inequality, consider a 3-coloring of $K_7^{(3)}$, which contains 35 hyperedges. Some color class (say, red) contains at most 11 hyperedges, otherwise, there would have to be at least 3 hyperedges. The generalized Turán number is known to satisfy $T(7) = 12$ [143]. Hence, there exists a 3-coloring of $K_7^{(3)}$ in which some $K_4^{(3)}$-subhypergraph does not contain any red hyperedges. It follows that $r_2^3(K_4^{(3)};3) \leq 7$. □

Theorem 3.33 ([49]) $r_3^4(K_4^{(3)};3) = 5$.

Proof If we give each of the four hyperedges in $K_4^{(3)}$ their own color, we obtain the lower bound $r_3^4(K_4^{(3)};3) \geq 5$. To prove the reverse inequality, suppose that $r_3^4(K_4^{(3)};3) > 5$. Then there exists a 4-coloring of $K_5^{(3)}$ in which every $K_4^{(3)}$ is rainbow (i.e., no two edges are the same color). There are five $K_4^{(3)}$-subhypergraphs and each hyperedge appears in exactly two of them. If two hyperedges have the same color, then they have exactly one vertex in common. Suppose that hyperedges vwx and xyz are both the same color (say, red). Then the only other hyperedges that intersect vwx at exactly one vertex are vyz and wyz, both of which intersect xyz at two vertices. It follows that there are at most two hyperedges of a given color. As there are four colors, this only allows for eight hyperedges in total, leaving two uncolored. So, no such 4-coloring of $K_5^{(3)}$ exists, and it follows that $r_3^4(K_4^{(3)};3) \leq 5$. □

For more general weakened hypergraph Ramsey numbers of the form $r_k^{k+1}(K_{k+1}^{(k)};k)$, Gyárfás [114] also proved the following theorem.

Theorem 3.34 ([114]) *For all* $k \geq 1$, $r_k^{k+1}(K_{k+1}^{(k)};k) \leq 2k+1$.

Proof Consider a $(k+1)$-coloring of $K_{2k+1}^{(k)}$ and select a set S of $k-1$ vertices. There are $r+2$ hyperedges that include the vertices in S (corresponding with each of the vertices in $V(K_{2k+1}^{(k)}) \setminus S$). By the pigeonhole principle, at least two of these hyperedges must receive the same color. Let e_1 and e_2 be two distinct hyperedges that include S and are the same color. Assume that $e_1 \setminus S = \{x\}$ and $e_2 \setminus S = \{y\}$. Then the subhypergraph induced by $S \cup \{x,y\}$ is a $K_{k+1}^{(k)}$ spanned by hyperedges using at most k colors. □

Relating the value of $r_k^{k+1}(K_{k+1}^{(k)};k)$ to the existence problem of large sets of t-designs, Gyárfás [114] showed that $r_4^5(K_5^{(4)};4) = 7$, $r_6^7(K_7^{(6)};6) > 9$, and

$$r_k^{k+1}(K_{k+1}^{(k)};k) = \begin{cases} k+2 & \text{if } k \text{ is odd} \\ k+3 & \text{if } k \equiv 2 \pmod{6}. \end{cases}$$

For $1 \leq c < k \leq n$, define the *k-uniform star* $S_{n,c}^{(k)}$ to be the k-uniform hypergraph of order n and size $\binom{n-c}{k-c}$ in which the hyperedge set consists of all hyperedges that share a common c-element set of vertices, called the *center* of $S_{n,c}^{(k)}$. The other $n - c$ vertices are called the *leaves* of $S_{n,c}^{(k)}$. For example, the 4-uniform stars $S_{7,3}^{(4)}$ and $S_{5,2}^{(4)}$ are shown in Figure 3.23. When $k = 2$, note that $S_{n,1}^{(2)} = K_{1,n-1}$.

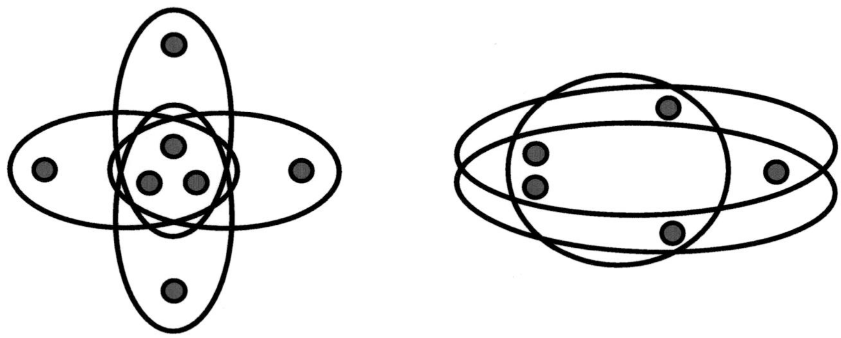

FIGURE 3.23: The 4-uniform stars $S_{7,3}^{(4)}$ and $S_{5,2}^{(4)}$.

Given a t-coloring $f : E(K_n^{(k)}) \longrightarrow \{1, 2, \ldots, t\}$, where $1 \leq k < n$, define the *complementary t-coloring* $f^* : E(K_n^{(n-k)}) \longrightarrow \{1, 2, \ldots, t\}$ by

$$f^*(e) = f(V(K_n^{(k)}) \setminus e), \quad \text{where } e \in E(K_n^{(n-k)}).$$

For a complete subhypergraph $K_m^{(k)}$ of $K_n^{(k)}$, the hyperedges in the $K_m^{(k)}$ correspond with the hyperedges in an $S_{n,n-m}^{(n-k)}$ in $K_n^{(n-k)}$ under a complementary coloring. The vertices that make up the center of the star are precisely those not included in the $K_m^{(k)}$. It follows that a t-coloring f of $K_n^{(k)}$ contains a $K_m^{(k)}$-subhypergraph spanned by hyperedges using at most s colors if and only if the complementary coloring f^* of $K_n^{(n-k)}$ contains an $S_{n,n-m}^{(n-k)}$-subhypergraph spanned by hyperedges using at most s colors. The following theorem immediately follows.

Theorem 3.35 *For all* $1 \leq c < k \leq n$, *if* $r_s^t(K_m^{(k)}; k) = n$, *then* $r_s^t(S_{n,n-m}^{(n-k)}; n-k) = n$.

From this theorem, Theorems 3.32 and 3.33 imply that

$$r_2^3(S_{7,3}^{(4)}; 4) = 7 \quad \text{and} \quad r_3^4(S_{5,1}^{(2)}; 2) = r_3^4(K_{1,4}) = 5,$$

respectively. Note that the second result agrees with Theorem 3.9. The result $r_4^5(K_5^{(4)}; 4) = 7$ proved in [114] implies that $r_4^5(S_{7,2}^{(3)}; 3) = 7$.

4

Weakened Ramsey Theory on the Integers

In this chapter, we consider weakened versions of a few numbers that arise in Ramsey theory involving sets of integers. Specifically, the weakened versions of Schur numbers, Rado numbers, and rainbow Schur numbers are defined, and several exact values of each one are determined.

4.1 Weakened Schur Numbers

Recall the following systems of equations/inequalities that were defined in Section 1.3:

$$\mathcal{E}_m: \ x_1 + x_2 + \cdots + x_{m-1} = x_m \qquad \text{and} \qquad \mathcal{E}_m^*: \begin{cases} x_1 + x_2 + \cdots + x_{m-1} = x_m \\ x_1 < x_2 < \cdots < x_m. \end{cases}$$

When considering a weakened version of a Schur number, we run into an issue with the equation $\mathcal{E}_m$. Specifically, the solution

$$\underbrace{1 + 1 + \cdots + 1}_{m-1 \text{ terms}} = m - 1$$

to $\mathcal{E}_m$ uses at most two colors, for all $t \geq 2$. To address this issue, we must use the system $\mathcal{E}_m^*$ instead. So, the weakened Schur numbers we will define are really a weakened version of a strict Schur number. It is also worth noting that we consider exact t-coorings for a similar reason.

Define the *weakened Schur number* $\widehat{S}_s^t(m)$ to be the least $n \in \mathbb{N}$ such that every exact t-coloring of $[1, n]$ contains a solution to $\mathcal{E}_m^*$ that uses at most s of the t colors, where $1 \leq s < \min\{m, t\}$. If we let $1 \leq s_1 \leq s_2 < \min\{m, t\}$, then every solution to $\mathcal{E}_m^*$ that uses at most s_1 of the colors, necessarily uses at most s_2 of the colors. It follows that $\widehat{S}_{s_1}^t(m) \geq \widehat{S}_{s_2}^t(m)$. We begin with a general result.

Theorem 4.1 (**[38]**) *For all* $t \geq 3$,

$$\widehat{S}_{t-1}^t(t) = \frac{t(t-1)}{2} + 2.$$

Proof To prove the lower bound, we provide an exact t-coloring of $[1, t(t-1)/2 + 1]$ that lacks a solution to $\mathcal{E}_t^*$ that uses at most $t-1$ colors. The only equations in $\mathcal{E}_t^*$ we must consider are

$$1 + 2 + \cdots + (t-2) + (t-1) = \frac{t(t-1)}{2} \tag{4.1}$$

DOI: 10.1201/9781003498865-4

and

$$1 + 2 + \cdots + (t-2) + t = \frac{t(t-1)}{2} + 1. \tag{4.2}$$

Consider the t-coloring in which $1, 2, \ldots, t$ receive distinct colors, $\frac{t(t-1)}{2}$ receives the same color as t, $\frac{t(t-1)}{2} + 1$ receives the same color as $t-1$, and all other numbers receive any of the t colors. Inequalities (4.1) and (4.2) are both rainbow, from which it follows that

$$\widehat{S}^t_{t-1}(t) \geq \frac{t(t-1)}{2} + 2.$$

To prove the reverse inequality, consider an exact t-coloring of $[1, t(t-1)/2+2]$. Besides Inequalities (4.1) and (4.2), we must also consider the inequalities

$$1 + 2 + \cdots + (t-2) + (t+1) = \frac{t(t-1)}{2} + 2 \tag{4.3}$$

and

$$1 + 2 + \cdots + (t-3) + (t-1) + t = \frac{t(t-1)}{2} + 2. \tag{4.4}$$

It every solution to $\mathcal{E}^*_t$ is rainbow, then Inequality (4.1) implies that $1, 2, \ldots, t-1$ must all receive distinct colors. By (4.2) and (4.4), the color of t is distinct from the colors assigned to $1, 2, \ldots, t-1$. Equations (4.3) and (4.4) then force $t-2$ and $\frac{t(t-1)}{2} + 2$ to receive the same color. It follows that

$$\widehat{S}^t_{t-1}(t) \leq \frac{t(t-1)}{2} + 2.$$

□

When $t = 3$, Theorem 4.1 implies that $\widehat{S}^3_2(3) = 5$. Next, we wish to fix $s = 2$ and $m = 3$, but to allow $t \geq 3$ to vary. The evaluation of $\widehat{S}^t_2(3)$ was completed in [38]. First, we need the following lemma.

Lemma 4.1 ([38]) *Let $n \geq 3$. If a coloring of $[1, n]$ does not contain a solution to $\mathcal{E}^*_3$ that uses at most 2 colors, then at least $n - \lfloor \frac{n}{2} \rfloor + \lfloor \frac{n}{3} \rfloor$ colors must be used.*

Proof Let $f : [1, n] \longrightarrow \{1, 2, \ldots, t\}$ be an exact t-coloring that does not contain a strict 3-Schur solution that uses at most 2 colors (i.e., every solution to $\mathcal{E}^*_3$ is rainbow). Such a coloring is guaranteed to exist since assigning unique colors to each number in $[1, n]$ gives one such coloring. Because of the equation $1+2 = 3$, each of 1, 2, and 3 must receive distinct colors. For any even $i > 3$, the equations

$$1 + (i-1) = i, \quad 2 + (i-2) = i, \quad \ldots, \quad \frac{i-2}{2} + \frac{i+2}{2} = i$$

force i to receive a different color than that of j, where $1 \leq j < i$, except possibly for $\frac{i}{2}$. From the equation $\frac{i}{2} + i = \frac{3i}{2}$, it follows that i requires a color diferent from that of $\frac{i}{2}$ if and only if $\frac{3i}{2} \leq n$.

For any odd $i > 3$, the equations

$$1 + (i-1) = i, \quad 2 + (i-2) = i, \quad \ldots, \quad \frac{i-2}{2} + \frac{i+2}{2} = i$$

force i to receive a different color than that of j, where $1 \leq j < i$. So, the only numbers that are allowed to repeat colors are those in the set

$$S := \{i \in [1, n] \mid 2i \in [1, n] \text{ and } 3i \notin [1, n]\}.$$

Note that $|S| = \lfloor \frac{n}{2} \rfloor - \lfloor \frac{n}{3} \rfloor$. The statement of the lemma follows from the fact that all elements in $[1, n] \setminus S$ must have distinct colors. □

Theorem 4.2 ([38]) *For all $n \geq 3$,*

$$\widehat{S}_2^t(3) = \begin{cases} \frac{6t+5}{5} & \text{if } t \equiv 0 \pmod 5 \\ \frac{6t+9}{5} & \text{if } t \equiv 1 \pmod 5 \\ \frac{6t+3}{5} & \text{if } t \equiv 2 \pmod 5 \\ \frac{6t+7}{5} & \text{if } t \equiv 3 \pmod 5 \\ \frac{6t+6}{5} & \text{if } t \equiv 4 \pmod 5). \end{cases}$$

Proof We separate the proof into cases, based on the value of n modulo 5.

Case 1: Assume that $t \equiv 0 \pmod 5$ and write $t = 5q$. In order to prove the lower bound, we provide an exact t-coloring of $[1, t+q] = [1, 6q]$ that lacks a solution to $\mathcal{E}_3^*$ that uses at most 2 colors. For each $i \in [1, 6q]$ such that $2i \in [1, 6q]$ and $3i \notin [1, 6q]$, both i and $2i$ may receive the same color. This is true for all i such that $2q < i \leq 3q$. It follows that q colors can be repeated without forcing a solution to $\mathcal{E}_3^*$ that is not rainbow. It follows that $\widehat{S}_2^t(3) \geq t + q + 1 = \frac{6t+5}{5}$.

To prove the reverse inequality, consider an exact t-coloring of $[1, t+q+1]$. If we let $n = t + q + 1 = 6q + 1$, then

$$n - \left\lfloor \frac{n}{2} \right\rfloor + \left\lfloor \frac{n}{3} \right\rfloor = 5q + 1 = t + 1 > t.$$

By the contrapositive statement to Lemma 4.1, there exists a solution to $\mathcal{E}_3^*$ that uses at most 2 colors. It follows that $\widehat{S}_2^t(3) \leq t + q + 1 = \frac{6t+5}{5}$.

Case 2: Assume that $t \equiv 1 \pmod 5$ and write $t = 5q+1$. We provide an exact t-coloring of $[1, t+q+1] = [1, 6q+2]$ that lacks a solution to $\mathcal{E}_3^*$ that uses at most 2 colors. For each $i \in [1, 6q+2]$ such that $2i \in [1, 6q+2]$ and $3i \notin [1, 6q+2]$, both i and $2i$ may receive the same color. This is true for all i such that $2q < i \leq 3q+1$. It follows that $q+1$ colors can be repeated without forcing a solution to $\mathcal{E}_3^*$ that is not rainbow. It follows that $\widehat{S}_2^t(3) \geq t + q + 2 = \frac{6t+9}{5}$.

To prove the reverse inequality, consider an exact t-coloring of $[1, t+q+2]$. If we let $n = t + q + 2 = 6q + 3$, then

$$n - \left\lfloor \frac{n}{2} \right\rfloor + \left\lfloor \frac{n}{3} \right\rfloor = 5q + 3 = t + 2 > t.$$

By the contrapositive statement to Lemma 4.1, there exists a solution to $\mathcal{E}_3^*$ that uses at most 2 colors. It follows that $\widehat{S}_2^t(3) \leq t + q + 2 = \frac{6t+9}{5}$.

Case 3: Assume that $t \equiv 2 \pmod 5$ and write $t = 5q+2$. We provide an exact t-coloring of $[1, t+q] = [1, 6q+2]$ that lacks a solution to $\mathcal{E}_3^*$ that uses at most 2 colors. For each $i \in [1, 6q+2]$ such that $2i \in [1, 6q+2]$ and $3i \notin [1, 6q+2]$, both i and $2i$ may receive the same color. This is true for all i such that $2q < i \leq 3q+1$. It follows that $q+1$ colors can be repeated without forcing a solution to $\mathcal{E}_3^*$ that is not rainbow. It follows that $\widehat{S}_2^t(3) \geq t + q + 1 = \frac{6t+3}{5}$.

To prove the reverse inequality, consider an exact t-coloring of $[1, t+q+1]$. If we let $n = t + q + 1 = 6q + 3$, then

$$n - \left\lfloor \frac{n}{2} \right\rfloor + \left\lfloor \frac{n}{3} \right\rfloor = 5q + 3 = t + 1 > t.$$

By the contrapositive statement to Lemma 4.1, there exists a solution to $\mathcal{E}_3^*$ that uses at most 2 colors. It follows that $\widehat{S}_2^t(3) \leq t + q + 1 = \frac{6t+3}{5}$.

Case 4: Assume that $t \equiv 3 \pmod 5$ and write $t = 5q+3$. We provide an exact t-coloring of $[1, t+q+1] = [1, 6q+4]$ that lacks a solution to $\mathcal{E}_3^*$ that uses at most 2 colors. For each $i \in [1, 6q+4]$ such that $2i \in [1, 6q+4]$ and $3i \notin [1, 6q+4]$, both i and $2i$ may receive the same color. This is true for all i such that $2q+1 < i \leq 3q+2$. It follows that $q+1$ colors can be repeated without forcing a solution to $\mathcal{E}_3^*$ that is not rainbow. It follows that $\widehat{S}_2^t(3) \geq t+q+2 = \frac{6t+7}{5}$.

To prove the reverse inequality, consider an exact t-coloring of $[1, t+q+2]$. If we let $n = t+q+2 = 6q+5$, then

$$n - \left\lfloor \frac{n}{2} \right\rfloor + \left\lfloor \frac{n}{3} \right\rfloor = 5q+4 = t+1 > t.$$

By the contrapositive statement to Lemma 4.1, there exists a solution to $\mathcal{E}_3^*$ that uses at most 2 colors. It follows that $\widehat{S}_2^t(3) \leq t+q+2 = \frac{6t+7}{5}$.

Case 5: Assume that $t \equiv 4 \pmod 5$ and write $t = 5q+4$. We provide an exact t-coloring of $[1, t+q+1] = [1, 6q+5]$ that lacks a solution to $\mathcal{E}_3^*$ that uses at most 2 colors. For each $i \in [1, 6q+5]$ such that $2i \in [1, 6q+5]$ and $3i \notin [1, 6q+5]$, both i and $2i$ may receive the same color. This is true for all i such that $2q+1 < i \leq 3q+2$. It follows that $q+1$ colors can be repeated without forcing a solution to $\mathcal{E}_3^*$ that is not rainbow. It follows that $\widehat{S}_2^t(3) \geq t+q+2 = \frac{6t+6}{5}$.

To prove the reverse inequality, consider an exact t-coloring of $[1, t+q+2]$. If we let $n = t+q+2 = 6q+6$, then

$$n - \left\lfloor \frac{n}{2} \right\rfloor + \left\lfloor \frac{n}{3} \right\rfloor = 5q+5 = t+1 > t.$$

By the contrapositive statement to Lemma 4.1, there exists a solution to $\mathcal{E}_3^*$ that uses at most 2 colors. It follows that $\widehat{S}_2^t(3) \leq t+q+2 = \frac{6t+6}{5}$. □

When $m = 4$, Theorem 4.1 implies that $\widehat{S}_3^4(4) = 8$. The next theorem handles another case where $m = 4$.

Theorem 4.3 ([38]) $\widehat{S}_2^3(4) = 11$.

Proof For the lower bound, consider the 3-coloring of $[1, 10]$ given by

$$\{1, 2, 3, 4, 5, 6, 7, 8, 9, 10\}.$$

We can confirm that all of the solutions to $\mathcal{E}_m^*$ within this set use all four colors by listing them:

$$\begin{array}{lll} 1+2+3=6, & 1+2+6=9, & 1+3+6=10, \\ 1+2+4=7, & 1+3+5=9, & 1+4+5=10, \\ 1+2+5=8, & 2+3+4=9, & 2+3+5=10, \\ 1+3+4=8, & 1+2+7=10, & \end{array}$$

from which it follows that $\widehat{S}_2^3(4) \geq 11$.

To prove the reverse inequality, suppose that there exists an exact 3-coloring of $[1, 11]$ in which every solution to $\mathcal{E}_4^*$ uses all three colors. Since $1+2+3=6$ is one such equation, we consider cases based on which two numbers in this equation receive the same color.

Case 1: Assume that 1 and 2 receive the same color, and without loss of generality, suppose that they are both red. Since 3 and 6 must receive the other two colors, assume that 3 is blue and 6 is green. The equation $1+2+6=9$ must also use all three colors, so it follows that 9 must be blue. The equation $2+3+4=9$ implies that 4 is green and the equation $1+4+6=11$ implies that 11 is blue. Finally, the equation $1+3+7=11$ implies that 7 is green, and it follows that $1+2+4=7$ only uses two colors (red and green). So, this case cannot occur.

Case 2: Assume that 1 and 3 receive the same color, and without loss of generality, suppose that they are both red, 2 is blue, and 6 is green. The equation $1+3+6=10$ forces 10 to be blue. Then $1+2+7=10$ and $2+3+5=10$ force 7 and 5 to both be green. The equation $1+3+7=11$ implies that 11 is blue and $1+2+8=11$ implies that 8 is green. From $1+3+4=8$, it follows that 4 is blue. Finally, we see that $2+4+5=11$ only uses two colors (blue and green). So, this case cannot occur.

Case 3: Assume that 2 and 3 receive the same color, and without loss of generality, suppose that they are both red, 1 is blue, and 6 is green. The equation $2+3+6=11$ forces 11 to be blue. Then $1+2+8=11$ and $1+3+7=11$ imply that 8 and 7 are green. The equation $1+4+6=11$ then forces 4 to be red. Finally, regardless of what color 9 receives, the equation $2+3+4=9$ uses at most two colors (one of which is red). So, this case cannot occur.

Case 4: Assume that 1 and 6 receive the same color, and without loss of generality, suppose that they are both red, 2 is blue, and 3 is green. The equation $1+2+6=9$ forces 9 to be green, and $1+3+6=10$ forces 10 to be blue. Then $2+3+4=9$ and $2+3+5=10$ imply that 4 and 5 are both red. It follows that $1+4+5=10$ uses two colors (red and blue). So, this case cannot occur.

Case 5: Assume that 2 and 6 receive the same color, and without loss of generality, suppose that they are both red, 1 is blue, and 3 is green. The equation $1+2+6=9$ forces 9 to be green, and $2+3+6=11$ forces 11 to be blue. Then $2+3+4=9$ implies that 4 is blue, and we see that $1+4+6=11$ uses two colors (red and blue). So, this case cannot occur.

Case 6: Assume that 3 and 6 receive the same color, and without loss of generality, suppose that they are both red, 1 is blue, and 2 is green. The equation $1+3+6=10$ forces 10 to be green, and $2+3+5=10$ forces 5 to be blue. Then $1+3+5=9$ implies that 9 is green, and $2+3+4=9$ implies that 4 is blue. It follows that the equation $1+4+5=10$ uses two colors (blue and green). So, this case cannot occur.

We have shown that none of the above cases can occur, so it follows that no exact 3-coloring of $[1,11]$ exists in which every solution to $\mathcal{E}_m^*$ uses all three colors. Therefore, we obtain the inequality $\widehat{S}_2^3(4) \leq 11$. $\square$

When $m=5$, Theorem 4.1 implies that $\widehat{S}_4^5(5)=12$. The next theorem covers another case where $m=5$.

Theorem 4.4 ([38]) $\widehat{S}_3^4(5)=14$.

Proof To obtain the lower bound for $\widehat{S}_3^4(5)$, consider the exact 4-coloring of $[1,13]$ given by

$$\{1,2,3,4,5,6,7,8,9,10,11,12,13\}.$$

We can confirm that all of the solutions to $\mathcal{E}_m^*$ in this set use all four colors by listing them:

$$1+2+3+4=10, \quad 1+2+4+5=12, \quad 1+2+4+6=13,$$
$$1+2+3+5=11, \quad 1+2+3+7=13, \quad 1+3+4+5=13,$$
$$1+2+3+6=12,$$

from which it follows that $\widehat{S}_3^4(5) \geq 14$.

To prove the upper bound, assume that there exists an exact 4-coloring of $[1,14]$ in which every solution to $\mathcal{E}_5^*$ uses all four colors. Since $1+2+3+4=10$ is one such solution, the proof can be separated into 10 cases, based on which two numbers in this equation receive the same color.

Case 1: Assume that 1 and 2 receive the same color, and without loss of generality, suppose that they are both red. Since 3, 4, and 10 must receive the other three colors, assume that 3 is blue, 4 is green, and 10 is purple. The equation $1 + 2 + 3 + 5 = 11$ must also use all four colors, so it follows that $\{5, 11\}$ is colored with green and purple. However, the equation $1 + 2 + 4 + 5 = 12$ prevents 5 from being colored green. So, 5 must be purple, 11 must be green, and 12 must be blue. Then regardless of the color of 6, the equation $1 + 2 + 3 + 6 = 12$ uses at most three colors.

Case 2: Assume that 1 and 3 receive the same color, and without loss of generality, suppose that they are both red, 2 is blue, 4 is green, and 10 is purple. The equation $1 + 2 + 3 + 5 = 11$ implies that $\{5, 11\}$ is colored with green and purple. However, the equation $1 + 3 + 4 + 5 = 13$ prevents 5 from being colored green. So, 5 must be purple, 11 must be green, and 13 must be blue. Regardless of the color of 7, the equation $1+2+3+7 = 13$ uses at most three colors.

Case 3: Assume that 1 and 4 receive the same color, and without loss of generality, suppose that they are both red, 2 is blue, 3 is green, and 10 is purple. The equation $1 + 2 + 4 + 5 = 12$ implies that $\{5, 12\}$ is colored with green and purple. However, the equation $1 + 3 + 4 + 5 = 13$ prevents5 from being colored green. So, 5 must be purple, 12 is green, and 13 is blue. Regardless of the color of 6, the equation $1 + 2 + 4 + 6 = 13$ uses at most three colors.

Case 4: Assume that 1 and 10 receive the same color, and without loss of generality, suppose that they are both red, 2 is blue, 3 is green, and 4 is purple. The equation $2 + 3 + 4 + 5 = 14$ implies that $\{5, 14\}$ must include the color red. If 5 is red, then the equations $1 + 2 + 4 + 5 = 12$, $1 + 2 + 3 + 6 = 12$, and $1 + 3 + 4 + 6 = 14$ imply that 12 is green, 6 is purple, and 14 is blue. It follows that $1 + 2 + 5 + 6 = 14$ uses at most three colors. If 14 is red, then the equation $1 + 3 + 4 + 6 = 14$ forces 6 to be blue, and regardless of the color of 5, the equation $1 + 2 + 5 + 6 = 14$ uses at most three colors.

Case 5: Assume that 2 and 3 receive the same color, and without loss of generality, suppose that they are both red, 1 is blue, 4 is green, and 10 is purple. The equation $1 + 2 + 3 + 5 = 11$ implies that $\{5, 11\}$ is colored with green and purple. However, the equation $2+3+4+5 = 14$ prevents 5 from being colored green. So, 5 is purple, 11 is green, and 14 is blue. Regardless of the color of 8, the equation $1 + 2 + 3 + 8 = 14$ uses at most three colors.

Case 6: Assume that 2 and 4 receive the same color, and without loss of generality, suppose that they are both red, 1 is blue, 3 is green, and 10 is purple. The equation $1 + 2 + 4 + 5 = 12$ implies that $\{5, 12\}$ is colored with green and purple. However, the equation $2 + 3 + 4 + 5 = 14$ prevents 5 from being green. So, 5 is purple, 12 is green, and 14 is blue. Regardless of the color of 7, the equation $1 + 2 + 4 + 7 = 14$ uses at most three colors.

Case 7: Assume that 2 and 10 receive the same color, and without loss of generality, suppose that they are both red, 1 is blue, 3 is green, and 4 is purple. The equation $2 + 3 + 4 + 5 = 14$ implies that $\{5, 14\}$ must include the color blue. If 5 is blue, then the equations $1 + 2 + 4 + 5 = 12$, $1 + 2 + 3 + 6 = 12$, and $1 + 3 + 4 + 5 = 13$ imply that 12 is green, 6 is purple, and 13 is red. It follows that $1 + 2 + 4 + 6 = 13$ uses at most three colors. If 14 is blue, then the equation $1 + 3 + 4 + 6 = 14$ forces 6 to be red, and regardless of the color of 5, the equation $1 + 2 + 5 + 6 = 14$ uses at most three colors.

Case 8: Assume that 3 and 4 receive the same color, and without loss of generality, suppose that they are both red, 1 is blue, 2 is green, and 10 is purple. The equation $2 + 3 + 4 + 5 = 14$ implies that $\{5, 14\}$ is colored with blue and purple. However, the equation $1 + 3 + 4 + 5 = 13$ prevents 5 from being blue. So, 5 is purple, 14 is blue, and 13 is green. Regardless of the color of 6, the equation $1+3+4+6=14$ uses at most three colors.

Case 9: Assume that 3 and 10 receive the same color, and without loss of generality, suppose that they are both red, 1 is blue, 2 is green, and 4 is purple. The equation $2+3+4+5=14$ implies that $\{5,14\}$ includes the color blue. If 5 is blue, then the equations $1+3+4+5=13$, $1+2+4+6=13$, and $1+2+3+6=12$ imply that 13 is green, 6 is red, and 12 is purple. It follows that $1+2+4+5=12$ uses at most three colors. If 14 is blue, then the equation $1+3+4+6=14$ forces 6 to be green, and regardless of the color of 5, the equation $1+2+5+6=14$ uses at most three colors.

Case 10: Assume that 4 and 10 receive the same color, and without loss of generality, suppose that they are both red, 1 is blue, 2 is green, and 3 is purple. The equation $2+3+4+5=14$ implies that $\{5,14\}$ includes the color blue. If 5 is blue, then the equations $1+3+4+5=13$, $1+2+4+6=13$, $1+2+3+7=13$, and $1+3+4+6=14$ imply that 13 is green, 6 is purple, 7 is red, and 14 is green. It follows that $1+2+4+7=14$ uses at most three colors. If 14 is blue, then regardless of the color of 5, the equation $1+2+4+7=14$ uses at most three colors.

We have shown that none of the above cases can occur, so it follows that no exact 4-coloring of $[1,14]$ exits in which every solution to $\mathcal{E}_m*$ uses all four colors. Therefore, we obtain the inequality $\widehat{S}_3^4(5)\le 14$. □

Theorems 4.3 and 4.4 do not easily generalize to weakened Schur numbers of the form $\widehat{S}_{t-1}^t(t+1)$ due to the large number of cases one must consider. Based on the results of these theorems, and the known strict Schur number $\widehat{S}_1^2(3)=9$ [2], the authors of [38] conjectured that

$$\widehat{S}_{t-1}^t(t+1)=\frac{t^2-t+16}{2},$$

for all $t\ge 2$ (see Problem 26 in Section 5.2).

4.2 Weakened Rado Numbers

Rado numbers can be viewed as a generaliation of Schur numbers. For any system $\mathcal{E}$ of equations/inequalities, define the *weakened Rado number* $R_s^t(\mathcal{E})$ to be the least $n\in\mathbb{N}$ such that every exact t-coloring of $[1,n]$ contains a solution to $\mathcal{E}$ that uses at most s colors, where $1\le s<\min\{m,t\}$. Note that $R_s^t(\mathcal{E}_m^*)=\widehat{S}_s^t(m)$. If we let $1\le s_1\le s_2<\min\{m,t\}$, then every solution to $\mathcal{E}$ that uses at most s_1 of the colors, necessarily uses at most s_2 of the colors. It follows that $R_{s_1}^t(\mathcal{E})\ge R_{s_2}^t(\mathcal{E})$.

Recall the system

$$\mathcal{E}_m^\ddagger:\begin{cases} x_1+x_2+\cdots+x_{m-1}<x_m\\ x_1<x_2<\cdots<x_m,\end{cases}$$

from Subsection 1.4.1. Weakened Rado numbers for $\mathcal{E}_m^\ddagger$ were first studied in 2022 in [39]. We start by proving the following general result.

Theorem 4.5 ([39]) *For all $t\ge 3$,*

$$R_{t-1}^t(\mathcal{E}_t^\ddagger)=\frac{t(t-1)}{2}+3.$$

Proof To prove the lower bound, we give an exact t-coloring of $[1,t(t-1)/2+2]$ that avoids a solution to $\mathcal{E}_t^\ddagger$ that uses at most $t-1$ colors. Observe that the only inequalities that we must consider are

$$1+2+\cdots+(t-1) < \frac{t(t-1)}{2}+1,$$
$$1+2+\cdots+(t-1) < \frac{t(t-1)}{2}+2,$$
$$1+2+\cdots+(t-2)+t < \frac{t(t-1)}{2}+2.$$

If we color $[1, t(t-1)/2+2]$ according to the partition

$$\{1\} \cup \{2\} \cup \cdots \cup \{t-2\} \cup \{t-1, t\} \cup \left\{t+1, t+2, \ldots, \frac{t(t-1)}{2}+2\right\},$$

then all of the solutions to $\mathcal{E}_t^\ddagger$ are rainbow. It follows that $R_{t-1}^t(\mathcal{E}_t^\ddagger) \geq \frac{t(t-1)}{2}+3$.

To prove the reverse inequality, consider an exact t-coloring of $[1, t(t-1)/2+3]$. Consider the inequalities

$$1+2+\cdots+(t-1) < \frac{t(t-1)}{2}+2, \tag{4.5}$$
$$1+2+\cdots+(t-2)+t < \frac{t(t-1)}{2}+2, \tag{4.6}$$
$$1+2+\cdots+(t-3)+(t-1)+t < \frac{t(t-1)}{2}+3. \tag{4.7}$$

If a solution to $\mathcal{E}_t^\ddagger$ that uses at most $t-1$ colors is avoided, then all of the numbers in Inequality (4.5) must receive distinct colors. From Inequality (4.6), it follows that t must receive the same color as $t-1$. However, this leads to a contradiction since Inequality (4.7) uses at most $t-1$ colors. It follows that $R_{t-1}^t(\mathcal{E}_t^\ddagger) \leq \frac{t(t-1)}{2}+3$. □

In the next theorem, we consider the case where $m=3$ and $s=2$.

Theorem 4.6 ([39]) *For all* $t \geq 3$,

$$R_2^t(\mathcal{E}_3^\ddagger) = \begin{cases} \frac{4t+6}{3} & \text{if } t \equiv 0 \pmod 3 \\ \frac{4t+5}{3} & \text{if } t \equiv 1 \pmod 3 \\ \frac{4t+4}{3} & \text{if } t \equiv 2 \pmod 3. \end{cases}$$

Proof We consider three cases, based on the value of t modulo 3.

<u>Case 1:</u> Assume that $t \equiv 0 \pmod 3$. Color the numbers in the set $[1, (4t+6)/3-1]$ according to the partition

$$\{1\} \cup \{2\} \cup \cdots \cup \left\{\frac{2t-3}{3}\right\} \cup \left\{\frac{2t}{3}, \frac{2t+3}{3}\right\} \cup \left\{\frac{2t+6}{3}, \frac{2t+9}{3}\right\} \cup \cdots \cup \left\{\frac{4t}{3}, \frac{4t+3}{3}\right\}.$$

Note that this partition consists of $\frac{2t-3}{3}$ sets with cardinality 1 and $\frac{t+3}{3}$ sets with cardinality 2, giving us a t-coloring. Every solution to $\mathcal{E}_3^\ddagger$ has the form $x_1+x_2<x_3$, where $x_1<x_2<\frac{4t+3}{3}$, from which it follows that

$$(x_2-1)+x_2 < \frac{4t+3}{3} \quad \Longrightarrow \quad x_2 < \frac{2t+3}{3}.$$

So, every solution to $\mathcal{E}_3^\ddagger$ is rainbow, implying that $R_2^t(\mathcal{E}_3^\ddagger) \geq \frac{4t+6}{3}$.

To prove the reverse inequality, consider an exact t-coloring of $[1, (4t+6)/3]$ that avoids a solution to $\mathcal{E}_3^{\ddagger}$ that uses at most two colors. Then all of the numbers in $[1, (4t+6)/6]$ must receive distinct colors and the inequality $1 + x_2 < x_3$ prevents any two numbers that differ by more than 1 from receiving the same color. So, at most two of $\frac{2t+6}{3}, \frac{2t+9}{3}, \ldots, \frac{4t+6}{3}$ can receive the same color. It follows that at least $t+1$ colors are required, and hence, $R_2^t(\mathcal{E}_3^{\ddagger}) \leq \frac{4t+6}{3}$.

Case 2: Assume that $t \equiv 1 \pmod 3$. Color the numbers in the set $[1, (4t+5)/3 - 1]$ according to the partition

$$\{1\} \cup \{2\} \cup \cdots \cup \left\{\frac{2t-2}{3}\right\} \cup \left\{\frac{2t+1}{3}, \frac{2t+4}{3}\right\} \cup \left\{\frac{2t+7}{3}, \frac{2t+10}{3}\right\} \cup \cdots \cup \left\{\frac{4t-1}{3}, \frac{4t+2}{3}\right\}.$$

Note that this partition consists of $\frac{2t-2}{3}$ sets with cardinality 1 and $\frac{t+2}{3}$ sets with cardinality 2, giving us a t-coloring. Every solution to $\mathcal{E}_3^{\ddagger}$ has the form $x_1 + x_2 < x_3$, where $x_1 < x_2 < \frac{4t+2}{3}$, from which it follows that

$$(x_2 - 1) + x_2 < \frac{4t+2}{3} \implies x_2 < \frac{4t+5}{6} < \frac{2t+4}{3}.$$

So, every solution to $\mathcal{E}_3^{\ddagger}$ is rainbow, implying that $R_2^t(\mathcal{E}_3^{\ddagger}) \geq \frac{4t+5}{3}$.

To prove the reverse inequality, consider an exact t-coloring of $[1, (4t+5)/3]$ that avoids a solution to $\mathcal{E}_3^{\ddagger}$ that uses at most two colors. Then all of the numbers in $[1, (4t+5)/6]$ must receive distinct colors and the inequality $1 + x_2 < x_3$ prevents any two numbers that differ by more than 1 from receiving the same color. So, at most two of $\frac{2t+7}{3}, \frac{2t+10}{3}, \ldots, \frac{4t+5}{3}$ can receive the same color. It follows that at least $t+1$ colors are required, and hence, $R_2^t(\mathcal{E}_3^{\ddagger}) \leq \frac{4t+5}{3}$.

Case 3: Assume that $t \equiv 2 \pmod 3$. Color the numbers in the set $[1, (4t+4)/3 - 1]$ according to the partition

$$\{1\} \cup \{2\} \cup \cdots \cup \left\{\frac{2t-1}{3}\right\} \cup \left\{\frac{2t+2}{3}, \frac{2t+5}{3}\right\} \cup \left\{\frac{2t+8}{3}, \frac{2t+11}{3}\right\} \cup \cdots \cup \left\{\frac{4t-2}{3}, \frac{4t+1}{3}\right\}.$$

Note that this partition consists of $\frac{2t-1}{3}$ sets with cardinality 1 and $\frac{t+1}{3}$ sets with cardinality 2, giving us a t-coloring. Every solution to $\mathcal{E}_3^{\ddagger}$ has the form $x_1 + x_2 < x_3$, where $x_1 < x_2 < \frac{4t+1}{3}$, from which it follows that

$$(x_2 - 1) + x_2 < \frac{4t+1}{3} \implies x_2 < \frac{2t+2}{3}.$$

So, every solution to $\mathcal{E}_3^{\ddagger}$ is rainbow, implying that $R_2^t(\mathcal{E}_3^{\ddagger}) \geq \frac{4t+4}{3}$.

To prove the reverse inequality, consider an exact t-coloring of $[1, (4t+4)/3]$ that avoids a solution to $\mathcal{E}_3^{\ddagger}$ that uses at most two colors. Then all of the numbers in $[1, (2t+2)/3]$ must receive distinct colors and the inequality $1 + x_2 < x_3$ prevents any two numbers that differ by more than 1 from receiving the same color. So, at most two of $\frac{2t+8}{3}, \frac{2t+11}{3}, \ldots, \frac{4t+4}{3}$ can receive the same color. It follows that at least $t+1$ colors are required, and hence, $R_2^t(\mathcal{E}_3^{\ddagger}) \leq \frac{4t+4}{3}$. □

In the next theorem, we consider the case where $m = 4$ and $s = 3$.

Theorem 4.7 ([39]) *For all $t \geq 4$,*

$$R_3^t(\mathcal{E}_4^\ddagger) = \begin{cases} 8\lfloor \frac{t}{5} \rfloor + 3 & \text{if } t \equiv 0 \pmod 5 \\ 8\lfloor \frac{t}{5} \rfloor + 5 & \text{if } t \equiv 1 \pmod 5 \\ 8\lfloor \frac{t}{5} \rfloor + 6 & \text{if } t \equiv 2 \pmod 5 \\ 8\lfloor \frac{t}{5} \rfloor + 7 & \text{if } t \equiv 3 \pmod 5 \\ 8\lfloor \frac{t}{5} \rfloor + 9 & \text{if } t \equiv 4 \pmod 5. \end{cases}$$

Proof Let $t = 5q + r$, where $q, r \in \mathbb{Z}$ and $0 \leq r < 5$ and note that $q = \lfloor \frac{t}{5} \rfloor$. Consider the following cases.

Case 1: Assume that $r = 0$ (i.e., $t = 5q$). To prove the lower bound, we consider the exact t-coloring of $[1, 8q + 2]$ defined by the following partition:

$$\{1\} \cup \{2\} \cup \cdots \cup \{4q - 1\} \cup \{4q, 4q + 1, 4q + 2\} \cup \{4q + 3, 4q + 4, 4q + 5, 4q + 6\} \\ \cup \{4q + 7, 4q + 8, 4q + 9, 4q + 10\} \cup \cdots \cup \{8q - 1, 8q, 8q + 1, 8q + 2\}.$$

If $x_2 \geq 4q$, then $x_3 \geq 4q+1$ and we find that the inequality $1+x_2+x_3 \geq 8q+2$ prevents us from having a solution to $\mathcal{E}_4^\ddagger$. Hence, every solution to $x_1 + x_2 + x_3 < x_4$ satisfies $x_2 < 4q$, from which we see that it must be rainbow. It follows that $R_3^t(\mathcal{E}_4^\ddagger) \geq 8q + 3$.

To prove the reverse inequality, consider an exact t-coloring of $[1, 8q + 3]$. If a rainbow solution to $\mathcal{E}_4^\ddagger$ is avoided, then the inequality $1 + 2 + x_3 < x_4$ implies that at most four consecutive numbers can receive the same color. The inequality $1 + (x_3 - 1) + x_3 < 8q + 3$ implies that $x_3 \leq 4q + 1$, from which it follows that $1, 2 \ldots, 4q$ must receive distinct colors that do not appear on any of the numbers greater than $4q$. So, at least

$$4q + \left\lceil \frac{(8q + 3) - 4q}{4} \right\rceil = 5q + 1 = t + 1$$

colors are required to avoid a rainbow solution to $\mathcal{E}_4^\ddagger$. It follows that $R_3^t(\mathcal{E}_4^\ddagger) \leq 8q + 3$.

Case 2: Assume that $r = 1$ (i.e., $t = 5q + 1$). Consider the exact t-coloring of $[1, 8q + 4]$ defined by the following partition:

$$\{1\} \cup \{2\} \cup \cdots \cup \{4q\} \cup \{4q + 1, 4q + 2, 4q + 3, 4q + 4\} \\ \cup \{4q + 5, 4q + 6, 4q + 7, 4q + 8\} \cup \cdots \cup \{8q + 1, 8q + 2, 8q + 3, 8q + 4\}.$$

If $x_2 \geq 4q+1$, then $x_3 \geq 4q+2$ and we find that the inequality $1+x_2+x_3 \geq 8q+4$ prevents us from having a solution to $\mathcal{E}_4^\ddagger$. Hence, every solution to $x_1+x_2+x_3 < x_4$ satisfies $x_2 < 4q+1$, from which we see that it must be rainbow. It follows that $R_3^t(\mathcal{E}_4^\ddagger) \geq 8q + 5$.

To prove the reverse inequality, consider an exact t-coloring of $[1, 8q + 5]$. If a rainbow solution to $\mathcal{E}_4^\ddagger$ is avoided, then the inequality $1 + 2 + x_3 < x_4$ implies that at most four consecutive numbers can receive the same color. The inequality $1 + (x_3 - 1) + x_3 < 8q + 5$ implies that $x_3 \leq 4q + 2$, from which it follows that $1, 2 \ldots, 4q + 1$ must receive distinct colors that do not appear on any of the numbers greater than $4q + 1$. So, at least

$$4q + 1 + \left\lceil \frac{(8q + 5) - (4q + 1)}{4} \right\rceil = 5q + 2 = t + 1$$

colors are required to avoid a rainbow solution to $\mathcal{E}_4^\ddagger$. It follows that $R_3^t(\mathcal{E}_4^\ddagger) \leq 8q + 5$.

Case 3: Assume that $r = 2$ (i.e., $t = 5q + 2$). Consider the exact t-coloring of $[1, 8q + 5]$ defined by the following partition:

$$\{1\} \cup \{2\} \cup \cdots \cup \{4q+1\} \cup \{4q+2, 4q+3, 4q+4, 4q+5\} \\ \cup \{4q+6, 4q+7, 4q+8, 4q+9\} \cup \cdots \cup \{8q+2, 8q+3, 8q+4, 8q+5\}.$$

If $x_2 \geq 4q+2$, then $x_3 \geq 4q+3$ and we find that the inequality $1+x_2+x_3 \geq 8q+6$ prevents us from having a solution to $\mathcal{E}_4^{\ddagger}$. Hence, every solution to $x_1+x_2+x_3 < x_4$ satisfies $x_2 < 4q+2$, from which we see that it must be rainbow. It follows that $R_3^t(\mathcal{E}_4^{\ddagger}) \geq 8q+6$.

To prove the reverse inequality, consider an exact t-coloring of $[1, 8q + 6]$. If a rainbow solution to $\mathcal{E}_4^{\ddagger}$ is avoided, then the inequality $1 + 2 + x_3 < x_4$ implies that at most four consecutive numbers can receive the same color. The inequality $1 + (x_3 - 1) + x_3 < 8q + 6$ implies that $x_3 \leq 4q + 2$, from which it follows that $1, 2 \ldots, 4q + 1$ must receive distinct colors that do not appear on any of the numbers greater than $4q + 1$. So, at least

$$4q + 1 + \left\lceil \frac{(8q+6) - (4q+1)}{4} \right\rceil = 5q + 3 = t + 1$$

colors are required to avoid a rainbow solution to $\mathcal{E}_4^{\ddagger}$. It follows that $R_3^t(\mathcal{E}_4^{\ddagger}) \leq 8q + 6$.

Case 4: Assume that $r = 3$ (i.e., $t = 5q + 3$). Consider the exact t-coloring of $[1, 8q + 6]$ defined by the following partition:

$$\{1\} \cup \{2\} \cup \cdots \cup \{4q+2\} \cup \{4q+3, 4q+4, 4q+5, 4q+6\} \\ \cup \{4q+7, 4q+8, 4q+9, 4q+10\} \cup \cdots \cup \{8q+3, 8q+4, 8q+5, 8q+6\}.$$

If $x_2 \geq 4q+3$, then $x_3 \geq 4q+4$ and we find that the inequality $1+x_2+x_3 \geq 8q+8$ prevents us from having a solution to $\mathcal{E}_4^{\ddagger}$. Hence, every solution to $x_1+x_2+x_3 < x_4$ satisfies $x_2 < 4q+3$, from which we see that it must be rainbow. It follows that $R_3^t(\mathcal{E}_4^{\ddagger}) \geq 8q+7$.

To prove the reverse inequality, consider an exact t-coloring of $[1, 8q + 7]$. If a rainbow solution to $\mathcal{E}_4^{\ddagger}$ is avoided, then the inequality $1 + 2 + x_3 < x_4$ implies that at most four consecutive numbers can receive the same color. The inequality $1 + (x_3 - 1) + x_3 < 8q + 7$ implies that $x_3 \leq 4q + 3$, from which it follows that $1, 2 \ldots, 4q + 2$ must receive distinct colors that do not appear on any of the numbers greater than $4q + 2$. So, at least

$$4q + 2 + \left\lceil \frac{(8q+7) - (4q+2)}{4} \right\rceil = 5q + 4 = t + 1$$

colors are required to avoid a rainbow solution to $\mathcal{E}_4^{\ddagger}$. It follows that $R_3^t(\mathcal{E}_4^{\ddagger}) \leq 8q + 7$.

Case 5: Assume that $r = 4$ (i.e., $t = 5q + 4$). Consider the exact t-coloring of $[1, 8q + 7]$ defined by the following partition:

$$\{1\} \cup \{2\} \cup \cdots \cup \{4q+2\} \cup \{4q+3, 4q+4\} \cup \{4q+5, 4q+6, 4q+7, 4q+8\} \\ \cup \{4q+9, 4q+10, 4q+11, 4q+12\} \cup \cdots \cup \{8q+5, 8q+6, 8q+7, 8q+8\}.$$

If $x_2 \geq 4q+3$, then $x_3 \geq 4q+4$ and we find that the inequality $1+x_2+x_3 \geq 8q+8$ prevents us from having a solution to $\mathcal{E}_4^{\ddagger}$. Hence, every solution to $x_1+x_2+x_3 < x_4$ satisfies $x_2 < 4q+3$, from which we see that it must be rainbow. It follows that $R_3^t(\mathcal{E}_4^{\ddagger}) \geq 8q+9$.

To prove the reverse inequality, consider an exact t-coloring of $[1, 8q + 9]$. If a rainbow solution to $\mathcal{E}_4^{\ddagger}$ is avoided, then the inequality $1 + 2 + x_3 < x_4$ implies that at most four consecutive numbers can receive the same color. The inequality $1 + (x_3 - 1) + x_3 < 8q + 9$

implies that $x_3 \le 4q+4$, from which it follows that $1, 2 \ldots, 4q+3$ must receive distinct colors that do not appear on any of the numbers greater than $4q+3$. So, at least

$$4q+3+\left\lceil \frac{(8q+9)-(4q+3)}{4} \right\rceil = 5q+5 = t+1$$

colors are required to avoid a rainbow solution to $\mathcal{E}_4^{\ddagger}$. It follows that $R_3^t(\mathcal{E}_4^{\ddagger}) \le 8q+7$. □

It is unclear what the values of $R_{m-1}^t(\mathcal{E}_m^{\ddagger})$ are for $t \ge m \ge 5$, but based on Theorems 4.6 and 4.7, perhaps the values are dependent on t modulo $2m-3$.

4.3 Weakened Rainbow Schur Numbers

For $m \ge 3$ and $2 \le s \le m$, define the *weakened rainbow Schur number* $RS_{s,m}(n)$ to be the minimum number of colors such that every exact $RS_{s,m}(n)$-coloring of $[1,n]$ contains a solution to $\mathcal{E}_m$ that uses at least s of the colors. Note that $RS_{m,m}(n) = RS_m(n)$. If $s_1 \le s_2$, then every solution to $\mathcal{E}_m$ that uses at least s_2 colors is a solution that uses at least s_1 colors. Therefore,

$$RS_{s_1,m}(n) \le RS_{s_2,m}(n),$$

for all $m \ge 3$. It follows that $RS_{s,m}(n)$ is defined for all $2 \le s < m$ since it is bounded above by $RS_m(n)$.

In 2024, the following two theorems were proved in [44].

Theorem 4.8 ([44]) *For all $m \ge 3$ and $n \ge 2m-4$, $RS_{2,m}(n) = 2$.*

Proof Since two colors are required in order to have a solution to $\mathcal{E}_m$ that uses at least two colors, $RS_{2,m}(n) \ge 2$. To prove the reverse inequality, consider an exact 2-coloring of $[1,n]$. Without loss of generality, suppose that 1 is colored red and $i \in [2,n]$ is the least natural number that is colored blue. Consider two cases.

Case 1: Suppose that $i \le n-m+2$. Then the equation

$$\underbrace{1+1+\cdots+1}_{m-2 \text{ terms}}+i = i+m-2 \le n$$

is in $\mathcal{E}_m$ and uses at least two colors.

Case 2: Suppose that $i > n-m+2$. Then the equation

$$\underbrace{1+1+\cdots+1}_{m-2 \text{ terms}}+(i-(m-2)) = i$$

is in $\mathcal{E}_m$ at uses at least two colors whenever $i \ge m-1$. In this case, this inequality is satisfied when $n-m+3 \ge m-1$, which is equivalent to $n \ge 2m-4$.

In both cases, we see that there exists a solution to $\mathcal{E}_m$ that uses at most two colors. It follows that $RS_{2,m}(n) \le 2$. □

Now we consider the values of $RS_{s,m}(n)$ when $s \ge 3$.

Theorem 4.9 ([44]) *Let $m \ge 4$ and $3 \le s \le m$. For all $n \ge \frac{s(s-1)}{2}+m-s$,*

$$RS_{s,m}(n) = \left\lceil \frac{(s-3)n+\frac{s(s-1)}{2}+m-s}{s-2} \right\rceil.$$

Proof To simplify notation, define

$$k = k(n,m,s) := \left\lceil \frac{(s-3)n + \frac{s(s-1)}{2} + m - s}{s-2} \right\rceil,$$

and note that

$$k(s-2) \leq (s-3)n + \frac{s(s-1)}{2} + m - s. \tag{4.8}$$

Consider the exact $(k-1)$-coloring of $[1,n]$ defined by the partition

$$\{1, 2\ldots, n+2-k\} \cup \{n+3-k\} \cup \{n+4-k\} \cup \cdots \cup \{n\}.$$

Suppose that

$$a_1 + a_2 + \cdots + a_{m-1} = a_m, \quad \text{where } a_1 \leq a_2 \leq \cdots \leq a_{m-1},$$

is a solution to $\mathcal{E}_m$ that uses at least s colors. Then Inequality (4.8) implies that a_m satisfies

$$\begin{aligned} a_m &\geq \underbrace{1 + 1 + \cdots + 1}_{m-s+1 \text{ terms}} + (n+3-k) + (n+4-k) + \cdots + (n+s-k) \\ &= (m-s+1) + 3 + 4 + \cdots + s + n(s-2) - k(s-2) \\ &= (m-s-2) + \frac{s(s+1)}{2} + n(s-2) - k(s-2) \\ &\geq (m-s-2) + \frac{s(s+1)}{2} + n(s-2) - \left((s-3)n + \frac{s(s-1)}{2} + m - s\right) \\ &= n + s - 2 \\ &\geq n + 1, \end{aligned}$$

which is a contradiction since $a_m \in [1,n]$. It follows that $RS_{s,m}(n) \geq k(n,m,s)$.

It remains to be shown that

$$RS_{s,m}(n) \leq k(n,m,s), \tag{4.9}$$

which we will accomplish by induction on $m+n$, where $m \geq 4$ and $n \geq \frac{s(s-1)}{2} + m - s$. First, we establish the base cases:

(*a*) $m = 4$, $3 \leq s \leq m$, and $n \geq \frac{s(s-1)}{2} + m - s$,

(*b*) $n = \frac{s(s-1)}{2} + m - s$, $m \geq 4$, and $3 \leq s \leq m$.

To prove that Inequality (4.9) holds in Base Case (*a*), note that if $m = 4$, $3 \leq s \leq m$, and $n \geq \frac{s(s-1)}{2} + m - s$, then $s = 3$ or $s = 4$. When $m = s = 4$, Inequality (4.9) follows from Equation (1.3). So, assume that $m = 4$ and $s = 3$. We will prove that $RS_{3,4}(n) \leq k(n,4,3) = 4$ by using induction on $n \geq 4$. Whne $n = 4$, an exact 4-coloring of $[1,4]$ necessarily has every number receiving its own unique color. Hence, $1 + 1 + 2 = 4$ is a solution to $\mathcal{E}_4$ that uses at least three colors, and it follows that $RS_{3,4}(4) \leq 4$. Now assume that $RS_{3,4}(n-1) \leq 4$, where $n > 4$ and consider an exact 4-coloring f of $[1,n]$. If the color of n is used on some other number $j < n$, then $[1, n-1]$ uses all four colors, and by the inductive hypothesis, there exists a solution to $\mathcal{E}_4$ that uses at least three colors. If n recieves its own unique color, then $[1, n-1]$ uses three colors. Let $i \in [2,n]$ be the least integer such that $f(i) \neq f(1)$. Then the equation

$$1 + i + (n - i - 1) = n$$

uses at least three colors, implying that Inequality (4.9) holds.

To prove that Inequality (4.9) holds in Base Case (*b*), observe that

$$\begin{aligned} k(n,m,s) &= \left\lceil \frac{(s-3)\left(\frac{s(s-1)}{2}+m-s\right)+\frac{s(s-1)}{2}+m-s}{s-2} \right\rceil \\ &= \left\lceil \frac{(s-2)\left(\frac{s(s-1)}{2}+m-s\right)}{s-2} \right\rceil \\ &= \frac{s(s-1)}{2}+m-s \\ &= n. \end{aligned}$$

So, in any exact $\left(\frac{s(s-1)}{2}+m-s\right)$-coloring of $[1,n]$, each element is the only member of its color class. Hence, every solution to $\mathcal{E}_m$ is a rainbow solution, implying that Inequality (4.9) holds.

Now let $m \geq 5$, $s \geq 3$, and $n \geq \frac{s(s-1)}{2}+m-s$. Assume that for all $m' \geq 4$ and $n' \geq \frac{s(s-1)}{2}+m-s$ with $m'+n' < m+n$,

$$RS_{s,m'}(n') \leq k(n',m',s),$$

for all $3 \leq s \leq m'$. The theorem will follow from proving that for all s such that $3 \leq s \leq m$, every exact $k(m,n,s)$-coloring of $[1,n]$ contains a solution to $\mathcal{E}_m$ that uses at least s colors. Let f be an exact $k(n,m,s)$-coloring of $[1,n]$. Write

$$(s-3)n+\frac{s(s-1)}{2}+m-s=(s-2)q+j, \tag{4.10}$$

where $q,j \in \mathbb{Z}$ and $0 \leq j < t-2$. We must consider two cases.

<u>Case 1:</u> Suppose that $1 \leq j < t-2$. Equation (4.10) implies that

$$k(n,m,s) = \left\lceil \frac{(s-3)n+\frac{s(s-1)}{2}+m-s}{s-2} \right\rceil = \left\lceil q+\frac{j}{t-2} \right\rceil = q+1.$$

So, the set $[1,n-1]$ uses at leasst q colors under the coloring f. By the inductive hypothesis,

$$\begin{aligned} RS_{s,m}(n-1) &\leq k(n-1,m,s) \\ &= \left\lceil \frac{(s-3)n+\frac{s(s-1)}{2}+m-s}{s-2} - \frac{s-3}{s-2} \right\rceil \\ &= \left\lceil q - \frac{(s-3)-j}{s-2} \right\rceil \\ &= q. \end{aligned}$$

So, in $[1,n-1]$ (and therefore, in $[1,n]$), there exists a solution to $\mathcal{E}_m$ that uses at least s colors.

<u>Case 2:</u> Suppose that $j=0$. Then $k(n,m,s)=q$ and Equation (4.10) implies that

$$k(n-1,m,s) = \left\lceil \frac{(s-3)(n-1)+\frac{s(s-1)}{2}+m-s}{s-2} \right\rceil = \left\lceil q-\frac{s-3}{s-2} \right\rceil = q.$$

If $[1,n-1]$ uses all q colors of f, then by the inductive hypothesis, there exists a solution to $\mathcal{E}_m$ that uses at least s colors, and we are done. So, assume that $[1,n-1]$ uses $q-1$ colors and n is the only number in its color class. We may assume that $s \le m-1$ since $s=m$ corresponds with Equation (1.3). Applying the inductive hypothesis again, we find that

$$\begin{aligned}
RS_{s,m-1}(n-1) &\le k(n-1,m-1,s)\\
&= \left\lceil \frac{(s-3)(n-1)+\frac{s(s-1)}{2}+m-1-s}{s-2} \right\rceil\\
&= \left\lceil q-\frac{s-3}{s-2}-\frac{1}{s-2} \right\rceil\\
&= q-1.
\end{aligned}$$

It follows that there exists a solution $a_1+a_2+\cdots+a_{m-2}=a_{m-1}$ to $\mathcal{E}_{m-1}$ that uses at least s colors, where $a_{m-1} \le n-1$. From this solution, we find that the equation

$$a_1+a_2+\cdots+a_{m-1}+(n-(a_1+a_2+\cdots+a_{m-2}))=n$$

is a solution to $\mathcal{E}_m$ that uses at least s colors.

In both cases, we find that f contains a solution to $\mathcal{E}_m$ that uses at least s colors. It follows that $RS_{s,m} \le k(n,m,s)$. □

5

Going Forward

In this chapter, we provide a list of research problems motivated by the topics covered in the preceding chapters. At present, the problems listed here have not be thoroughly studied. The chapter concludes with a collection of tables containing some of the main results known in weakened Ramsey theory.

5.1 Problems in Weakened Ramsey Theory for Graphs and Hypergraphs

In this section, numerous problems in weakened Ramsey theory for graphs and hypergraphs are discussed.

Problem 1 (3-Color 2-Chromatic Ramsey Numbers) By Theorems 2.11, 2.12, and 3.1 , we have that $r_2^3(K_3, K_4, K_5) = 8$, $r(K_3, K_5, K_5) = 9$, and $r_2^3(K_4) = 10$, respectively. Theorem 2.4 then leads to the inequalities

$$r_2^3(K_4, K_4, K_5) \leq \left\lceil \frac{1}{2}\left(r_2^3(K_3, K_4, K_5) + r_2^3(K_4, K_3, K_5) + r_2^3(K_4)\right)\right\rceil = 13,$$

$$r_2^3(K_4, K_5, K_5) \leq \left\lceil \frac{1}{2}\left(r_2^3(K_3, K_5, K_5) + r_2^3(K_4, K_4, K_5) + r_2^3(K_4, K_5, K_4)\right)\right\rceil \leq 18,$$

and

$$r_2^3(K_5) \leq \left\lceil \frac{1}{2}\left(r_2^3(K_4, K_5, K_5) + r_2^3(K_5, K_4, K_5) + r_2^3(K_5, K_5, K_4)\right)\right\rceil \leq 27.$$

Theorem 2.3 and Table 3.1 then gives us the ranges

$$10 \leq r_2^3(K_4, K_4, K_5) \leq 13, \quad 13 \leq r_2^3(K_4, K_5, K_5) \leq 18, \quad \text{and} \quad 20 \leq r_2^3(K_5) \leq 27.$$

Can these bounds be improved upon?

Problem 2 (t-Color 2-Chromatic Ramsey Numbers) In Theorem 2.1, it was shown that for all $s, t \geq 2$ and $m = r(K_s, K_t)$, if T_m is any tree of order m and $p \geq 2$, then

$$r_2^3(T_m, K_{p(s-1)+1}, K_{p(t-1)+1}) \geq p(m-1) + 1.$$

A 4-color version of this lower bound was also proved in Theorem 2.2. In both proofs, a lexicographic construction was used. Can these bounds be generalized to t-colorings, for $t \geq 5$? Also, are there any values of the parameters in which these inequalities are tight?

Problem 3 (3-Color (1,2)-Chromatic Ramsey Numbers Involving Stars) In Subsection 2.3.2, we considered chromatic Ramsey numbers of the form

$$r_{1,2}^3(K_3, K_3, K_3, K_{1,i}, K_{1,j}, K_{1,k}),$$

DOI: 10.1201/9781003498865-5

determining numerous values when $2 \leq i \leq j \leq k$, where $j \leq 6$. These numbers were originally considered in Jacobson's dissertation [132]. The proofs contained there had some mistakes and were mostly incomplete. Besides the cases considered in Subsection 2.3.2, what are the values for larger i, j, and k?

Problem 4 (Star-Critical Chromatic Ramsey Numbers) In a very recent preprint, Yi and Luo [212] introduced the concept of a star-critical chromatic Ramsey number. Given $t \geq 3$ colors, let $n, s_1, s_2, \ldots, s_n \in \mathbb{N}$ satisfy $1 \leq s_1 < s_2 < \cdots < s_n < t$. Order the $\binom{t}{s_1}$ subsets of s_1 of the colors, the $\binom{t}{s_2}$ subsets of s_2 of the colors, $\ldots$, the $\binom{t}{s_n}$ subsets of s_n of the colors, and let $b := \sum_{i=1}^{n} \binom{t}{s_i}$. For graphs $G_1, G_2, \ldots, G_b$, define the *star-critical t-color $(s_1, s_2, \ldots, s_n)$-chromatic Ramsey number* $r_*^{s_1,s_2,\ldots,s_n,t}(G_1, G_2, \ldots, G_b)$ to be the least k such that every t-coloring of $K_{r^t_{s_1,s_2,\ldots,s_n}(G_1,G_2,\ldots,G_b)-1} \sqcup K_{1,k}$ contains a subgraph isomorphic to G_i whose edges only use colors in the ith subset of colors, for some $i \in \{1, 2, \ldots, b\}$.

The star-critical chromatic Ramsey numbers determined by Yi and Luo [212] were weakened Ramsey numbers for stars (see Theorem 3.12 in Section 3.2). At present, nondiagonal star-critical chromatic Ramsey numbers have not been studied. In the case where all of the arguments are complete graphs with order at least 2, a proof similar to that of Theorem 3.8 in Section 3.1 can be used to show that

$$r_*^{s_1,s_2,\ldots,s_n,t}(K_{n_1}, K_{n_2}, \ldots, K_{n_b}) = r^t_{s_1,s_2,\ldots,s_n}(K_{n_1}, K_{n_2}, \ldots, K_{n_b}) - 1.$$

What are the values of such numbers for graphs that are not all complete?

Problem 5 (Weakened Chromatic Gallai-Ramsey Numbers) As we saw in Sections 3.3 and 3.4, a fair amount of research has been done on weakened Gallai-Ramsey numbers. At present, chromatic Gallai-Ramsey numbers have not been studied in the nondiagonal case. For example, in the case where $t = 3$, one could define the *3-color 2-chromatic Gallai-Ramsey number* $gr_2^3(G_1, G_2, G_3)$ to be the least $n \in \mathbb{N}$ such that every Gallai 3-coloring of K_n contains a red/blue G_1, a red/green G_2, or a blue/green G_3.

Problem 6 (Weakened Connected Ramsey Numbers) In 1978, David Sumner [201] introduced the concept of a connected Ramsey number. We know from Theorem 1.3 that in every 2-coloring of a complete graph, at least one of the subgraphs spanned by edges in a single color is connected. Sumner's idea was to consider Ramsey numbers for colorings in which the subgraphs spanned by edges in each of the colors are connected. Such colorings are called *connected 2-colorings*.

The *connected Ramsey number* $r_c^2(G_1, G_2)$ is defined to be the least $p \in \mathbb{N}$ such that every connected 2-coloring of K_p contains a red copy of G_1 or a blue copy of G_2. These numbers were also studied in [34], [52], [94], [95], and [117]. Fundamental to the determination of connected Ramsey numbers is the observation that if every connected 2-coloring of K_p contains a red G_1 or a blue G_2, then the same is true for every connected 2-coloring of K_n, when $n \geq p$. This observation follows from Theorem 11 of [27] (and the discussion immediately following Theorem 11), where it was shown that every connected 2-coloring of K_p, where $p \geq 5$, contains a *removable vertex*. (i.e., a vertex whose deletion results in a connected 2-coloring of K_{p-1}).

For $t \geq 3$, define a *connected t-coloring* of K_p to be a t-coloring in which the subgraphs spanned by edges in each color are connected. Unfortunately, when $t \geq 3$, not every connected t-coloring contains a removable vertex (for example, see the connected coloring of K_7 given in Figure 5.1). If one were to define a connected Ramsey number in this setting, it would lack the fundamental property that allows one to obtain an upper bound for such a

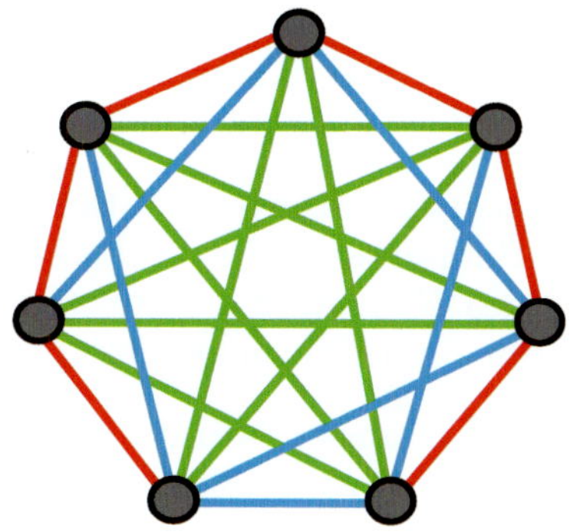

FIGURE 5.1: A connected 3-coloring of K_7 in which the deletion of any vertex results in a 3-coloring of K_6 in which the subgraph spanned by edges in one of the colors is disconnected.

number. To get around this issue, in [46], the authors introduced the concept of a strongly connected t-coloring. When $p > 2t$, a *strongly connected t-coloring* of K_p is a connected t-coloring $f : E(K_p) \longrightarrow \{1, 2, \ldots, t\}$ that contains a sequence of vertices $x_1, x_2, \ldots, x_k$, where $k = p - 2t$, such that the restricted colorings

$$f_\ell : (E(K_p - \{x_1, x_2, \ldots, \ell\}) \longrightarrow \{1, 2, \ldots, t\}$$

given by $f_\ell = f|_{K_p - \{x_1, x_2, \ldots, x_\ell\}}$, are connected for all $1 \leq \ell \leq k$. When $p = 2t$, by convention, we say that every connected t-coloring of K_{2t} is a strongly connected t-coloring. Theorem 11 of [27] implies that a 2-coloring of K_p is a connected 2-coloring if and only if it is a strongly connected 2-coloring.

For $t \geq 3$, define the *connected Ramsey number* $r_c^t(G_1, G_2, \ldots, G_t)$ to be the least $p \in \mathbb{N}$ such that every strongly connected t-coloring of K_p contains a copy of G_i in color i, for some $i \in \{1, 2, \ldots, t\}$. When $G_1 = G_2 = \cdots = G_t$, we shorten the notation to $r_c^t(G_1)$. In [46], the authors evaluated $r_c^t(K_{1,n})$, $r_c^3(K_{1,3}, K_{1,4}, K_{1,4})$, $r_c^3(P_4)$, $r_c^3(P_5)$, and $r_c^3(P_6)$.

Now that a multicolor version of connected Ramsey numbers has been introduced, for $t \geq 3$ and $1 \leq s < t$, one can define the *weakened connected Ramsey number* $r_c^{s,t}(G)$ to be the least $p \in \mathbb{N}$ such that every strongly connected t-coloring of K_p contains a copy of G that is spanned by edges using at most s colors. Of course, one could also define a chromatic version of these numbers in which different graphs make up the arguments. At present, neither of these numbers have been studied.

Problem 7 (Star-Critical Weakened Connected Ramsey Numbers) This problem builds off of the background given in Problem 6. In [180], the *star-critical connected Ramsey number* $r_c^*(G_1, G_2)$ was defined to be the least k such that every connected 2-coloring of $K_{r_c(G_1,G_2)-1} \sqcup K_{1,k}$ contains a red copy of G_1 or a blue copy of G_2. Several cases involving stars and paths were determined. Multicolor star-critical connected Ramsey numbers have not yet been studied, let alone their weakened analogues.

Problem 8 (Weakened Bipartite Ramsey Numbers) The 2-color bipartite Ramsey number was first introduced by Beineke and Schwenk [19] in 1975 (see also [93]). In a more general form, the *bipartite Ramsey number* $br^t(G_1, G_2, \ldots, G_t)$ is defined to be the least $p \in \mathbb{N}$ such that every t-coloring of $K_{p,p}$ contains a copy of G_i in color i, for some $i \in \{1, 2, \ldots, t\}$. Here, it is assumed that G_i is bipartite for all $i \in \{1, 2, \ldots, t\}$. When $G_1 = G_2 = \cdots = G_t$, the notation can be shortened to $br^t(G)$.

For $1 \leq s < t$ and bipartite graph G, one can define the *weakened bipartite Ramsey number* $br_s^t(G)$ to be the least $p \in \mathbb{N}$ such that every t-colorong of $K_{p,p}$ contains a copy of G spanned by edges using at most s colors. These numbers have not yet been studied, nor have their chromatic versions, in which different graphs can be used for the arguments.

Problem 9 (Weakened Hypergraph Ramsey Numbers) As we saw in Section 3.6, the only weakened hypergraph Ramsey numbers $r_s^t(H; k)$ that have been considered when $s \geq 2$ and $k \geq 3$ are those in which H is either a complete hypergraph or a k-uniform star. These numbers have not yet been studied for hypergraph paths, cycles, matchings, etc.

Problem 10 (Weakened Gallai-Ramsey Numbers for Cycles) In Section 3.4, Inequality (3.4) provided the following upper bounds:

$$gr_2^t(C_n) \leq \begin{cases} \frac{n-4}{2} \cdot t + \frac{n+2}{2} & \text{if } n \text{ is even} \\ \frac{n-3}{2} \cdot t + \frac{n+1}{2} & \text{if } n \text{ is odd,} \end{cases}$$

for all $t \geq 3$ and $n \geq 7$. In Theorems 3.19 and 3.21, $gr_2^t(C_5)$ and $gr_2^t(C_6)$ were evaluated. What are the values of $gr_2^t(C_n)$ for $n \geq 7$? What are the values of the corresponding star-critical weakened Gallai-Ramsey numbers $gr_*^{2,t}(C_n)$ for $n \geq 6$? Also, the Gallai-Ramsey number $gr_3^t(C_n)$ and its star-critical analogue have not been studied.

Problem 11 (Star-Critical Weakened Gallai-Ramsey Numbers for Fans) In [31], weakened Gallai-Ramsey numbers for fans were considered (see Subsection 3.4.2). There, it was shown that

$$gr_2^t(F_2) = t + 3, \quad \text{for all } t \geq 3,$$
$$gr_2^3(F_3) = 9,$$

and

$$gr_{2n-1}^t(F_n) = 2n + 1, \quad \text{for all } n \geq 2 \text{ and } t \geq 2n.$$

What are the values of the corresponding star-critical weakened Gallai-Ramsey numbers? What are the values of $gr_2^t(F_3)$ when $t > 3$?

Problem 12 (Weakened Gallai-Ramsey Numbers Wheels) In 2026, Jakhar and Moun [136] proved that

$$gr_2^t(W_5) = t + 3, \quad \text{for all } t \geq 3$$

and

$$gr_2^t(W_6) = 2t + 2, \quad \text{for all } t \geq 3$$

(cf. Theorems 3.28 and 3.29 in Subsection 3.4.3). At present, $gr_2^t(W_n)$ is unknown when $t \geq 3$ and $n \geq 7$. The star-critical weakened Gallai-Ramsey numbers $gr_*^{2,t}(W_n)$ are also unknown for all $t \geq 3$ and $n \geq 5$.

Problem 13 (Weakened Gallai-Ramsey Numbers for Books) The smallest book athat is not a complete graph is $B_2 = K_4 - e$. In Corollary 3.2 of Subsection 3.4.4, it was shown that $gr_2^3(B_2) = 4$. Here, it does not make sense to consider more than three colors since every Gallai coloring of K_4 uses at most three colors by Lemma 1.1. Jakhar and Moun [136] considered larger books, proving that

$$gr_2^3(B_3) = 6 \quad \text{and} \quad gr_2^3(B_4) = 7 \tag{5.1}$$

(see Theorems 3.28 and 3.29 in Subsection 3.4.4). The numbers $gr_2^t(B_3)$ and $gr_2^t(B_4)$ have not yet been determined when $t > 3$, nor have the star-critical analogues been determined when $t \geq 3$.

Since $K_{2,n}$ is a subgraph of B_n, one immediately obtains the inequality

$$gr_2^t(K_{2,n}) \leq gr_2^t(B_n),$$

for all $t \geq 3$ and $n \geq 3$. Motivated by this inequality and the values in (5.1), Jakhar and Moun [136] also proved that

$$gr_2^3(K_{2,3}) = 6 \quad \text{and} \quad gr_2^3(K_{2,4}) = 7$$

(see Corollaries 3.3 and 3.4 in Subsection 3.4.4). Similar to the case of books, the numbers $gr_2^t(K_{2,3})$ and $gr_2^t(K_{2,4})$ have not been determined when $t > 3$, nor have their star-critical analogues been determined when $t \geq 3$.

Problem 14 (Weakened Gallai-Ramsey Numbers for Generalized Fans) In [31], it was suggested that one consider weakened Gallai-Ramsey numbers for generalized fans. The *generalized fan* $F_{k,n}$ is defined by $F_{k,n} := K_1 + nK_k$. For example, $F_{3,2}$ and $F_{3,3}$ are shown in shown in Figure 5.2. The generalized fan $F_{k,n}$ has order $|V(F_{k,n})| = kn + 1$ and

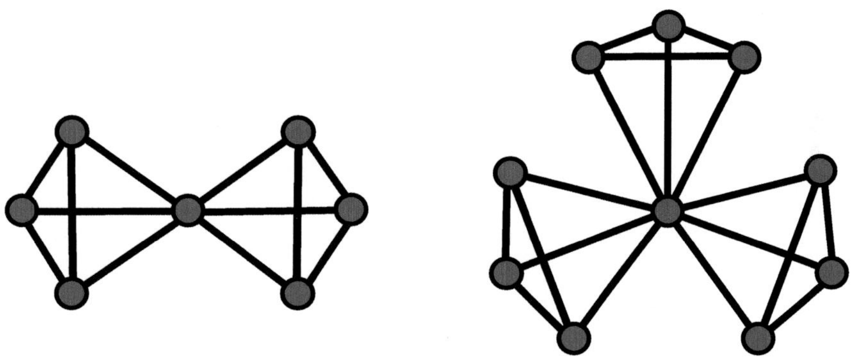

FIGURE 5.2: The generalized fans $F_{3,2}$ and $F_{3,3}$.

size $|E(F_{k,n})| = \frac{nk(k-1)}{2}$. Note that when $k = 2$, we have $F_n = F_{2,n}$. Ramsey and Gallai-Ramsey numbers involving generalized fans were studied in [47], [48], [118], [119], and [205]. The weakened Gallai-Ramsey number $gr_s^t(F_{k,n})$ (and its star-critical analogue) have not yet been studied.

Problem 15 (Determination of $gr_2^3(K_6)$) In Theorem 3.16, it was shown that $gr_2^3(K_6) \geq 12$. Based on this construction, Jakhar and Moun [136] stated the following conjecture.

Conjecture 5.1 ([136]) $gr_2^3(K_6) = 12$.

In fact, Theorem 3.16 provided lower bounds for $gr_2^3(K_n)$, for all $n \geq 6$. Are any of these bounds tight?

Problem 16 (Weakened Rainbow Numbers) In Section 4.3, we considered a weakened version of Rainbow Schur numbers. From a graph-theoretic perspective, these numbers are closely related to rainbow numbers (see Section 11.4 of [59]). For a graph G that lacks isolated vertices and $n \geq |V(G)|$, define the *rainbow number* $rb_n(G)$ to be the least t such that every exact t-coloring of K_n contains a rainbow subgraph that is isomorphic to G (i.e., one in which all of its edges have different colors).

Here, we are fixing the number of vertices and determining the number of colors required to force a rainbow subgraph. In comparison, with Ramsey numbers, one fixes the number of colors, then determines the number of vertices required to have a monochromatic subgraph. Due to the interchange of the parameters in these two problems, this topic is referred to as "anti-Ramsey theory" (see [86]). One can define the *anti-Ramsey number* $ar_n(G)$ to be the maximum number of colors that can be used in a coloring of K_n without producing a rainbow copy of G. It follows that

$$ar_n(G) = rb_n(G) - 1,$$

so the determinations of these two numbers are equivalent. For that reason, we focus on the case of rainbow numbers.

For a graph G without isolated vertices, let $2 \leq s \leq |E(G)|$ and $n \geq |V(G)|$. Define the *weakened rainbow number* $rb_{s,n}(G)$ to be the least t such that every t-coloring of K_n contains a subgraph that is isomorphic to G whose edges use at least s colors. From this definition, we see that

$$rb_{s,n}(G) \leq rb_{|E(G)|,n}(G) = rb_n(G).$$

A weakened anti-Ramsey number could be defined analogously. At present, weakened rainbow numbers have have not been studied.

Besides the case of graphs, rainbow numbers have also been studied within the setting of uniform hypergraphs [50, 97, 113, 183]. The weakened version of such numbers has not yet been considered.

Problem 17 (Weakened Rainbow Numbers for Various Host Graphs) Let G be a graph that lacks isolated vertices and let $\mathcal{F}$ be a nonempty collection of graphs that all contain a graph G as a subgraph. The *rainbow number* $rb(\mathcal{F}, G)$ is defined to be the least $t \in \mathbb{N}$ such that every exact t-coloring of a graph $F \in \mathcal{F}$ contains a rainbow subgraph that is isomorphic to G. When $\mathcal{F} = \{F\}$, we simplify the notation to $rb(F, G)$. Note that $rb(K_n, G) = rb_n(G)$. Common host graphs that have been considered include bipartite graphs [11, 12, 161], planar graphs [128, 135, 137, 139, 152, 153, 154, 155, 185, 186, 187], and Kneser graphs [138].

For a graph G that lacks isolated vertices, $2 \leq s \leq |E(G)$, and a nonempty collection $\mathcal{F}$ of graphs that contain G as a subgraph, define the *weakened rainbow number* $rb_s(\mathcal{F}, G)$ to be the least $t \in \mathbb{N}$ such that every exact t-coloring of a graph $F \in \mathcal{F}$ contains a subgraph whose edges use at least s colors. Other than the case $s = |E(G)|$, these numbers have not yet been studied.

Problem 18 (Weakened Ramsey Multiplicity) In 1974, Harary and Prins [122] introduced the concept of Ramsey multiplicity, building off of the observation that every 2-coloring of K_6 contains at least two monochromatic K_3-subgraphs (see [107]). For graphs $G_1, G_2, \ldots, G_t$, the *Ramsey multiplicity* $RM(G_1, G_2, \ldots, G_t)$ is defined to be least possible total number of G_1 in color 1, G_2 in color 2, ..., G_t in color t among all t-colorings of $K_{r(G_1,G_2,\ldots,G_t)}$. The 2-color Ramsey multiplicity for many small graphs were studied in [77], [76], [56], [122], [131], and [184]. The t-color Ramsey multiplicity for stars was determined in [134]. When $G_1 = G_2 = \cdots = G_t$, we write $RM^t(G_1)$ for the corresponding Ramsey multiplicity.

Define the *weakened Ramsey multiplicity* $RM_s^t(G)$ to be the least possible total number copies of G spanned by edges that use at most s colors, among all possible t-colorings of $K_{r_s^t(G)}$. At present, these numbers have not been studied. One could also define an anaologous chromatic (non-diagonal) version of the Ramsey multiplicity.

Problem 19 (The Diagonal Conjecture) In 2008, R. Wang [206] claimed that $r(K_s, K_t) > r(K_{s-1}, K_{t+1})$ is true for $s \leq t$. While this statement appears to be true based on known Ramsey numbers (see [189]), Wang's proof was not correct. In 2019, Liang,

Radziszowski, and Xu [163] revisited this concept, referring to the claim as the "Diagonal Conjecture" since it states that given a table of values for $r(K_s, K_t)$, the values should decrease as one moves away from the diagonal while keeping $s+t$ fixed. They generalized this conjecture to more than two colors.

Conjecture 5.2 ([163]) *Let $n_i \geq 3$, for all $i \in \{1, 2, \ldots, t\}$, where $t \geq 2$. If $n_{t-1} \leq n_t$, then*

$$r(K_{n_1}, K_{n_2}, \ldots, K_{n_t}) \geq r(K_{n_1}, K_{n_2}, \ldots, K_{n_{t-2}}, K_{n_{t-1}-1}, K_{n_t+1}).$$

The authors in [163] presented some evidence for the Diagonal Conjecture and proved several implications that follow from its validity. When $t \geq 3$, one can ask whether or not the chromatic Ramsey number satisfies

$$r_{t-1}^t(K_{n_1}, K_{n_2}, \ldots, K_{n_t}) \geq r_{t-1}^t(K_{n_1}, K_{n_2}, \ldots, K_{n_{t-2}}, K_{n_{t-1}-1}, K_{n_t+1}).$$

If true, what other implications follow from this inequality?

Problem 20 (Weakened Regular Ramsey Numbers) Recently, Zhidan Luo [166] introduced the concept of a regular Ramsey number. For graphs $G_1, G_2, \ldots, G_t$, the *regular Ramsey number* $rr(G_1, G_2, \ldots, G_t)$ is defined to be the least $k \in \mathbb{N}$ such that every t-coloring of a k-regular graph G of order $r(G_1, G_2, \ldots, G_t)$ contains a copy of G_i in color i, for some $i \in \{1, 2, \ldots, t\}$. Since $K_{r(G_1, G_2, \ldots, G_t)}$ is $(r(G_1, G_2, \ldots, G_t) - 1)$-regular, the regular Ramsey number is guaranteed to exist. Since every k-regular graph of order $r(G_1, G_2, \ldots, G_t)$ is a subgraph of $K_{r(G_1, G_2, \ldots, G_t)-1} \sqcup K_{1,k}$, it follows that

$$r_*(G_1, G_2, \ldots, G_t) \leq rr(G_1, G_2, \ldots, G_t) \leq r(G_1, G_2, \ldots, G_t) - 1.$$

When $G_1 = G_2 = \cdots = G_t$, we write $rr^t(G_1)$ in place of $rr(G_1, G_2, \ldots, G_t)$.

For $1 \leq s < t$, define the *weakened regular Ramsey number* $rr_s^t(G)$ to be the least $k \in \mathbb{N}$ such that every t-coloring of of a k-regular graph of order $r^t(G)$ contains a copy of G whose edges use at most s colors. Similar to its non-weakened counterpart, it is easy to confirm that

$$r_*^{s,t}(G) \leq rr_s^t(G) \leq r_s^t(G) - 1.$$

These numbers have not currently been studied, but they may be useful in the determination of certain weakened star-critical Ramsey numbers.

Problem 21 (Weakened Size Ramsey Theory) In Section 3.5, both the weakened size Ramsey number $\hat{r}_s^t(G)$ and the weakened size Gallai-Ramsey number $\widehat{gr}_s^t(G)$ (see [208]) were defined. In the proof of Theorem 3.31, and the discussion immediately following it, it was shown that

$$\hat{r}_s^t(K_n) = \binom{r_s^t(K_n)}{2} \quad \text{and} \quad \widehat{gr}_s^t(K_n) = \binom{gr_s^t(K_n)}{2},$$

for all $t \geq 2$, $1 \leq s < t$, and $n \geq 2$. Currently, $\hat{r}_s^t(G)$ and $\widehat{gr}_s^t(G)$ have not been studied when $s \geq 2$ and G is not a complete graph.

Problem 22 (Weakened Upper and Lower Size Ramsey Numbers) In 1992, Erdős and Faudree [83] introduced the concepts of lower and upper size Ramsey numbers. They are similar to the size Ramsey number, but focus on t-colorings of $K_{r(G_1, G_2, \ldots, G_t)}$. For graphs $G_1, G_2, \ldots, G_t$, the *upper size Ramsey number* $u(G_1, G_2, \ldots, G_t)$ is defined to be the minimum number such that if a subgraph G of $K_{r(G_1, G_2, \ldots, G_t)}$ has at least $u(G_1, G_2, \ldots, G_t)$ edges, then every t-coloring of G contains a monochromatic copy of G_i in color i, for some

$i \in \{1, 2, \ldots, t\}$. The *lower size Ramsey number* $\ell(G_1, G_2, \ldots, G_t)$ is defined to be the minimum number of edges in any subgraph G of $K_{r(G_1, G_2, \ldots, G_t)}$ such that every t-coloring of G contains a monochromatic copy of G_i in color i, for some $i \in \{1, 2, \ldots, t\}$. From their definitions, we see that these numbers are related to the star-critical Ramsey number in the following way:

$$\ell(G_1, G_2, \ldots, G_t) \leq \binom{r(G_1, G_2, \ldots, G_t) - 1}{2} + r_*(G_1, G_2, \ldots, G_t) \leq u(G_1, G_2, \ldots, G_t).$$

Neither Gallai nor weakened versions of the upper and lower size Ramsey numbers have been studied.

Problem 23 (Weakened Ordered Ramsey Numbers) The study of ordered Ramsey numbers was initiated by Erdős and Szekeres [87] in 1935, where they proved the existence of monotone sequences among collections of distinct real numbers. An *ordered graph* $H^<$ of order n is a graph H whose vertices have been distinctly labeled with the numbers in the set $\{1, 2, \ldots, n\}$. An ordered graph $G^<$ with vertex labels $\{1, 2, \ldots, N\}$ contains the ordered graph $H^<$ if there exists a mapping $\phi : [1, n] \longrightarrow [1, N]$ such that $\phi(i) < \phi(j)$ for $1 \leq i < j \leq n$ and $\phi(i)\phi(j)$ is an edge of $G^<$ whenever ij is an edge of $H^<$.

Given an ordered graph $H^<$, the *ordered Ramsey number* $r^t_<(H^<)$ is defined to be the least $p \in \mathbb{N}$ such that every t-coloring of a complete ordered graph of order p contains a monochromatic ordered copy of $H^<$. If H is the underlying graph of $H^<$ (i.e., withou the vertex labels), then note that

$$r^t(H) \leq r^t_<(H^<).$$

Ordered Ramsey numbers have been studied for various ordered graphs (e.g., see [13], [14], [15], [16], [60], and [75]), with an ordered Gallai-Ramsey number version being considered by Mao [171].

Given an ordered graph $H^<$ and $1 \leq s < t$, define the *weakened ordered Ramsey number* $r^{s,t}_<(H^<)$ to be the least $p \in \mathbb{N}$ such that every t-coloring of a complete ordered graph of order p contains an ordered copy of $H^<$ whose edges use at most s colors. At present, weakened ordered Ramsey numbers have not been studied, nor have their Gallai-colored counterparts.

Problem 24 (Weakened Ramsey Theory for ℓ-Connected Graphs) In 1983, David Matula [175] considered Ramsey numbers for ℓ-connected graphs. A graph G is called *ℓ-connected* if G has order greater than ℓ, and the deletion of $\ell - 1$ or fewer vertices results in a connected graph. In other words, you must remove at least ℓ vertices in order to disconnect G.

Define the Ramsey number $r^t(\mathcal{F}^\ell)$ to be the least $n \in \mathbb{N}$ such that every t-coloring of K_n contains a monochromatic ℓ-connected subgraph. Note that this definition is different from most of the Ramsey number definitions we have considered since it guarantees the existence of a graph from among a collection of graphs rather than a specific graph. Matula's main result was a proof that

$$2t(\ell - 1) + 1 \leq r^t(\mathcal{F}^\ell) < \frac{10}{3}t(\ell - 1) + 1.$$

In the special case where $t = 2$, he further proved that

$$4(\ell - 1) + 1 \leq r^2(\mathcal{F}^\ell) < (3 + \sqrt{11/3})(\ell - 1) + 1.$$

In general, if $\mathcal{H}$ is any collection of graphs and $1 \leq s < t$, one can consider the weakened Ramsey number $r^t_s(\mathcal{H})$ to be the least $n \in \mathbb{N}$ such that every t-coloring of K_n contains a subgraph from $\mathcal{H}$ spanned by edges that use at most s colors. The numbers $r^t_s(\mathcal{F}^\ell)$ have not yet been studied, although some cases will be trivial for small values of ℓ.

Problem 25 (Rainbow G-Free Colorings of Complete Graphs) Let G be a nonempty graph and let $\mathcal{H}$ be a collection of nonempty graphs. For all $t \geq |E(G)|$, define the *Gallai-Ramsey number* $gr^t(G : \mathcal{H})$ to be the least $n \in \mathbb{N}$ such that every t-coloring of K_n that avoids a rainbow copy of G contains a monochromatic copy of some graph in $\mathcal{H}$. When $s = 1$, we write $gr^t(G : \mathcal{H})$ in place of $gr_1^t(G : \mathcal{H})$. When $\mathcal{H} = \{H\}$, we shorten the notation to $gr^t(G : H)$. Note that by the notation introduced in Subsection 1.3.3, $gr^t(H) = gr^t(K_3 : H)$.

In order for K_n to have an exact k-coloring, it is necessary that

$$\binom{n}{2} = \frac{n(n-1)}{2} \geq k,$$

which is equivalent to $n^2 - n - 2k \geq 0$. Solving this inequality for n results in the inequality

$$gr^t(G : \mathcal{H}) \geq \left\lceil \frac{1 + \sqrt{1 + 8k}}{2} \right\rceil.$$

Similar to Theorem 1.22, which describes the structure of rainbow K_3-free colorings, the structure of rainbow G-free colorings are described in Theorem 1 of [202] when $G = P_4$, Theorem 2 of [202] when $G = P_5$, Theorem 3 of [17] when $G = K_{1,3}$, Theorem 4 of [17] when $G = P_4^+$, and Theorem 2.1 of [100] when $G = S_3^+$. Here, P_4^+ is the graph formed by adding a pendant edge onto one of the internal vertices of P_4, and S_3^+ is the graph formed by adding a pendant edge onto one of the vertices in K_3. For example, in [207], the cases where $G \in \{P_5, K_{1,3}\}$ and $\mathcal{H} = \mathcal{F}^\ell$ were considered, and in [159], the cases where $G \in \{K_{1,3}, P_5, P_4^+\}$ and $H = K_{m,n}$ were considered.

For $t \geq 3$ and $1 \leq s < t$, define the *weakened Gallai-Ramsey number* $gr_s^t(G : \mathcal{H})$ to be the least $n \in \mathbb{N}$ such that every t-coloring of K_n that avoids a rainbow copy of G contains a copy of some element in $\mathcal{H}$ whose edges use at most s colors. Currently, the only weakened Gallai-Ramsey numbers that have been studied are those in which $\mathcal{H} = \{K_3\}$.

5.2 Problems in Weakened Ramsey Theory Over the Integers

In this section, we list numerous problems in weakened Ramsey theory over the integers that have not been studied as of the writing of this text.

Problem 26 (Weakened Schur Numbers) Using the known strict Schur number $\widehat{S}_1^2(3) = 9$ [2] as well as the known weakened Schur numbers $\widehat{S}_2^3(4) = 11$ and $\widehat{S}_3^4(5) = 14$ (Theorems 4.3 and 4.4), the authors of [38] made the following conjecture.

Conjecture 5.3 (**[38]**) *For all* $t \geq 2$,

$$\widehat{S}_{t-1}^t(t+1) = \frac{t^2 - t + 16}{2}.$$

Unfortunately, the methods used in the proofs of Theorems 4.3 and 4.4 do not readily extend to this general statement due to the number of cases one must consider. An alternate method of proof will be needed to determine the validity of this conjecture.

Problem 27 ((Weakened) Gallai-Schur Numbers) A t-coloring of $[1, n]$ is called a *Gallai t-coloring* if it does not contain a rainbow solution to $\mathcal{E}_3$. Define the *Gallai-Schur* *number* $GS_m(t)$ to be the least $n \in \mathbb{N}$ such that every Gallai t-coloring of $[1, n]$ contains a monochromatic solution to $\mathcal{E}_m$. In the case where $m = 3$, it was shown in [32] that

$$GS_3(t) = \begin{cases} 5^k & \text{if } t = 2k \\ 2 \cdot 5^k & \text{if } t = 2k+1, \end{cases}$$

by relating these numbers to the known values of $gr^3(K_3)$ (see Theorem 1.23). Can $GS_m(t)$ be related in a similar manner to $gr^m(K_n)$, for $m \geq 3$?

More generally, for $1 \leq s < \min\{m, t\}$, define the *weakened Gallai-Schur number* $GS_{s,m}(t)$ to be the least natural number such that every Gallai t-coloring of $[1, GS_{s,m}(t)]$ contains a solution to $\mathcal{E}_m$ that uses at most s colors. These numbers have not currently been studied.

Problem 28 (Weakened (Gallai-)Rado Numbers) In Section 4.2, the weakened Rado number $R_s^t(\mathcal{E})$ was defined for any system of equations/inequalities $\mathcal{E}$. Besides the systems $\mathcal{E}_m$, $\mathcal{E}_m^*$, $\mathcal{E}_m^\dagger$, and $\mathcal{E}_m^\ddagger$ defined in Section 1.4, for what other systems can we determine $R_s^t(\mathcal{E})$? For example, in a recent preprint, Mao, Robertson, Wang, Yang, and Yang [172] also considered the system

$$\mathcal{E}_{b,3}: \; x_1 + x_2 + b = x_3, \quad \text{where } b \in \mathbb{N}.$$

For $1 \leq s < \min\{m, t\}$ and any system $\mathcal{E}$ of equations/inequalities, define the *weakened Gallai-Rado number* $GR_s^t(\mathcal{E})$ to be the least $n \in \mathbb{N}$ such that every Gallai t-coloring of $[1, n]$ contains a solution to $\mathcal{E}$ that uses at most s colors. Weakened Gallai-Rado numbers have not yet been studied.

Problem 29 (Weakened Rainbow Rado Numbers) Generalizing the concept of a weakened rainbow Schur number as introduced in Section 4.3, one can consider such numbers for systems of equations/inequalities other than $\mathcal{E}_m$. If $\mathcal{E}$ is any system of equations/inequalities in m variables, and if $m \geq 3$ and $2 \leq s \leq m$, define the *weakened rainbow Rado number* $RR_{s,\mathcal{E}}(n)$ to be the minimum number of colors such that every exact $RR_{s,\mathcal{E}}(n)$-coloring of $[1, n]$ contains a solution to $\mathcal{E}$ that uses at least s of the colors. These numbers have not yet been studied for systems other than $\mathcal{E}_m$.

Problem 30 (Weakened Rainbow Schur Numbers on a Rectangular Grid) In 2023, Fallon, Manhart, Miller, Rehm, Warnberg, and Zinnel [91] considered rainbow solutions to $\mathcal{E}_3$ in the rectangular grid $[1, m] \times [1, n]$, with coordinate-wise addition. It was suggested in [44] that these numbers be considered for $\mathcal{E}_m$, where $m \geq 4$, and that weakened rainbow Schur numbers also be considered in this setting.

Problem 31 (Weakened van der Waerden Numbers) If $a, b, k \in \mathbb{N}$, then the sequence of k integers

$$a, \; a+b, \; a+2b, \; \ldots, \; a+(k-1)b$$

is called an *arithmetic progression of length* k. In particular, any two consecutive numbers in an arithmetic progression have the same difference. If $t \geq 2$ is an integer, then the *van der Waerden number* $w^t(k)$ is defined to be the least $n_0 \in \mathbb{N}$ such that for any $n \geq n_0$, every t-coloring of $[1, n]$ contains a monochromatic arithmetic progression of length k. Van der Waerden's Theorem guarantees that the number $w^t(k)$ is finite (see Theorem 2.1 in [156]). Known values of $w^t(k)$ are discussed in Section 2.3 of [156].

If $1 \leq s < t$, define the *weakened var der Waerden number* $w_s^t(k)$ to be the least $n_0 \in \mathbb{N}$ such that for any $n \geq n_0$, every t-coloring of $[1, n]$ contains an arithmetic progression of length k that uses at most s colors. At present, these numbers have not been studied.

Problem 32 (Weakened Continuous Rado Numbers) In 2005, Brady and Haas [29] considered Rado numbers for the real number line. If $y, c \in \mathbb{R}$ and $m, t \in \mathbb{N}$ such that $m \geq 3$ and $t \geq 2$, define the *continuous Rado number* $R^t(y, m, c)$ to be the least real number such that every t-coloring of the real numbers in the interval $[y, R^t(y, m, c)]$ contains a monochromatic solution to the equation

$$\mathcal{E}_{c,m} : x_1 + x_2 + \cdots + x_{m-1} + c = x_m.$$

Note that this is the only place in this book where interval notation is used to indicate an interval on the real number line, rather than an interval of integers.

In addition to the parameters above, if $1 \leq s < \min\{m, t\}$, define the *weakened continuous Rado number* $R^t_s(y, m, c)$ to be the least real number such that every t-coloring of the real numbers in the interval $[y, R^t_s(y, m, c)]$ contains a solution to $\mathcal{E}_{c,m}$ that uses at most s colors. These numbers have not yet been studied. In a recent preprint, Vestal and Sax [204] considered an off-diagonal version of continuous Rado numbers, which may also be generalized to a weakened version.

5.3 Tables of Some Known Results

In this section, we compile many of the results discussed in this book into tables for quick reference.

TABLE 5.1: Known values for $r^3_2(G_1, G_2, G_3)$. Here, it is assumed that $2 \leq i \leq j \leq k$ and $S = \{(i, j, k) \mid i + j + k \equiv 3 \pmod 4$ and not all of i, j, and k are odd$\}$. Proofs of these numbers can be found in Subsections 2.2.1, 2.3.1, and 2.3.3.

(G_1, G_2, G_3)	$r^3_2(G_1, G_2, G_3)$	Source
(K_2, K_j, K_k)	j	[63]
(K_3, K_3, K_k)	$\begin{cases} 5 & \text{if } k \leq 5 \\ 6 & \text{if } k \geq 6 \end{cases}$	[63]
(K_3, K_4, K_k)	$\begin{cases} 7 & \text{if } k = 4 \\ 8 & \text{if } k \in \{5, 6\} \\ 9 & \text{if } k \geq 7 \end{cases}$	[63]
(K_3, K_5, K_k)	$\begin{cases} 9 & \text{if } k = 5 \\ 11 & \text{if } k = 6 \\ 12 & \text{if } k \in \{7, 8\} \\ 13 & \text{if } k \in \{9, 10, 11, 12, 13\} \\ 14 & \text{if } k \geq 14 \end{cases}$	[63]
$(K_{1,i}, K_{1,j}, K_{1,k})$	$\begin{cases} i + j - 1 & \text{if } k \geq i + j \text{ and } i \text{ and } j \text{ are both even} \\ i + j & \text{if } k \geq i + j \text{ and } i \text{ is odd or } j \text{ is odd} \\ \lceil (i + j + k)/2 \rceil - 1 & \text{if } k < i + j \text{ and } (i, j, k) \in S \\ \lceil (i + j + k)/2 \rceil & \text{if } k < i + j \text{ and } (i, j, k) \notin S \end{cases}$	[62]
(P_i, P_j, P_k)	$\begin{cases} j + \lfloor i/2 \rfloor - 1 & \text{if } k \geq j + \lfloor i/2 \rfloor - 1 \\ \lfloor (i + 2j + 4k - 2)/6 \rfloor & \text{if } K < j + \lfloor i/2 \rfloor - 1 \end{cases}$	[179]

TABLE 5.2: Known values for $r^3_{1,2}(K_i, K_j, K_k, K_m, K_n, K_p)$. These numbers were proved by Liu and Chung [164] and can be found in Subsection 2.2.2.

i	j	k	m	n	p	$r^3_2(K_i, K_j, K_k, K_m, K_n, K_p)$
3	3	3	3	3	≤ 5	5
3	3	3	3	3	≥ 6	6
3	3	3	3	4	≥ 4	6
3	3	3	4	4	4	8
3	3	4	4	4	4	8
3	4	4	4	4	4	9
3	3	3	4	≥ 4	≥ 5	9
3	3	3	5	≥ 5	≥ 5	14
3	3	3	≥ 6	≥ 6	≥ 6	17

TABLE 5.3: Known values for $r^4_{1,2}(K_3, K_3, K_3, K_3, K_i, K_j, K_k, K_m, K_n, K_p)$. These numbers are due to Jacobson [132, 133], and their proofs can be found in Subsection 2.2.3.

i	j	k	m	n	p	$r^4_{1,2}(K_3, K_3, K_3, K_3, K_i, K_j, K_k, K_m, K_n, K_p)$
3	3	3	3	3	3	5
3	3	3	3	3	≥ 4	6
3	3	≥ 3	≥ 4	≥ 4	≥ 4	9
3	4	3	4	3	4	8
3	5	≥ 3	≥ 5	≥ 3	≥ 5	14
3	5	≥ 4	≥ 4	≥ 4	≥ 4	14
3	6	≥ 3	≥ 6	≥ 3	≥ 6	17
4	4	≥ 5	≥ 5	≥ 5	≥ 5	18

TABLE 5.4: Known values for $r^4_{1,3}(K_3, K_3, K_3, K_3, K_4 K_i, K_j, K_k)$. These numbers were determined by Liu and Chung [164], and their proofs can be found in Subsection 2.2.4.

i	j	k	$r^4_{1,3}(K_3, K_3, K_3, K_3, K_4, K_i, K_j, K_k)$
4	4	4	7
4	4	≥ 5	8
≥ 4	≥ 5	≥ 5	9

TABLE 5.5: Known values for $r^3_{1,2}(K_3, K_3, K_3, K_{1,i}, K_{1,j}, K_{1,k})$ when $2 \leq i \leq j \leq k$. These numbers were evaluated by Jacobson [132] and their proofs can be found in Subsection 2.3.2.

i	j	k	$r^3_{1,2}(K_3, K_3, K_3, K_{1,i}, K_{1,j}, K_{1,k})$
2	2	≥ 2	3
2	3	3	4
2	≥ 3	≥ 4	5
3	≥ 3	≤ 4	5
3	3	≥ 5	6
3	4	≥ 5	6
3	5	5	6
3	≥ 5	≥ 6	7
4	4	4	6
4	4	≥ 5	7
4	5	≥ 5	7
4	6	6	7
4	≥ 6	≥ 7	9
5	5	5	8
5	5	6	8
5	5	7	9
5	6	7	9
5	6	6	9
5	7	7	9
5	7	≥ 8	10
5	≥ 8	≥ 8	11
6	6	6	9
6	6	7	9

TABLE 5.6: Known weakened Ramsey numbers of the form $r^3_2(G)$. These numbers follow from Theorems 2.10, 3.1, and 3.9.

G	$r^3_2(G)$	Source
K_3	5	[63]
K_4	10	[63]
$K_{1,m}$	$\begin{cases} m & \text{if } m = 3q+1 \text{ with } m \text{ and } q \text{ both odd} \\ m+1 & \text{otherwise} \end{cases}$	[150]

TABLE 5.7: Known weakened Gallai-Ramsey numbers of the form $gr_2^3(G)$. Proofs of these numbers can be found in Sections 3.3 and 3.4.

G	$gr_2^3(G)$	Source
K_4	5	[51]
K_5	9	[160]
C_4	4	[51]
C_5	6	[51]
C_6	7	Theorem 3.21
F_2	6	[31]
F_3	9	[31]
W_5	6	[136]
W_6	8	[136]
B_2	4	Corollary 3.2
B_3	6	[136]
B_4	7	[136]
$K_{2,3}$	6	[136]
$K_{2,4}$	7	[136]
$K_{3,3}$	8	[136]

Bibliography

[1] T. Ahmed, L. Boza, M. Revuelta, and M. Sanz. Exact values and lower bounds on the n-color weak Schur numbers for $n = 2, 3$. *Ramanujan J.*, 62:347–363, 2023.

[2] T. Ahmed, M. Eldridge, J. Marler, and H. Snevily. Strict Schur numbers. *Integers*, 13:#A22, 2013.

[3] T. Ahmed and D. Schaal. On generalized Schur numbers. *Exp. Math.*, 25:213–218, 2016.

[4] J.L. Ramírez Alfonsín and B. Reed, editors. *Perfect Graphs*. John Wiley & Sons, Inc., Chichester, England, 2001.

[5] N. Alon, P. Erdős, D. Gunderson, and M. Molloy. A Ramsey-type problem and the Turán numbers. *J. Graph Theory*, 40(2):120–129, 2002.

[6] M. Alspach. Cayley graphs. In J. Gross and J. Yellen, editors, *Handbook of Graph Theory*, Discrete Mathematics and its Applications. CRC Press, Boca Raton, FL, 2004.

[7] V. Angeltveit and M. McKay. $R(5,5) \leq 46$. *preprint: arXiv:2409.15709v2.*

[8] L. Aragão, M. Collares, J.P. Marciano, T. Martins, and R. Morris. A lower bound for set-coloring Ramsey numbers. *Random Struct. Alg.*, 64:157–169, 2024.

[9] M. Artin. *Algebra*. Pearson Education, Inc., second edition, 2011.

[10] R. Atanasov, M. Budden, J. Lambert, K. Murphy, and A. Penland. On certain induced subgraphs of Paley graphs. *Acta Univ. Apulensis Math. Inform.*, 40:51–65, 2014.

[11] M. Axenovich, T. Jiang, and A. Kündgen. Bipartite anti-Ramsey numbers of cycles. *J. Graph Theory*, 47(1):9–28, 2004.

[12] D. Bal and P. Bennett. Edge-coloring $K_{n,n}$ with no 2-colored C_{2k}. *preprint: arXiv:2507.13329v1.*

[13] M. Balko, J. Cibulka, K. Král, and J. Kynčl. Ramsey numbers of ordered graphs. *Electron. Notes Discrete Math.*, 49:419–424, 2015.

[14] M. Balko, V. Jelínek, and P. Valtr. On ordered Ramsey numbers of bounded-degree graphs. *J. Combin. Theory Ser. B*, 134:179–202, 2019.

[15] M. Balko and M. Polijak. On off-diagonal ordered Ramsey numbers of nested matchings. *Discrete Math.*, 346(2):#113223, 2023.

[16] M. Balko and M. Polijak. On ordered Ramsey numbers of matchings versus triangles. *Electron. J. Combin.*, 31(2):#P2.23, 2024.

[17] R. Bass, C. Magnant, K. Ozeki, and B. Pyron. Characterizations of edge-colorings of complete graphs that forbid certain rainbow subgraphs. *preprint.*

[18] G. Beam and M. Budden. Weakened Gallai-Ramsey numbers. *Surv. Math. Appl.*, 13:131–145, 2018.

[19] L. Beineke and A. Schwenk. On a bipartite form of the Ramsey problem. In *Proceedings of the Fifth British Combinatorial Conference*, volume XV of *Congr. Numer.*, pages 17–22. Univ. Aberdeen, 1975.

[20] C. Berge. *Graphs and Hypergraphs*, volume 6 of *North-Holland Mathematical Library.* North-Holland Publishing Company, Amsterdam, 1973.

[21] A. Beutelspacher and W. Brestovansky. Generalized Schur numbers. *Lecture Notes in Math.*, 969:30–38, 1982.

[22] A. Bialostocki, P. Dierker, and W. Voxman. Either a graph or its complement is connected: a continuing saga. *Unpublished manuscript*, 2001.

[23] A. Bialostocki and D. Schaal. On a variation of Schur numbers. *Graphs Combin.*, 16:139–147, 2000.

[24] A. Bialostocki and W. Voxman. On monochromatic-rainbow generalizations of two Ramsey-type theorems. *Ars Combin.*, 68:131–142, 2003.

[25] P. Blanchard, F. Harary, and R. Reis. Partitions into sum-free sets. *Integers*, 6:#A7, 2006.

[26] J. Bondy and P. Erdős. Ramsey numbers for cycles in graphs. *J. Combin. Theory Ser. B*, 14:46–54, 1973.

[27] J. Bosák, A. Rosa, and Š. Znám. On decompositions of complete graphs into factors with given diameters. In *Theory of Graphs*, pages 37–56. Academic Press, New York, 1969.

[28] L. Boza, M. Revuelta, and M. Sanz. A general lower bound on the weak Schur number. *Electron. Notes Discrete Math.*, 68:137–142, 2018.

[29] C. Brady and R. Haas. Rado numbers for the real line. *Congr. Numer.*, 177:109–114, 2005.

[30] M. Bucić and A. Khamseh. The $(t-1)$-chromatic Ramsey number for paths. *J. Graph Theory*, 102(4):784–796, 2023.

[31] M. Budden. Weakened Gallai-Ramsey numbers for fans to appear in *Util. Math.*

[32] M. Budden. Schur numbers involving rainbow colorings. *Ars Math. Contemp.*, 18:281–288, 2020.

[33] M. Budden. The Gallai-Ramsey number for a tree versus complete graphs. *Integers*, 21A:#A7, 2021.

[34] M. Budden. Ramsey numbers for connected 2-colorings of complete graphs. *Theory Appl. Graphs*, 10:Article No. 7 (10 pages), 2023.

[35] M. Budden. *Star-Critical Ramsey Numbers for Graphs.* SpringerBriefs in Mathematics, Springer, Cham, 2023.

[36] M. Budden. The star-critical weakened Gallai-Ramsey number $gr_*^{2,t}(C_5)$. *Filomat*, 39(24):8545–8549, 2025.

[37] M. Budden, N. Calkins, W. Hack, J. Lambert, and K. Thompson. Enumeration of triangles in quartic residue graphs. *Integers*, 11:#A48 (16 pages), 2011.

[38] M. Budden and J. Clifton. A weakened version of strict Schur numbers. *Acta Univ. Apulensis Math. Inform.*, 65:111–122, 2021.

[39] M. Budden and B. Collins. Some variations of multicolor Rado numbers. *Acta Univ. Apulensis Math. Inform.*, 69:93–105, 2022.

[40] M. Budden, L. Goch, and J. Hiller. Computational lower bounds for weakened Ramsey numbers from strongly regular t-colorings of complete graphs. *to appear in Bull. Inst. Combin. Appl.*

[41] M. Budden, J. Hiller, and A. Penland. Constructive methods in Gallai-Ramsey theory for hypergraphs. *Integers*, 20A:#A4, 2020.

[42] M. Budden, J. Hiller, and A. Rapp. Generalized Ramsey theorems for r-uniform hypergraphs. *Australas. J. Combin.*, 63:142–152, 2015.

[43] M. Budden, Y. Khobragade, and S. Sarkar. Lower bounds for multicolor star-critical Ramsey numbers. *preprint: arXiv:2407.00872v1.*

[44] M. Budden and B. Landman. Rainbow numbers for the generalized Schur equation $x_1 + x_2 + \cdots + x_{m-1} = x_m$. *Australas. J. Combin.*, 90:66–79, 2024.

[45] M. Budden, M. Moun, and J. Jakhar. Star-critical weakened Ramsey numbers. *Theory Appl. Math.*, 12(2):Article #4 (16 pages), 2025.

[46] M. Budden, M. Moun, J. Jakhar, and R. Prange. Multicolor connected Ramsey numbers. *under review.*

[47] M. Budden and R. Prange. Ramsey theory for a generalized fan versus triangles. *Util. Math.*, 123:165–178, 2025.

[48] M. Budden and R. Prange. The star-critical Gallai-Ramsey number $gr_*^t(F_{3,2}, K_3, K_3)$ and related critical colorings. *Congr. Numer.*, 236:115–122, 2025.

[49] M. Budden, M. Stender, and Y. Zhang. Weakened Ramsey numbers and their hypergraph analogues. *Integers*, 17:#A23, 2017.

[50] M. Budden and W. Stiles. Anti-Ramsey hypergraph numbers. *Electron. J. Graph Theory Appl. (EJGTA)*, 9:397–407, 2021.

[51] M. Budden and T. Wimbish. Subgraphs of Gallai-colored complete graphs spanned by edges using at most two colors. *Australas. J. Combin.*, 84:375–387, 2022.

[52] M. Budden and T. Wimbish. Connected Ramsey numbers involving trees. *Electron. J. Math.*, 9:31–37, 2025.

[53] S. Burr. Ramsey numbers involving graphs with long suspended paths. *J. London Math. Soc. (2)*, 24:405–413, 1981.

[54] S. Burr, P. Erdős, and L. Lovasz. On graphs of Ramsey type. *Ars Combin.*, 1(1):167–190, 1976.

[55] S. Burr and J. Roberts. On Ramsey numbers for stars. *Util. Math.*, 4:217–220, 1973.

[56] S. Burr and V. Rosta. On the Ramsey multiplicities of graphs - problems and recent results. *J. Graph Theory*, 4:347–361, 1980.

[57] S. Bustmante and M. Stein. Monochromatic tree covers and Ramsey numbers for set-coloured graphs. *Discrete Math.*, 341:266–276, 2018.

[58] G. Chartrand and A. Polimeni. Ramsey theory and chromatic numbers. *Pacific J. Math.*, 55(1):39–43, 1974.

[59] G. Chartrand and P. Zhang. *Chromatic Graph Theory.* Discrete Mathematics and its Applications. CRC Press, Boca Raton, Florida, 2009.

[60] S.A. Choudum and B. Ponnusamy. Ordered Ramsey numbers. *Discrete Math.*, 247:79–92, 2002.

[61] F. Chung and R. Graham. Edge-colored complete graphs with precisely colored subgraphs. *Combinatorica*, pages 315–324, 1983.

[62] K. Chung, M. Chung, and C. Liu. A generalization of Ramsey theory for graphs - with stars and complete graphs as forbidden subgraphs. *Congr. Numer.*, 19:155–161, 1977.

[63] K. Chung and C. Liu. A generalization of Ramsey theory for graphs. *Discrete Math.*, 21:117–127, 1978.

[64] M. Chung. A Generalization of Ramsey Number for Graphs. Master's thesis, University of Illinois at Urbana-Champaign, 1978.

[65] V. Chvátal. Tree-complete graph Ramsey numbers. *J. Graph Theory*, 1:93, 1977.

[66] V. Chvátal and F. Harary. Generalized Ramsey theory for graphs. *Bull. Amer. Math. Soc.*, 78(3):423–426, 1972.

[67] V. Chvátal and F. Harary. Generalized Ramsey theory for graphs, II. Small off-diagonal numbers. *Proc. Amer. Math. Soc.*, 32(2):389–394, 1972.

[68] V. Chvátal and F. Harary. Generalized Ramsey theory for graphs, III. Small off-diagonal numbers. *Pacific J. Math.*, 41:335–345, 1972.

[69] V. Chvátal and F. Harary. Generalized Ramsey theory for graphs, I. Diagonal numbers. *Period. Math. Hungar.*, 3(1–2):115–124, 1973.

[70] C. Clapham, A. Flockhart, and J. Sheehan. Graphs without four-cycles. *J. Graph Theory*, 13(1):29–47, 1989.

[71] E. Cockayne and P. Lorimer. The Ramsey number for stripes. *J. Austral. Math. Soc.*, 19(Series A):252–256, 1975.

[72] M. Codish, M. Frank, A. Itzhakov, and A. Miller. Computing the number $R(4,3,3)$ using abstraction and symmetry breaking. *Constraints*, 21:375–393, 2016.

[73] C. Colbourn and J. Dinitz, editors. *Handbook of Combinatorial Designs.* Discrete Mathematics and its Applications. Taylor & Francis Group, LLC, second edition, 2007.

[74] D. Conlon, J. Fox, X. He, D. Mubayi, A. Suk, and J. Verstraëte. Set-coloring Ramsey numbers via codes. *Studia Sci. Math. Hungar.*, 61(1):1–15, 2024.

[75] D. Conlon, J. Fox, C. Lee, and B. Sudakov. Ordered Ramsey numbers. *J. Combin. Theory Ser. B*, 122:353–383, 2017.

[76] D. Conlon, J. Fox, B. Sudakov, and F. Wei. Threshold Ramsey multiplicity for odd cycles. *Rev. Un. Mat. Argentina*, 64(1):49–68, 2022.

[77] D. Conlon, J. Fox, B. Sudakov, and F. Wei. Threshold Ramsey multiplicity for paths and even cycles. *European J. Combin.*, 107:Article #103612, 2023.

[78] R. Diestel. *Graph Theory*, volume 173 of *Graduate Texts in Mathematics.* Springer, Berlin, third edition, 2005.

[79] G. Dirac. Some theorems on abstract graphs. *Proc. London Math. Soc.*, 2:69–81, 1952.

[80] T. Dzido. A note on Turán numbers for even wheels. *Graphs Combin.*, 29:1305–1309, 2013.

[81] S. Eliahou, J. Marín, M. Revuelta, and M. Sanz. Weak Schur numbers and the search for G.W. Walker's lost partitions. *Comput. Math. Appl.*, 63:175–182, 2012.

[82] P. Erdős. Problems and results on finite and infinite graphs. In *Recent Advances in Graph Theory*, pages 183–192, Prague, 1975. Proc. Second Czechoslovak Sympos., Prague, 1974, Adademia.

[83] P. Erdős and R. Faudree. Size Ramsey functions. In *Sets, Graphs and Numbers*, volume 60, pages 219–238. Colloq. Math. Soc. Bolyai, 1992.

[84] P. Erdős, R. Faudree, C. Rousseau, and R. Schelp. The size Ramsey number. *Period. Math. Hungar.*, 9:145–161, 1978.

[85] P. Erdős, A. Hajnal, and R. Rado. Partition relations for cardinal numbers. *Acta Math. Acad. Sci. Hungar.*, 16:93–196, 1965.

[86] P. Erdős, M. Simonovits, and V. Sós. Anti-Ramsey theorems. *Coll. Math. Soc. J. Bolyai*, 10:633–643, 1973.

[87] P. Erdős and G. Szekeres. A combinatorial problem in geometry. *Compos. Math.*, 2:463–470, 1935.

[88] P. Erdős and A. Szemerédi. On a Ramsey type theorem. *Period. Math. Hungar.*, 2:295–299, 1972.

[89] G. Exoo. A lower bound for $R(5,5)$. *J. Graph Theory*, 13:97–98, 1989.

[90] K. Fallon, C. Giles, H. Rehm, S. Wagner, and N. Warnberg. Rainbow numbers of $[n]$ for $\sum_{i=1}^{k-1} x_i = x_k$. *Australas. J. Combin.*, 77(1):1–8, 2020.

[91] K. Fallon, E. Manhart, J. Miller, H. Rehm, N. Warnberg, and L. Zinnel. Rainbow numbers of $[m] \times [n]$ for $x_1 + x_2 = x_3$. *Integers*, 23:#A47, 2023.

[92] R. Faudree, R. Gould, M. Jacobson, and C. Magnant. Ramsey numbers in rainbow triangle free colorings. *Australas. J. Combin.*, 46:269–284, 2010.

[93] R. Faudree and R. Schelp. Path-path Ramsey-type numbers for the complete bipartite graph. *J. Combin. Theory Ser. B*, 19:161–173, 1975.

[94] R. Faudree and R. Schelp. Connected Ramsey numbers for paths and stars. *Colloq. Math. Soc. János Bolyai*, 18:307–313, 1978.

[95] R. Faudree and R. Schelp. Some connected Ramsey numbers. *J. Graph Theory*, 2(3):119–128, 1978.

[96] J. Fox, A. Grinshpun, and J. Pach. The Erdős-Hajnal conjecture for rainbow triangles. *J. Combin. Theory Ser. B*, 111:75–125, 2015.

[97] P. Frankl and A. Kupavskii. Two problems on matchings in set families - in the footsteps of Erdős and Kleitman. *J. Combin. Theory Ser. B*, 138:286–313, 2019.

[98] A. Fröhlich and M. Taylor. *Algebraic Number Theory*, volume 27 of *Cambridge Studies in Advanced Mathematics*. Cambridge University Press, Cambridge, 1991.

[99] S. Fujita and C. Magnant. Gallai-Ramsey numbers for cycles. *Discrete Math.*, 311:1247–1254, 2011.

[100] S. Fujita and C. Magnant. Extensions of Gallai-Ramsey results. *J. Graph Theory*, 70(4):404–426, 2012.

[101] S. Fujita, C. Magnant, and K. Ozeki. Rainbow generalizations of Ramsey theory - a dynamic survey. *Theory Appl. Graphs*, 0(1):Article 1, 2014.

[102] T. Gallai. Transitiv orientierbare graphen. *Acta Math. Acad. Sci. Hungar.*, 18:25–66, 1967.

[103] J. Gallian. *Contemporary Abstract Algebra*. Brooks/Cole, Cengage Learning, seventh edition, 2010.

[104] C.F. Gauss. *Disquisitiones Arithmeticae*, volume Articles 125–145. 1801.

[105] L. Gerencsér and A. Gyárfás. On Ramsey-type problems. *Ann. Univ. Sci. Budapest. Eötvös Sect. Math.*, 10:167–170, 1967.

[106] S. Golomb and L. Baumert. Backtrack programming. *J. ACM*, 12(4):516–524, 1965.

[107] A. Goodman. On sets of acquaintances and strangers at any party. *Amer. Math. Monthly*, 66:778–783, 1959.

[108] R. Graham and S. Butler. *Rudiments of Ramsey Theory*, volume 123 of *Regional Conference Series in Mathematics*. American Mathematical Society, Providence, Rhode Island, second edition, 2015.

[109] R. Graham, B. Rothschild, and J. Spencer. *Ramsey Theory*. Wiley-Interscience Series in Discrete Mathematics and Optimization. John Wiley & Sons, Inc., second edition, 1990.

[110] J. Graver and J. Yackel. Some graph theoretic results associated with Ramsey's theorem. *J. Combin. Theory*, 4:125–175, 1968.

[111] R. Greenwood and A. Gleason. Combinatorial relations and chromatic graphs. *Canadian J. Math.*, 7:1–7, 1955.

[112] C. Grinstead and S. Roberts. On the Ramsey numbers $R(3,8)$ and $R(3,9)$. *J. Combin. Theory Ser. B*, 33:27–51, 1982.

[113] R. Gu, J. Li, and Y. Shi. Anti-Ramsey number of paths in hypergraphs. *SIAM J. Discrete Math.*, 34(1):271–307, 2020.

[114] A. Gyárfás. Large sets of t-designs and a Ramsey-type problem. *Integers*, 18:#A53, 2018.

[115] A. Gyárfás, G. Sárközy, A. Sebö, and S. Selkow. Ramsey-type results for Gallai colorings. *J. Graph Theory*, 64(3):233–243, 2010.

[116] A. Gyárfás and G. Simonyi. Edge colorings of complete graphs without tricolored triangles. *J. Graph Theory*, 46(3):211–216, 2004.

[117] R. Häggkvist, R. Faudree, and R. Schelp. Pancyclic graphs - connected Ramsey number. *Ars Combin.*, 11(37–49), 1981.

[118] Y. Hao and Q. Lin. Ramsey number of K_3 versus $F_{3,n}$. *Discrete Appl. Math.*, 251:345–348, 2018.

[119] Y. Hao and Q. Lin. Star-critical Ramsey numbers for large generalized fans and books. *Discrete Math.*, 341:3385–3393, 2018.

[120] F. Harary. *Graph Theory.* Addison Wesley, Reading, Mass., 1969.

[121] F. Harary. Recent results on generalized Ramsey theory for graphs. *Lecture Notes in Math.*, 303:125–138, 1972.

[122] F. Harary and G. Prins. Generalized Ramsey theory for graphs, IV. The Ramsey multiplicity of a graph. *Networks*, 4:163–173, 1974.

[123] H. Harborth and M. Möller. Weakened Ramsey numbers. *Discrete Appl. Math.*, 95:279–284, 1999.

[124] M. Heule. Schur number five. *Proceedings of the Thirty-Second AAAI Conference on Artificial Intelligence and Thirtieth Innovative Applications of Artificial Intelligence Conference and Eighth AAAI Symposium on Educational Advances in Artificial Intelligence*, pages 6598–6606, 2018.

[125] J. Hook. *The Classification of Critical Graphs and Star-Critical Ramsey Numbers.* PhD thesis, Lehigh University, 2010.

[126] J. Hook. Recent developments of star-critical Ramsey numbers. In *Combinatorics, Graph Theory and Computing*, volume 448 of *Springer Proc. Math. Stat.*, Cham, 2024. Springer.

[127] J. Hook and G. Isaak. Star-critical Ramsey numbers. *Discrete Appl. Math.*, 159:328–334, 2011.

[128] M. Horňák, S. Jendrol', I. Schiermeyer, and R. Soták. Rainbow numbers for cycles in plane triangulations. *J. Graph Theory*, 78(4):248–257, 2015.

[129] K. Ireland and M. Rosen. *A Classical Introduction to Modern Number Theory.* Springer-Verlag, New York, second edition, 1998.

[130] R. Irving. An extension of Schur's theorem on sum-free partitions. *Acta Arith.*, 25:55–64, 1973/74.

[131] M. Jacobson. A note on Ramsey multiplicity. *Discrete Math.*, 29:201–203, 1980.

[132] M. Jacobson. *On Various Extensions of Ramsey Theory.* PhD thesis, Emory University, 1980.

[133] M. Jacobson. On a generalization of Ramsey theory. *Discrete Math.*, 38:191–195, 1982.

[134] M. Jacobson. On the Ramsey multiplicity for stars. *Discrete Math.*, 42:63–66, 1982.

[135] J. Jakhar, M. Budden, and M. Moun. Rainbow numbers in planar host graphs. *Filomat*, 39(8):2695–2710, 2025.

[136] J. Jakhar and M. Moun. Weakened Gallai-Ramsey number for various graphs of order up to six. *Discrete Appl. Math.*, 378:426–435, 2026.

[137] S. Jendrol', I. Schiermeyer, and J. Tu. Rainbow numbers for matchings in plane triangulations. *Discrete Math.*, 331:158–164, 2014.

[138] Z. Jin, F. Wang, H. Wang, and B. Lv. Rainbow triangles in edge-colored Kneser graphs. *Appl. Math. Comput.*, 365:#124724, 2020.

[139] Z. Jin and K. Ye. Rainbow number of matchings in planar graphs. *Discrete Math.*, 342(10):2846–2858, 2018.

[140] V. Jungić. *Basics of Ramsey Theory.* CRC Press, Boca Raton, Florida, 2023.

[141] J. Kalbfleisch. Construction of special edge-chromatic graphs. *Canad. Math. Bull.*, 8:575–584, 1965.

[142] J. Kalbfleisch. *Chromatic Graphs and Ramsey's Theorem.* PhD thesis, University of Waterloo, 1966.

[143] G. Katona, T. Nemetz, and M. Simonovits. On a problem of Turán in the theory of graphs (in Hungarian). *Mat. Lapok*, 15:228–238, 1964.

[144] G. Kéry. On a theorem of Ramsey (in Hungarian). *Mat. Lapok*, 15:204–224, 1964.

[145] A. Khamseh. On $(t-1)$-colored paths in t-colored complete graphs. *Bull. Belg. Math. Soc. Simon Stevin*, 25(2):197–207, 2018.

[146] A. Khamseh. A note on $(t-1)$-chromatic Ramsey number of linear forests. *Int. J. Comput. Math. Comput. Syst. Theory*, 5(2):92–97, 2020.

[147] A. Khamseh. A short note on the $(t-1)$-chromatic Ramsey number of linear forests. *Int. J. Comput. Math. Comput. Syst. Theory*, 10(3):117–125, 2025.

[148] A. Khamseh and G. Omidi. A generalization of Ramsey theory for linear forests. *Int. J. Comput. Math.*, 89(10):1303–1310, 2012.

[149] A. Khamseh and G. Omidi. A generalization of Ramsey theory for stars and one matching. *Math. Reports*, 19(69):85–92, 2017.

[150] A. Khamseh and G. Omidi. Large stars with few colors. *Math. Rep. (Bucur.)*, 24(74):363–375, 2022.

[151] D. König. *Theory of Finite and Infinite Graphs.* Translated by R. McCoart with commentary by W.T. Tutte. Birkhäuser, 1990.

[152] Y. Lan and Y. Shi. Planar Turán numbers of short paths. *Graphs Combin.*, 35(1035–1049), 2019.

[153] Y. Lan, Y. Shi, and Z.-X. Song. Extremal theta-free planar graphs. *Discrete Math.*, 342(12):#111610, 2019.

[154] Y. Lan, Y. Shi, and Z.-X. Song. Planar anti-Ramsey numbers of paths. *Discrete Math.*, 342(11):2316–3224, 2019.

[155] Y. Lan, Y. Shi, and Z.-X. Song. Planar Turán number and planar anti-Ramsey number of graphs. *Oper. Res. Trans.*, 25(3):200–216, 2021.

[156] B. Landman and A. Robertson. *Ramsey Theory on the Integers*, volume 73 of *Student Mathematical Library.* American Mathematical Society, Providence, Rhode Island, second edition, 2014.

[157] H. Lei, Y. Shi, Z.-X. Song, and J. Zhang. Gallai-Ramsey numbers of C_{10} and C_{12}. *Australas. J. Combin.*, 79(3):380–400, 2021.

[158] F. Lemmermeyer. *Reciprocity Laws: From Euler to Eisenstein.* Springer-Verlag, Berlin, 2000.

[159] L. Li, X. Li, Y. Mao, and Y. Si. Gallai-Ramsey numbers for rainbow trees and monochromatic complete bipartite graphs. *Discuss. Math. Graph Theory*, 45(3):823–844, 2025.

[160] X. Li, H. Broersma, and L. Wang. The Erdős-Gyárfás function with respect to Gallai-colorings. *J. Graph Theory*, 101(2):242–264, 2022.

[161] X. Li, J. Tu, and Z. Jin. Bipartite rainbow numbers of matchings. *Discrete Math.*, 309(8):2575–2578, 2009.

[162] Y. Li and Q. Lin. *Elementary Methods of Graph Ramsey Theory*, volume 211 of *Applied Mathematical Sciences.* Springer, Cham, 2022.

[163] M. Liang, S. Radziszowski, and X. Xu. On a diagonal conjecture for classical Ramsey numbers. *Discrete Appl. Math.*, 267:195–200, 2019.

[164] C. Liu and K. Chung. A Generalization of Ramsey Theory for Graphs. University of Illinois at Urbana-Champaign, Department of Computer Science, 1977.

[165] H. Liu, C. Magnant, A. Saito, I. Schiermeyer, and Y. Shi. Gallai-Ramsey number for K_4. *J. Graph Theory*, 94:192–205, 2020.

[166] Z. Luo. Star-critical Ramsey numbers and regular Ramsey numbers for stars. *Discuss. Math. Graph Theory*, 45:755–762, 2025.

[167] C. Magnant and P. Salehi-Nowbandegani. *Topics in Gallai-Ramsey Theory.* SpringerBriefs in Mathematics, Springer, Cham, 2020.

[168] C. Magnant and I. Schiermeyer. Gallai-Ramsey number for K_5. *J. Graph Theory*, 101:455–492, 2022.

[169] B. Maheswari and M. Lavaku. Enumeration of triangles and hamilton cycles in quadratic residue Cayley graphs. *Chamchuri J. Math.*, 1(1):95–103, 2009.

[170] Y. Mao. Size Gallai-Ramsey number. *Graphs Combin.*, 39:Article #126, 2023.

[171] Y. Mao. Ordered Gallai-Ramsey numbers. *Discrete Appl. Math.*, 376(62–71), 2025.

[172] Y. Mao, A. Robertson, J. Wang, C. Yang, and G. Yang. Gallai-Schur triples and related problems. *preprint: arXiv:2502.2122v1.*

[173] Y. Mao, Z. Wang, C. Magnant, and I. Schiermeyer. Ramsey and Gallai-Ramsey number for wheels. *Graphs Combin.*, 38:Article #42, 2022.

[174] Y. Mao, Z. Wang, C. Magnant, and I. Schiermeyer. Gallai-Ramsey numbers for fans. *Discrete Math.*, 346:Article #113338, 2023.

[175] D. Matula. Ramsey theory for graph connectivity. *J. Graph Theory*, 7:95–103, 1983.

[176] B. McKay and S. Radziszowski. $R(4,5) = 25$. *J. Graph Theory*, 19:309–322, 1995.

[177] B. McKay and S. Radziszowski. Subgraph counting identities and Ramsey numbers. *J. Combin. Theory Ser. B*, 69(2):193–209, 1997.

[178] B. McKay and K. Zhang. The value of the Ramsey number $R(3,8)$. *J. Graph Theory*, 16:99–105, 1992.

[179] R. Meenakshi and P. Sundararaghavan. Generalized Ramsey numbers for paths in 2-chromatic graphs. *Internat. J. Math. & Math. Sci.*, 9:273–276, 1986.

[180] M. Moun, J. Jakhar, and M. Budden. Star-critical connected Ramsey numbers for 2-colorings of complete graphs. *Trans. Comb.*, 14:211–222, 2025.

[181] A. Munemasa and M. Shinohara. Complementary Ramsey numbers and Ramsey graphs. *J. Indones. Math. Soc.*, 25(02):146–153, 2019.

[182] J. Nešetřil and V. Rödl, editors. *Mathematics of Ramsey Theory*, volume 5 of *Algorithms and Combinatorics*. Springer-Verlag, Berlin, 1990.

[183] L. Özkahya and M. Young. Anti-Ramsey number of matchings in hypergraphs. *Discrete Math.*, 313(20):2359–2364, 2013.

[184] K. Piwakowski and S. Radziszowski. The Ramsey multiplicity of K_4. *Ars Combin.*, 60:131–135, 2001.

[185] Z. Qin, Y. Lan, and Y. Shi. Improved bounds for rainbow numbers of matchings in plane triangulations. *Discrete Math.*, 342(1):221–225, 2019.

[186] Z. Qin, Y. Lan, Y. Shi, and J. Yue. Exact rainbow numbers for matchings in plane triangulations. *Discrete Math.*, 344(4):#112301, 2021.

[187] Z. Qin, H. Lei, and S. Li. Rainbow numbers for small graphs in planar graphs. *Appl. Math. Comput.*, 371:#124888, 2020.

[188] R. Rado. Studien zur kombinatorik (in German). *Math. Z.*, 36(1):424–470, 1933.

[189] S. Radziszowski. Small Ramsey numbers–revision 17. *Electron. J. Combin.*, DS1.17: 1–133, 2024.

[190] F. Ramsey. On a problem of formal logic. *Proc. London Math. Soc.*, 30:264–286, 1929.

[191] D. Ray-Chaudhuri and R. Wilson. Solution of Kirkman's schoolgirl problem. In *Combinatorics*, volume XIX of *Proc. Sympos. Pure Math.*, pages 187–203. Amer. Math. Soc., 1971.

[192] A. Robertson. *Fundamentals of Ramsey Theory.* Discrete Mathematics and its Applications. CRC Press, Boca Raton, Florida, 2021.

[193] A. Robertson and D. Schaal. Off-diagonal generalized Schur numbers. *Adv. Appl. Math.*, 26:252–257, 2001.

[194] M. Sanz. *Números de Schur y Rado.* PhD thesis, Universidad de Sevilla, 2010.

[195] I. Schur. Über die kongruenz $x_m + y_m \equiv z_m (\text{mod } p)$ (in German). *Jahresber. Dtsch. Math.-Ver.*, 25:114–117, 1916.

[196] A. Soifer. *The Mathematical Coloring Book: Mathematics of Coloring and the Colorful Life of Its Creators.* Springer, New York, 2009.

[197] A. Soifer, editor. *Ramsey Theory: Yesterday, Today, and Tomorrow*, volume 285 of *Progress in Mathematics.* Birkhäuser/Springer, New York, 2011.

[198] Z.-X. Song, B. Wei, F. Zhang, and Q. Zhao. A note on Gallai-Ramsey number of even wheels. *Discrete Math.*, 343:Article #111725, 2020.

[199] Z.-X. Song and J. Zhang. A conjecture on Gallai-Ramsey numbers of even cycles and paths. *Australas. J. Combin.*, 75(3):296–308, 2019.

[200] W. Su, Q. Li, H. Luo, and G. Li. Lower bounds of Ramsey numbers based on cubic residues. *Discrete Math.*, 250:197–209, 2002.

[201] D. Sumner. The connected Ramsey number. *Discrete Math.*, 22:49–55, 1978.

[202] A. Thomason and P. Wagner. Complete graphs with no rainbow path. *J. Graph Theory*, 54(3):261–266, 2007.

[203] P. Turán. Eine extremalaufgabe aus der graphentheorie. *Mat. Fiz. Lapok*, 48:436–452, 1941.

[204] D. Vestal and J. Sax. Off-diagonal continuous Rado numbers for $x_1+x_2+\cdots+x_k = x_0$. *preprint: arXiv:2511.20528v1.*

[205] M. Wang and J. Qian. Ramsey numbers for complete graphs versus generalized fans. *Graphs Combin.*, 38:#186 (8 pages), 2022.

[206] R. Wang. Another definition for Ramsey numbers. *IEEE Int. Sump. Inf. Sci. Eng.*, 2:405–409, 2008.

[207] Z. Wang, L. Zhang, M. Wei, and M. Budden. Gallai-Ramsey numbers for ℓ-connected graphs. *preprint: arXiv:2601.13944v1.*

[208] Z. Wang and Y. Zhang. Weakened size Gallai-Ramsey numbers for K_5. *Bull. Iranian Math. Soc.*, 51:Article #72, 2025.

[209] A. Weil. *Basic Number Theory*, volume 144 of *Die Grundlehren der mathematischen Wissenschaften.* Springer-Verlag, New York, 1967.

[210] X. Xu, M. Liang, and H. Luo. *Ramsey Theory: Unsolved Problems and Results.* De Gruyter, Berlin, 2018.

[211] X. Xu, Z. Shao, W. Su, and Z. Li. Set-coloring of edges and multigraph Ramsey numbers. *Graphs Combin.*, 25:863–870, 2009.

[212] A. Yi and Z. Luo. The d-chromatic Ramsey number for stars. *preprint: arXiv:2512.16599v1.*

[213] F. Zhang, Z.-X. Song, and Y. Chen. Gallai-Ramsey number of even cycles with chords. *Discrete Math.*, 345:Article #112738, 2022.

[214] F. Zhang, Z.-X. Song, and Y. Chen. Gallai-Ramsey number of odd cycles with chords. *European J. Combin.*, 107:Article #103598, 2023.

[215] J. Zou, Y. Mao, C. Magnant, Z. Wang, and C. Ye. Gallai-Ramsey numbers for books. *Discrete Appl. Math.*, 268:164–177, 2019.

Index

For Product Safety Concerns and Information please contact our EU representative GPSR@taylorandfrancis.com Taylor & Francis Verlag GmbH, Kaufingerstraße 24, 80331 München, Germany

Batch number: 10740923

Printed by Printforce, the Netherlands